Drug Discrimination
and
State Dependent Learning

Drug Discrimination and State Dependent Learning

EDITED BY

BENG T. HO

*Texas Research Institute of Mental Sciences
and The University of Texas
Health Science Center at Houston*

DANIEL W. RICHARDS, III

Houston Community College

DOUGLAS L. CHUTE

University of Houston

ACADEMIC PRESS New York San Francisco London

A Subsidiary of Harcourt Brace Jovanovich, Publishers

ACADEMIC PRESS, INC.
111 Fifth Avenue, New York, New York 10003

United Kingdom Edition published by
ACADEMIC PRESS, INC. (LONDON) LTD.
24/28 Oval Road, London NW1 7DX

Library of Congress Cataloging in Publication Data
Main entry under title:

Drug discrimination and state dependent learning.

Includes bibliographies.
1. Psychopharmacology. 2. Sensory discrimination.
3. Drugs. I. Ho, Beng T., Date II. Chute, Douglas L. III. Richards, Daniel W. [DNLM: 1. Psychopharmacology. 2. Learning—Drug effects. QV77 D788]
RM315.D79 615'.78 77-2027
ISBN 0-12-350250-0

PRINTED IN THE UNITED STATES OF AMERICA

Contents

II RESEARCH METHODS AND NEW TECHNIQUES

List of Contributors

Numbers in parentheses indicate the pages on which the authors' contributions begin.

HAROLD L. ALTSHULER (263), Neuropsychopharmacology Research, Texas Research Institute of Mental Sciences and Baylor College of Medicine, Texas Medical Center, Houston, Texas

JAMES B. APPEL (149), Behavioral Pharmacology Laboratory, Department of Psychology, University of South Carolina, Columbia, South Carolina

ROBERT L. BALSTER (131), Pharmacology Department, Medical College of Virginia, Virginia Commonwealth University, Richmond, Virginia

HERBERT BARRY, III (3, 175), Department of Pharmacology, School of Pharmacy, University of Pittsburgh, Pittsburgh, Pennsylvania

RONALD G. BROWNE (79), Psychiatry Department, School of Medicine, University of California, San Diego, La Jolla, California

JUDY BUELKE (47), Department of Pharmacology and Research, Institute of Pharmaceutical Sciences, School of Pharmacy, University of Mississippi, University, Mississippi

WILLIAM T. CHANCE (119), Department of Pharmacology, Medical College of Virginia, Virginia Commonwealth University, Richmond, Virginia

DOUGLAS L. CHUTE (345), Department of Psychology, University of Houston, Houston, Texas

ROBERT D. FORD* (131), Pharmacology Department, Medical College of Virginia, Virginia Commonwealth University, Richmond, Virginia

LEROY G. FREY (35), Department of Pharmacology and Therapeutics, School of Medicine, State University of New York at Buffalo, Buffalo, New York

JOEL E. GREENE (203), Department of Psychology, New Mexico Highlands University, Las Vegas, New Mexico

RONALD L. HAYES (193, 249), Neurobiological and Anesthesiological Branch, National Institute of Dental Research, NIH, Bethesda, Maryland

** Present address: Pharmacology Department, Michigan State University, East Lansing, Michigan.*

I. D. HIRSCHHORN* (163), Department of Pharmacology, New York Medical College, Valhalla, New York

BENG T. HO (67), Texas Research Institute of Mental Sciences and The University of Texas Health Sciences Center at Houston, Houston, Texas

FRANK A. HOLLOWAY (319), Department of Psychiatry and Behavioral Sciences, University of Oklahoma Health Sciences Center, Oklahoma City, Oklahoma

EDWARD C. KRIMMER (3, 175), Department of Pharmacology, School of Pharmacy, University of Pittsburgh, Pittsburgh, Pennsylvania

DONALD M. KUHN† (149), Behavioral Pharmacology Laboratory, Department of Psychology, University of South Carolina, Columbia, South Carolina

MARY McKENNA (67), Texas Research Institute of Mental Sciences and The University of Texas Health Sciences Center at Houston, Houston, Texas

DAVID J. MAYER (249), Department of Physiology, Virginia Commonwealth University, Richmond, Virginia

DONALD A. OVERTON (283), Temple University School of Medicine and Eastern Pennsylvania Psychiatric Institute, Philadelphia, Pennsylvania

PAUL E. PHILLIPS (263), Neuropsychopharmacology Research, Texas Research Institute of Mental Sciences, Houston, Texas

DANIEL W. RICHARDS, III (227), Division of Social Sciences, Houston Community College, Houston, Texas

JOHN A. ROSECRANS (119), Department of Pharmacology, Medical College of Virginia, Virginia Commonwealth University, Richmond, Virginia

MARTIN D. SCHECHTER (103), Department of Pharmacology, Eastern Virginia Medical School, Norfolk, Virginia

HERBERT WEINGARTNER (361), Adult Psychiatry Branch, National Institute of Mental Health, Bethesda, and University of Maryland, Baltimore County

FRANCIS J. WHITE (149), Behavioral Pharmacology Laboratory, Department of Psychology, University of South Carolina, Columbia, South Carolina

MARVIN C. WILSON (47), Department of Pharmacology, School of Pharmacy, University of Mississippi, University, Mississippi

J. C. WINTER (35), Department of Pharmacology and Therapeutics, School of Medicine, State University of New York at Buffalo, Buffalo, New York

HAROLD ZENICK (203), Department of Psychology, New Mexico Highlands University, Las Vegas, New Mexico

* *Present address: Department of Pharmacology, New York College of Podiatric Medicine, New York, New York.*

† *Present address: Department of Psychology, Princeton University, Princeton, New Jersey.*

Preface

The effects of drugs upon behavior have been traditionally investigated by the change in baseline response produced by specified drug administration. This book describes two extensions of this traditional approach. Drug discrimination deals with behavior when the drug is a cue to a choice response task, and state dependent learning concerns the behavior affected by a drug-induced change in condition between training and testing trials. These two rapidly growing fields of research aid in offering explanations as to the roles that interoceptive physiological conditions have upon behavior. The chapters of this volume are representative samples of input from the investigators active in the field today. In addition to the studies using drugs, investigators are currently attempting to produce the same phenomena by such means as electrical stimulation, ablation, and changes in endogenous rhythms.

This is the first volume devoted exclusively to the definition and explanation of existing relationships between drug discrimination and state dependent learning. It now appears that drug discriminative control and state dependency are no longer one entity, as some investigators once spoke of state dependent discriminative stimuli, but rather are two complementary functions served by alterations of the interoceptive milieu upon behavior. In this context, drug discrimination describes the means that interoceptive stimuli can elicit or guide organisms in their selection of responses. In contrast, state dependency describes the effects that the organism's physiological state has upon coding and availability of acquired information.

This book is intended for researchers in psychology, pharmacology, and physiology interested in relating the interoceptive state to behavior, as well as for new investigators interested in current findings and research problems in the fields of drug discrimination and state dependent learning. It is divided into three general sections: mechanisms of drugs as discriminative stimuli, research methods and new techniques, and state dependent phenomena. The chapters in the section on drug discrimination include general information on the pharmacological actions of the agents discussed. This provides a background for scientists of related disciplines. The methodological section familiarizes investigators with the statistical problems and treatments commonly applied to studies in this area, as well as behavioral data considerations for those of neighboring disciplines. In the state dependent learning section the role of the central nervous system's state in the acquisition of learned responses as well as disorders of memory are presented. Both provide evidence which, integrated into the models of learning and memory, will facilitate the understanding of those processes. Information on the failure of memory induced by a state change are as important as are descriptions of normal processes of memory.

This book was generated following the success of a symposium held by the Southwestern Psychological Association in Houston, Texas in April, 1975. It was after this conference that we felt there was a need for such a book, and the next year and a half was spent in planning and preparing the manuscripts for publication.

We would like to thank the Southwestern Psychological Association, the faculty and graduate students of the University of Houston, and the Texas Research Institute of Mental Sciences who facilitated the initial discussions leading to this book.

We are also grateful to Mrs. Lore Feldman and Ms. Meredith Riddell for their excellent help in editing the book, and to Mrs. Barbara Dufresne for her skillful coordination of all communications between the authors, editors, and publisher.

In addition, we would like to express our deep appreciation to the staff of Academic Press for their encouragement and cooperation.

Drug Discrimination and State Dependent Learning

Overview

1

Pharmacology of Discriminative Drug Stimuli[1]

Herbert Barry, III, and Edward C. Krimmer
Department of Pharmacology
School of Pharmacy
University of Pittsburgh

Discriminative drug stimuli are based on pharmacological and psychological attributes and are measured by pharmacological and psychological procedures. Therefore, the topic of drugs as discriminative stimuli is a part of the interdisciplinary field of behavioral pharmacology or psychopharmacology.

Most behavioral pharmacology procedures measure the drug effect as an unconditional stimulus, evoking strong physiological and behavioral changes. Studies of discriminable drug stimuli establish the drug effect as a conditional stimulus, signaling the occurrence of subsequent unconditional stimuli, such as food reward or painful electric shock. Usually, animals are trained for different but equivalent responses under the different conditions, such as choice of opposite directions in which to run or different levers to press. The unconditional drug stimuli provide the basis for learning the differential response but are subordinated to the other unconditional stimuli, such as food or shock.

The present chapter summarizes characteristics of discriminative drug stimuli, compared with related attributes of drugs, such as dissociative, punitive, and reinforcing drug effects. A change in central nervous system (CNS) function seems to be necessary for the discriminative drug stimulus to occur, but there is little information on the specific brain areas or changes in brain function involved. Most of the available information consists of tests with novel drug doses, time intervals, and, especially, different drugs in animals

[1] This chapter was prepared with the support of Public Health Service Research Scientist Development Award K2-MH-5921 (to H. B.), Post-Doctoral Research Fellowship DA-2376 (to E. C. K.), and Research Grant MH-13595 (to both authors). The chapter contains portions of the data reported in a dissertation (by E. C. K.) in partial fulfillment of the requirements for the Ph.D. degree in pharmacology, University of Pittsburgh.

trained to discriminate a particular drug from vehicle. These tests determine to what degree the discrimination generalizes to other doses, time intervals, and drugs. The findings on discriminable drug effects are building up a classification that can be compared with classification of the same drugs based on other attributes.

STIMULUS CHARACTERISTICS OF DRUGS

The sensory modalities of vision, hearing, smell, taste, and touch can differentiate among a tremendous variety of environmental conditions. Stimuli from each modality have almost infinite variations because each combination of conditions can provide a differential stimulus pattern. For example, in the visual modality, a single rectangle may vary in height, width, brightness, color, and duration of exposure.

The most prominent characteristic of these sensory stimuli is their changeability. A particular pattern of environmental events may be rapidly initiated, modified, and terminated, not only by changes in the environment but also by such selective responses as turning toward or away from the source of the stimulus. The discriminative response is almost always associated with onset, cessation, or other changes of the stimulus. Since the sensory stimuli generally have weak motivational properties and are important only as signals, the organism learns to ignore the environmental conditions when they remain unchanged for a long time. This is the process of sensory adaptation, which protects the individual from being overwhelmed by the huge complexity and variety of these environmental conditions.

In contrast to environmental events, drug effects generally are stable and persistent. Their onset and termination are gradual, and they are resistant to change caused by selective orienting or motor responses. The drug stimulus is usually defined in terms of its peak effect rather than the transition from the nondrug to drug condition. This separation from the nondrug stimulus sometimes leads to definition of the drug effect as a dissociative rather than discriminative stimulus.

Another important aspect of drug effects is that sufficiently high doses constitute strong unconditional stimuli that evoke drastic unconditional responses, including changes in physiological systems. These changes include the pleasurable effects that give rise to voluntary drug consumption and the aversive effects that provide the basis for learning to avoid consumption of a drug or of another substance associated with administration of the drug.

Dissociative or Discriminative

The discriminability of a drug from the nondrug condition has been attributed to a dissociation between the two conditions, which results in amnesia for a response learned under the other condition. The term "state-dependent learning" (Overton, 1966, 1968, 1973) implies that responses learned under one condition are specific to that condition and fail to transfer to the other condition. This concept has recently been discussed by Overton (1974) and Bliss (1974).

Contrary to the rapidly changing and highly specific sensory stimuli, many behaviorally active drugs produce long-lasting and pervasive effects on the CNS, which might be expected to result in a drug-specific condition dissociated from the nondrug condition. According to a number of reports reviewed by Overton (1968), a new response learned under the influence of a drug fails to occur in a subsequent nondrug test. Similarity to amnesia for events immediately preceding electroconvulsive shock is suggested by reports of amnesia for new responses learned immediately preceding pentobarbital injection (Chute & Wright, 1973; Wright & Chute, 1973; Wright, 1974).

A dissociative drug condition has generally been inferred when an active shock-avoidance response learned under one condition fails to be performed in an initial exposure to the other condition. This concept has serious limitations. The amnesic dissociation is generally limited to a novel drug effect at a high dose. When alternative, equivalent responses are trained, such as left-hand and right-hand directions of turn in a T-maze under the differential drug and nondrug conditions, animals occasionally make the response that is incorrect for their current condition. This is contrary to the uniform response that would be expected if completely dissociated habits had been acquired under the different conditions.

Bindra and Reichert (1967) have suggested that dissociation of movement initiation but not of response choice may occur under a novel drug condition. A more probable explanation is that the behavior of active shock avoidance is unstable and tends to be replaced by a passive, freezing tendency when the stimulus conditions are changed.

Overton (1968) has reviewed evidence for less transfer of a learned response from the drug to nondrug state than from the nondrug to drug state. This asymmetrical dissociation may be attributable to the greater familiarity and variety of the nondrug state, so that habits tend to generalize readily to differences in drug or other conditions. The novelty and distinctiveness of the drug state may cause the habits to be associated more specifically with that condition. A thorough discussion by Overton (1974) points out many alternative possible interpretations of results obtained when animals trained under the drug or nondrug condition are shifted to the other condition. Goodwin (1974) has summarized evidence that the alcoholic "blackout" in humans often constitutes an impairment in acquisition of the memory while intoxicated rather than an impairment in retrieval of the memory caused by the change from the intoxicated to the sober state.

Punishing or Reinforcing

A high dose of a behaviorally active drug is a strong unconditional stimulus that evokes strong unconditional responses, indicated by various physiological effects. According to Dollard and Miller (1950), any sufficiently strong stimulus is an aversive motivational condition. The aversive drug effect may be seen not only in avoidance of drug consumption but also in avoidance of an originally attractive conditional stimulus associated with the drug. An experimental model for demonstrating and testing this aversive conditioning is the "poisoning" procedure. When drinking of a saccharin solution is followed by administration of a drug, the aversive effect of the drug condition is demonstrated by avoidance of the saccharin solution in a subsequent test. The taste of saccharin is the conditional stimulus associated with the unconditional aversive drug stimulus.

This aversive drug effect could be the basis for learning to discriminate between drug and nondrug conditions. The aversive drug effect may be associated with a saccharin solution tasted many hours earlier (Garcia, Hankins, Robinson, & Vogt, 1972); similarly, the discriminative response may be established at the time of peak drug effect, an hour or longer after onset of the drug stimulus. Important differences suggest, however, that an aversive, unconditional stimulus is not the general basis for drug discrimination. Drugs with peripheral but not central actions, such as methyl atropine, are ineffective as discriminative stimuli (Barry, 1974), whereas they are highly effective as aversive, unconditional stimuli (Berger, 1972). Effective aversive conditioning seems to require gastric distress as the unconditional stimulus, gustatory conditional signals (Garcia & Koelling, 1966; Frumkin, 1976), and also a novel drug as the aversive, unconditional stimulus (Cappell & Le Blanc, 1975). The importance of novelty corresponds to the situation for establishing a dissociative but not discriminative drug stimulus.

An opposite type of strong stimulus is a reinforcing or pleasurable effect, indicated by repetition of a response that produces the stimulus. This is a strong unconditional stimulus, usually attributable to relief or reduction of an aversive drive, such as hunger or pain (Dollard & Miller, 1950). The reinforcing effect of many drugs is demonstrated by their repeated voluntary consumption. Overton (1973) has pointed out a general correspondence between effectiveness of drugs as discriminative signals, in laboratory animals, and frequency of addictive abuse by humans. Strong effects on the CNS generally characterize both the discriminable and reinforcing drug effects.

There are important differences between the reinforcing and discriminative drug effects. Reinforcing drug effects have been demonstrated only with the technique of self-administration. The same drug may be reinforcing or punishing, depending on whether the animals controls its administration (Steiner, Beer, & Shaffer, 1969; Wise, Yokel, & deWit, 1976). An important characteristic for reinforcing drug effects is immediacy of onset, indicated by the greater effectiveness of intravenous than of slower-acting routes of administration. In contrast to reinforcing effects of drugs, discriminative effects have generally been trained at fairly long intervals and with the drug administered by the experimenter rather than by the animal.

Classical Conditioning

The reinforcing or punishing effect of a drug is a strong unconditional stimulus that evokes strong unconditional responses. These responses become associated with various weak stimuli that regularly precede the unconditional stimulus. The previously weak stimuli become conditional stimuli by the process of classical conditioning originally described by Pavlov (1927). The strong unconditional responses occur as conditional responses to the conditional stimuli, before or in the absence of the unconditional stimulus. A recent discussion of conditioning of drug reactions (Lynch, Fertziger, Teitelbaum, Cullen, & Gantt, 1973) gives evidence that conditional responses can be established only if they are mediated by the CNS.

The unconditional responses include a variety of physiological and behavioral reactions to the reinforcing or punishing stimulus. The conditional response may be a direct expression of the unconditional aversive or rewarding drug effect. An unconditional punishing drug effect becomes associated with a preceding weak stimulus when nausea to an emetic drug is evoked as a conditional response to the taste of a saccharin solution, experienced for the first time shortly before the drug adminstration. The conditional aversive response is indicated by avoidance of the saccharin solution. A conditional rewarding drug effect, such as relief from the illness of morphine withdrawal, may occur as a conditional response before the drug is injected. A placebo often has been reported to produce the relief or pleasure caused by a drug. Roffman, Reddy, and Lal (1973) reported that in rats a bell associated with morphine injections prevented or reversed the hypothermia associated with morphine withdrawal.

An opposite unconditional response to the drug is a compensatory or inhibitory reaction, counteracting the direct effects in order to minimize the disturbance of normal functioning. This is the basis for acquiring tolerance to the drug effect and can become a conditional response to a stimulus that is repeatedly followed by the drug administration. Roffman and Lal (1974) found that the rate of hexobarbital metabolism was increased by a signal that preceded a physiological condition (hypoxia) inducing increased rate of metabolism of the drug. The increased metabolism was a conditional compensatory response to the conditional stimulus (airflow) for the unconditional stimulus (hypoxia) for the unconditional compensatory or adjustive response of enhanced metabolism. Siegel (1975) found that after developing tolerance to the analgesic effect of morphine, rats responded to the signals for drug administration by increased sensitivity to the painful stimulation when placebo was substituted for morphine. A similar type of compensatory response to a drug may explain observations summarized by Wikler and Pescor (1967) that a long time after cessation of the unconditional withdrawal illness, conditional stimuli associated with previous use of morphine or another narcotic drug can elicit a conditional withdrawal reaction and thus induce craving for the drug effect.

The classical conditioning of unconditional drug effects cannot explain learning of discriminative responses to differential drug conditions. The direct drug effect, when evoked as a conditional stimulus without drug administration, obscures the differences between the drug and nondrug condition. The inhibitory or compensatory response to the drug further detracts from the distinctiveness and consistency of the drug effect as a discriminative stimulus.

Attributes of the Discriminative Stimulus

A drug is an unconditional stimulus. The unconditional response to the drug includes the various physiological and behavioral effects that are consistently elicited by a sufficiently high dose. When the drug functions as a discriminative stimulus, however, it is associated with a different unconditional stimulus, such as delivery of a food pellet or escape from painful shock. The strong reinforcing nature of this stimulus provides the incentive for instrumental learning, whereby differential responses are made to obtain food or to escape from shock under the drug and control conditions.

In this situation, the drug acts as a conditional stimulus. The differential responses associated with food or escape from shock under the drug and control conditions are conditional responses, replacing the unconditional responses to these conditions. Therefore, the original physiological and behavioral effects of the drug are subordinated to the new function of the drug as a signal for the differential responses.

A strong reinforcing or punishing drug effect would be expected to retard the discrimination of drug from control condition because the strong unconditional response to the drug interferes with the differential conditional responses. Pavlov (1927, p. 29) reported on the use of painful electric shock as a conditional stimulus for food presented shortly afterward to dogs. The unconditional responses to shock, such as struggling and barking, were completely replaced by the conditional response to food. If the shock were severe, however, it would prevent the conditioning. This was demonstrated by Williams and Barry (1966) in rats trained to press a lever for food pellets delivered only after variable intervals. Punishment of lever pressing by painful shocks had less inhibitory effect when the food accompanied the shocks (counterconditioning) rather than occurring at different times, but only when the shocks were fairly mild. Appel (1968) demonstrated similar counterconditioning of the punishing effect of shocks when accompanied by food reward.

It seems likely that the discriminative stimulus attributes of drugs are not the strongly reinforcing or punishing effects. Dollard and Miller (1950) made the distinction between strong and distinctive stimuli. Discriminative sensory signals, such as visual or auditory stimuli, are generally distinctive but weak. Similarly, certain attributes of painful electric shock or of strongly reinforcing or punishing drugs may provide distinctive but weak signals, suitable for functioning as conditional stimuli.

In comparison with sensory signals, including painful electric shock, drug effects are more stable, characterized by gradual onset and prolonged persistence. This attribute of the drug effect might be expected to retard discriminative learning. Sensory signals are effective discriminative stimuli only when the onset or other change of the signal coincides closely with the opportunity to make the discriminative response. The process of adaptation eliminates the response to a sensory stimulus that remains unchanged for a prolonged duration.

Drug effects may function as effective discriminative stimuli an hour or longer after administration. Tolerance to a drug develops more slowly and to a lesser degree than the corresponding process of adaptation to sensory stimuli. The persistence of the discriminative drug stimulus may be due to the stable, slowly changing nature of the drug effect. Another possible reason is pervasiveness of the drug effect throughout the CNS, in which sensory signals are more localized. The strong reinforcing or punishing effects of many drugs might retard tolerance and enhance persistence of all the responses to the drugs. This effect might include the discriminative stimulus attributes even if these are not strongly reinforcing or punishing.

DISCRIMINABILITY OF DRUGS

Each study of drug discrimination has usually been concerned with a particular training technique, dosage of the drug, and time interval from drug administration to training. All three types of variation, however, are represented in the diverse studies by different experimenters. Comparisons among the studies lead to some general conclusions about the effects of different training techniques, dosages, and time intervals.

Techniques for Training

The drug effect becomes a conditional stimulus by consistently preceding an unconditional stimulus, such as food reward or escape from painful shock. The unconditional response to the unconditional stimulus includes such anticipatory motor responses as choice of one direction of turn in a T-maze or one lever to press in a chamber. These anticipatory motor responses to the conditional stimulus are established by instrumental learning, reinforced or punished by differential unconditional stimuli. Classical conditioning causes these motor responses to become differential conditional responses associated with the differential drug and control conditions.

In the majority of studies on drug discrimination, the instrumental learning consists of associating differential but equivalent responses with the drug and nondrug conditions. Overton (1961, 1964) required opposite turns in a T-maze, and Stewart (1962) required choice of opposite compartments from a central one in a box, escape from shock being the incentive for the correct choice. In subsequent studies, the choice in the T-maze has also been trained by food reward for the correct response, punishing the incorrect response by no food

(Barry, Koepfer, & Lutch, 1965), or in addition by painful electric shock (Schechter & Rosecrans, 1971a). Differential but equivalent responses have also been trained in other test environments, such as the Lashley jumping stand (Brown, Feldman, & Moore, 1968) and a two-lever test chamber (Kubena & Barry, 1969a). Food reward was the incentive for the correct response in these studies, but shock escape has also been used as the incentive for learning to choose the correct lever (Krimmer, 1974b).

Another type of training procedure is to establish differential rates or patterns of correct responses under the drug and nondrug conditions. Harris and Balster (1968) used a fixed-ratio (FR) schedule, which results in high response rates, and a schedule of differential reinforcement for low rate (DRL). These schedules were associated with the differential drug and nondrug conditions, so that the animals learned different rates of pressing two alternative levers or else a single lever.

Different types of responses, which express different motivations, can also be associated with the drug and nondrug conditions. Conger (1951) trained rats to run to a food cup in a straight alley. Discrimination between the alcohol and saline conditions was demonstrated by establishing an avoidance response under one condition in which painful electric shock was administered at the food cup (alcohol for half the animals and saline for the other half). Kubena and Barry (1969a) applied this technique to a lever-pressing response. The opposite approach and avoidance responses provide a measure of differential drug effects on the motivations for obtaining food and avoiding pain. Barry and Krimmer (1972) used this technique to demonstrate that shock avoidance is attenuated by barbiturates and chlordiazepoxide but not by alcohol and chlorpromazine. Krimmer and Barry (1976) trained animals in a two-compartment box to make an active avoidance response under one condition (pentobarbital or saline) and a passive avoidance response under the other condition. This technique was more effective when the animals were trained immediately after intravenous infusion than when they were trained 10 min after intraperitoneal injection because the discriminative response was impaired by a general passive-response tendency at the longest interval.

The choice between differential but equivalent responses in a T-maze, in the studies by Overton (1966), was trained in 10 successive trials in close succession on the same day. Only the first trial of each day can be used as the measure of drug discrimination because in the subsequent trials the correct choice might be attributable to short-term memory of the consequence of the preceding choice. An influence of short-term memory is suggested by the fact that the percentage of correct choices rapidly increases in the successive trials. Nevertheless, the learning of the discriminative responses seems to be facilitated by the repeated correct choices, associated with the drug or control condition, in the later trials of the same day. The animal's learning in these trials therefore is not limited to the habit of repeating the immediately preceding correct choice regardless of the drug or control condition with which it is associated. This is consistent with the hypothesis that discrimination learning is influenced by all the relevant signals associated with the differential responses rather than the alternative hypothesis that the animal attends only to a single, dominant signal (Mackintosh, 1975). Our experience indicates that the discrimination learning is retarded with a procedure of only two or three trials in close succession per day. One disadvantage of this schedule is that even when the choice on the first trial is correct, a tendency for alteration or variation of response often causes the animal to make the incorrect choice on the second or third trial. Therefore, a small number of trials per day results in a high percentage of incorrect choices.

Drug discrimination is often trained by an operant conditioning procedure in which the animal has continual access to the alternative, equivalent levers. The measure of drug discrimination is limited to the choice responses preceding the first reinforcement of the session. An intermittent schedule of reinforcement, such as variable interval, allows a prolonged choice interval and a quantitative measure of number of presses on the alternative levers at the beginning of the session (Kubena & Barry, 1969a). The addition of a longer, variable interval of no reinforcement at the beginning of the session allows this initial test to be prolonged. Krimmer (1974b) used a 10-sec fixed-interval schedule in 5-min sessions, preceded by a variable interval (0–60 sec) without food reinforcement. An advantageous schedule might be differential reinforcement of low rates (DRL) on both levers (Huang & Ho, 1974) because tests without food reward are minimally differentiated from training sessions.

The fixed-ratio schedule of reinforcement establishes repeated, consistent choice of one of the alternative levers. Colpaert, Niemegeers, and Janssen (1976a) have reported training of 100% correct choice in many animals with this schedule. The disadvantage is that in tests of drug doses or other conditions intermediate between the training drug and control conditions, the animal tends to continue its exclusive selection of one of the choices. A schedule that encourages the animals to sample both choices, such as fixed or variable interval, may provide a more accurate measure of intermediate conditions.

After the first reinforcement of the session, the animal can learn to repeat the correct choice. Nevertheless, the association of the drug or control condition with the differential response apparently continues to be learned during intermittent reinforcement. A schedule of food reinforcement for each correct lever press has been reported to retard learning of the drug discrimination (Harris & Balster, 1971) probably because of the drastic change in the situation produced by the frequent delivery and consumption of the food pellets.

There is some evidence for differential effects of procedural variations on the discriminable drug stimulus. Figure 1.1 shows that the pentobarbital response generalized to a lower dosage of the same drug in animals trained to discriminate the drug (10 mg/kg) from saline with shock escape rather than food pellets as the unconditional stimulus. The pentobarbital response also generalized to sufficiently high doses of alcohol in the animals trained with

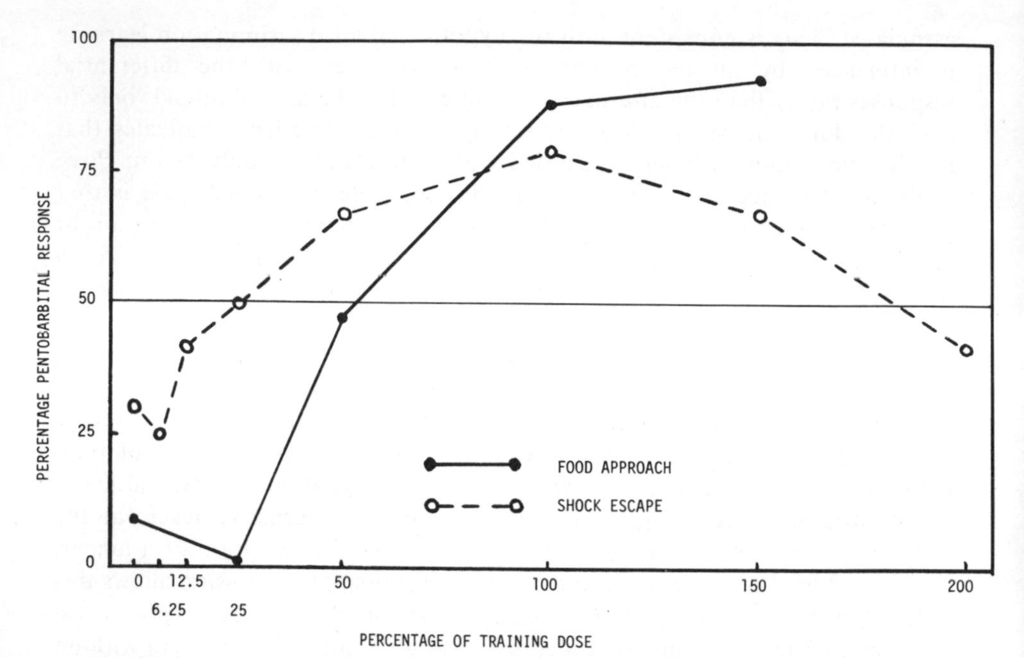

Figure 1.1. Percentage of pentobarbital response by groups of rats trained to discriminate pentobarbital (10 mg/kg) from saline with different techniques (food approach and shock escape) in tests under different doses, shown as percentage of training dose.

shock escape but not with food pellets. These comparisons were reported by Krimmer (1974b).

In general, remarkably consistent results have been obtained in studies of the same drugs with different unconditional stimuli, responses, and test situations. This fact, noted by Kubena and Barry (1969b) and by Barry (1974), gives evidence that the drug discrimination is not based on perception of the drug effect on a specific attribute of response, such as its coordination or vigor, which would be elicited only by certain test procedures.

In a study of monkeys trained to perform two different but equivalent responses under the drug and nondrug conditions, Trost and Ferraro (1974) demonstrated that drug discrimination was independent of drug effects on response rate. This does not exclude the possibility that the drug might cause other changes in performance not measured by response rate. Therefore, a more effective control for differential responses was an experiment by Barry (1968) in which food reward was associated with differential environmental conditions (lighted or dark) under the drug and nondrug condition. Rats were trained to press a lever that alternated their illumination condition (lighted or dark environment). The animals learned to select the illumination associated with food under their current drug or nondrug condition.

Different Dosages

Drug dosage determines not only the strength of the effects but also the distinctiveness of the discriminative stimulus characteristics. In general, a moderately high dose seems necessary to elicit the discriminative response, but not as high as the dose usually required for marked dissociation of responses from the nondrug condition when the animal is first exposed to the test situation in the drug condition. The early studies (Overton, 1966) were with doses of drugs so high that they would probably preclude responding for food reinforcement. Most of the subsequent studies reviewed by Barry (1974) have been with lower doses.

Direct relationships between increasing stimulus strength and increasing drug dosage are evident from dose-response curves reported by Overton (1974). He defined sessions to criterion as the beginning of a series of 10 consecutive training days in which the correct choice was made on the initial trial on at least 8 of the days. Table 1.1 summarizes the findings based on a different group of rats for each dosage of each drug trained in the T-maze shock-escape situation. Most of the drugs have a rather narrow range (less than threefold) between the dosages at which the discrimination is acquired slowly and rapidly. An exception is the antimuscarinic agent, scopolamine, which shows an almost 20-fold range. Dose-response curves for some additional drugs (Overton, 1969) indicate that for another antimuscarinic agent, atropine, the discrimination from saline is acquired slowly at a dose of 2.6 mg/kg and rapidly at about 20 mg/kg. This indicates much less potency than for scopolamine but also a wide range between the two dosage levels.

In a series of experiments (Barry & Krimmer, 1972; Krimmer, 1974b), dose-response functions for the training drugs and novel drugs indicated that the discriminability of a drug increases with the dosage. Above the optimal dosage

Table 1.1
Doses of Designated Drugs in mg/kg at which Discrimination from Saline Is Acquired Rapidly (10 Trials to Criterion) and Slowly (40 Trials to Criterion) according to Linear Functions Shown by Overton (1974)

	Dose (mg/kg) for acquisition of discrimination	
	Slowly	Rapidly
Alcohol	600	1000
Pentobarbital	3	8
Ketamine	6	12
Nicotine	.3	1
Scopolamine	.05	.90

level, behavioral toxicity is evidenced by a trend toward random responding or nonresponding at the highest doses. Figure 1.1 shows that under the training conditions (0 and 10 mg/kg of pentobarbital) the percentage of correct responses was higher with the food-approach than shock-escape procedure. An apparent greater sensitivity to pentobarbital with the shock-escape procedure is indicated by a higher percentage of drug responses at low dosages and by the deteriorating performance at higher dosages.

Barry (1974) has suggested a simplified method for calculating an ED_{50} (effective dose with 50% probability of eliciting the specified response) in tests with lower doses of a drug used for training the drug discrimination. Table 1.2 shows the percentage of the training dose at which the ED_{50} was found for several drugs. The measure shows some variations, which might be attributable to differences in training procedures, but the ED_{50} is fairly close to 50% of the training dose for each drug except for atropine. The generalization of the atropine response to much lower doses is consistent with the wide span of doses between slow and rapid acquisition of the discriminative response for the same drug (Overton, 1969) and for scopolamine, shown in Table 1.1.

Comparisons between groups of animals trained to discriminate different doses of a drug from the control condition might distinguish between two alternative possible attributes of the ED_{50}. One possible attribute is a constant dose indicating the minimal detectable drug effect regardless of the dose that the animal has been trained to discriminate from the control condition. The other possible attribute is a constant percentage of the training dose, intermediate between the training dose and no drug effect. These attributes might

Table 1.2
Percentage of Training Dose at which ED_{50} for Drug Response Occurs (ED_{50}%) in Designated Experiments

Drug	Training dose (mg/kg)	Reference	ED_{50} (% of training dose)
Pentobarbital	10	Barry and Krimmer (1972)	39
		Krimmer (1974b), Experiment 1	52
		Krimmer (1974b), Experiment 3	62
		Krimmer (1974b), Experiment 5	25
		Krimmer and Barry (1976)	35
Chlordiazepoxide	10	Krimmer (1974b), Experiment 4	67
	20	Krimmer and Barry (unpublished)	30
Alcohol	1200	Kubena and Barry (1969b)	47
	1000	Krimmer (1974b)	61
THC	4	Kubena and Barry (1972)	48
	5	Jarbe and Henriksson (1974)	34
	5	Bueno *et al.* (1976)	54
	2	Barry and Krimmer (1976)	41
Atropine	10	Kubena and Barry (1969b)	8

be combined or might characterize different drugs. In Table 1.2, the ED_{50} for chlordiazepoxide is about 7 mg/kg regardless of whether the animals were trained with 10 or 20 mg/kg, whereas the ED_{50} for Δ^9-tetrahydrocannabinol (THC) is close to 50% of the dosage at which the animals had been trained, whether the training was with 2, 4, or 5 mg/kg.

The data summarized in Table 1.2 were all obtained in rats, at 10–60 min after intraperitoneal (i.p.) injection. In the unpublished study by Krimmer and Barry (Table 1.2), the ED_{50} was lower (22% instead of 30% of the training dose) in a separate group of rats trained to discriminate chlordiazepoxide (20 mg/kg) from saline after oral administration. Krimmer and Barry (1976) found that the ED_{50} was higher (60% instead of 35%) at an earlier stage when the same rats were trained at 15 sec after intravenous (i.v.) infusion of 5 mg/kg instead of at 10 min after intraperitoneal injection of 10 mg/kg of pentobarbital.

Waters, Richards, and Harris (1972) reported evidence that the ED_{50} for a drug effect might be lower if animals are trained to discriminate a high dose from a very low dose of the same drug rather than from the nondrug condition. With amphetamine (2.5 mg/kg) as one of the training conditions, a test at 1 mg/kg elicited the correct response for this higher dose with 87% frequency in a group for which .3 mg/kg was the alternative condition and with only 56% frequency in a group for which saline was the alternative condition. A third group (0 versus .3 mg/kg amphetamine) made random choices for all doses including the training dosage, indicating that the low dose was a very weak discriminative stimulus. With a discrimination between two pentobarbital doses (2 and 10 mg/kg), Trost and Ferraro (1974) found that the ED_{50} for the high-dose response was substantially less than the median dose of 6 mg/kg for 2 of 3 rhesus monkeys. These findings may indicate that animals trained to discriminate two doses of the same drug learn to focus the discrimination on the quantitative dimension of magnitude of the drug effect, thus increasing sensitivity to effects of intermediate doses.

Time Intervals after Drug Administration

Drug discrimination has generally been trained and tested at what is believed to be the time of peak drug effect, at least 5 min and usually 15 min or longer after intraperitoneal administration. The few tests at different time intervals indicate rapid onset and long persistence of the drug stimulus.

Figure 1.2 summarizes data reported by Krimmer (1974b) on rats trained at 20 min after i.p. administration of pentobarbital (10 mg/kg) or saline. The maximal drug response in tests at the much shorter interval of 5 min, when drug absorption was probably incomplete, indicates that the short interval from the nondrug to drug condition probably compensated for that disadvantage. The results were similar for the different groups of animals trained with food reward and shock escape as the incentive. Similar peak discrimi-

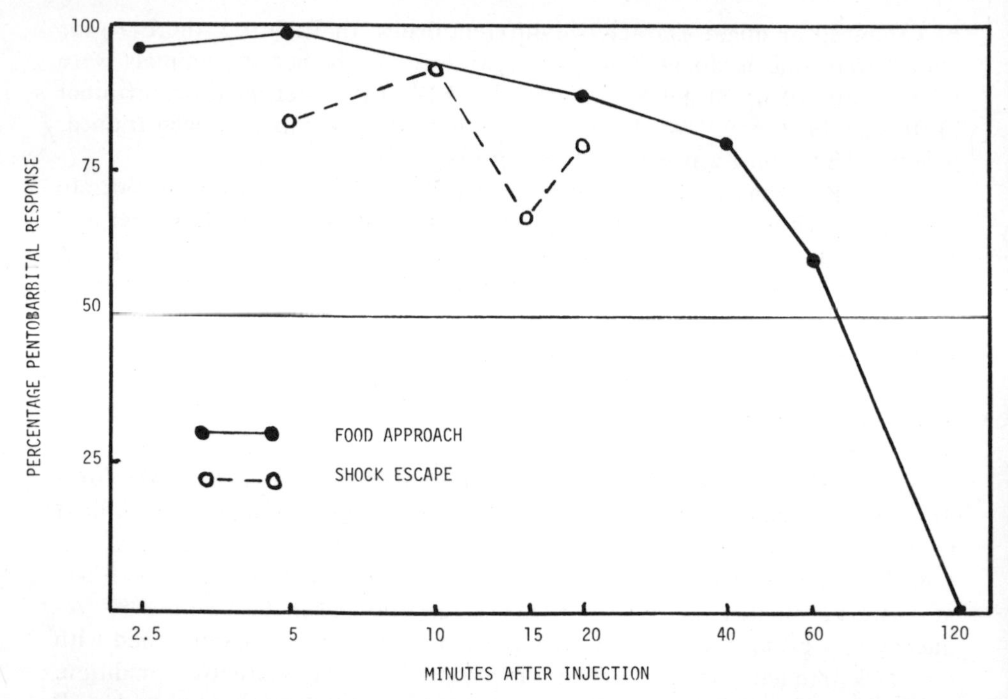

Figure 1.2. Percentage of pentobarbital response by the same two groups of rats shown in Figure 1.1 trained at 20 min after intraperitoneal injection, in tests at several time intervals.

nability in a test at 5 min was found for another group of animals trained to discriminate alcohol from saline at the same 20-min interval with food reward as the incentive (Krimmer, 1974b).

Persistence of the drug stimulus is indicated by the slow decline in drug response at longer intervals, shown in Figure 1.2. Drug responding remained about 50% at 1 hr after pentobarbital, whereas most behavioral effects of this moderately short-acting barbiturate would be expected to disappear before that interval. The alcohol stimulus seems to be even more persistent because the group trained to discriminate alcohol (1000 mg/kg) from saline (Krimmer, 1974b) continued to make predominantly the drug response in a test at 120 min. The linear, zero-order rate of alcohol metabolism (Wallgren & Barry, 1970, p. 44) might account for a more persistent effect, compared with the exponential rate of metabolism of pentobarbital and most other drugs.

An extremely short time for drug onset has been investigated by Krimmer and Barry (1976). They trained rats to discriminate pentobarbital (5 mg/kg) from saline at 15 sec after i.v. infusion through a chronically implanted venous cannula. Tests at longer time intervals showed that the drug stimulus persisted as long as 7.5 min after intravenous infusion of 5 mg/kg and 15 min after intraperitoneal injection of 10 mg/kg. The immediate stimulus induced by this drug was thereby shown to be qualitatively similar to the stimulus at substantially longer time intervals.

The discriminative-stimulus attributes of other drugs have been trained and tested at various time intervals. Schechter and Rosecrans (1971a) reported that rats trained at 5 min after administration of nicotine (0.4 mg/kg subcutaneously [s.c.]) or saline continued to make the nicotine response up to 20 min after injection but that they predominantly made the nondrug response at 30 min and longer intervals. Hirschhorn and Rosecrans (1974c) showed that in rats trained to discriminate nicotine (.2 and .4 mg/kg) from saline at the same 5-min interval, the drug response persisted longer, declining below chance after 80 min with .2 mg/kg and after 120 min with .4 mg/kg. Huang and Ho (1974) found that the amphetamine stimulus persisted for about 2 hr after .8 mg/kg, 5 hr after 4 mg/kg, and 8 hr after 8 mg/kg. Barry and Krimmer (1976) trained rats to discriminate 2 mg/kg of THC from vehicle at 30 min after i.p. injection. The drug response was present after 7.5 min and persisted for 2 hr but had disappeared at 4 hr. Ferraro, Gluck, and Morrow (1974), in a study of monkeys trained at 150 min after oral administration of 3 or 4 mg/kg of THC, found that the animals made the drug response in tests at all intervals between 90 and 330 min and the saline response before and after this time span. The later onset and longer duration of the drug stimulus might be explainable by the oral route of administration and the longer interval used in training.

RELATIONSHIPS AMONG DRUGS

Each drug might have a distinctive combination of effects, providing the basis for discriminating it from all other drugs. This is analogous to the individual's ability to discriminate a particular visual stimulus from all others on the basis of a unique combination of characteristics, such as size, brightness, and shape. This would constitute a high degree of specificity of the drug stimulus, so that the nondrug response would be elicited in tests with all other drugs or even with lower or higher doses of the same drug.

An alternative possibility is that the drug discrimination is based on a single preponderant action, so that the drug response is generalized to all other drugs with the same action. Many behavioral tests measure a single, general drug effect, such as depression or stimulation, so that the same change in performance is caused by a variety of different drugs. With high doses, many drugs have generalized toxic effects, such as loss of righting reflex or diminished intake of food and water. An extreme generalization of drug effects might consist of a single dissociated state, differentiating all drugs from the normal condition but not from each other.

In studies of animals trained to discriminate a drug from the control condition, the drug stimulus seems to be intermediate between these possible extremes of specificity and generalization. Several categories of drugs are differentiated from each other on the basis of their discriminative-stimulus effects. The number of categories seems to be fairly small, however, and the discriminative response learned with one drug generalizes to a number of other drugs with the same preponderant effect. Thus, discriminative responses seem to be learned on the basis of several fairly generalized effects of different drugs. Since most of the training has been between a drug and nondrug condition, it is possible that greater specificity could be established by training a discrimination between two drugs or two dosages of the same drug.

Sedative and Stimulant Agents

Sedatives have frequently been used as discriminative signals. Rats have been trained to discriminate pentobarbital (Overton, 1966), or alcohol (Kubena & Barry, 1969b), or chlordiazepoxide (Colpaert, Niemegeers, & Janssen, 1976; Colpaert, Desmedt, & Janssen, 1976) from the saline condition. The drug response is usually elicited by sufficiently high doses of any of the drugs in this general category of hypnotic–sedative agents. The specificity of the discriminative response trained with pentobarbital has been demonstrated by preponderantly nondrug responses in tests with various compounds (gallamine, lysergic acid diethylamide [LSD], physostigmine, chlorpromazine, and THC).

Within the broad category of sedatives or depressants, differential stimulus attributes have been demonstrated between pentobarbital and alcohol. Animals trained to discriminate pentobarbital from saline with a food-reward procedure make the saline response in tests with alcohol (Barry & Krimmer, 1972). Animals readily learn to discriminate the effects of pentobarbital (10 mg/kg) from alcohol (1000 mg/kg). At these doses, pentobarbital is the stronger discriminative stimulus, but a qualitative rather than merely quantitative difference has been indicated by tests with higher doses of both drugs (Krimmer & Barry, 1973) and by successful training of another group of rats to discriminate the same pentobarbital dose of 10 mg/kg from a higher alcohol dose of 1500 mg/kg (Krimmer, 1974b).

Table 1.3 summarizes the ED_{50} tested for three drugs (pentobarbital, chlordiazepoxide, and alcohol) in rats trained to discriminate one of these drugs from the saline condition. Some effects of different training procedures are suggested. The pentobarbital response generalized to alcohol in animals trained to press alternative levers for escape from shock (Krimmer, 1974b, Experiments 3 and 5) but not in animals trained to discriminate pentobarbital from saline by procedures involving food reward (Barry & Krimmer, 1972; Krimmer & Barry, 1973; Krimmer, 1974b, Experiment 1) or avoidance of painful shock by active or passive responses in a two-compartment box (Krimmer & Barry, 1976). In the groups trained to discriminate alcohol from the saline condition, a greater degree of generalization to the other two drugs (pentobarbital and chlordiazepoxide) after training with a lower dose is suggested

Table 1.3
Comparison of ED_{50} for Drug Response in Tests of Same Animals with Pentobarbital, Chlordiazepoxide, and Alcohol (Alc.) after Training with Designated Agent

Training drug	Dose (mg/kg)	Reference	ED_{50} (mg/kg)		
			Pentobarbitol	Chlordiazepoxide	Alcohol
Pentobarbital	10	Barry and Krimmer (1972)	3.9	2.5	—
		Krimmer (1974b), Experiment 1	5.2	3.8	—
		Krimmer (1974b), Experiment 3	6.2	4	790
		Krimmer (1974b), Experiment 5	2.5	1.3	845
		Krimmer and Barry (1976)	3.5	2.5	—
Chlordiazepoxide	10	Krimmer (1974b), Experiment 4	5.9	6.7	1300
	20	Krimmer and Barry (unpublished)	12.9	6.1	—
Alcohol	1200	Kubena and Barry (1969b)	6	5.4	566
	1000	Krimmer (1974b), Experiment 1	5.4	2.9	610

by lower ED_{50} values for these drugs in animals trained with 1000 mg/kg (Krimmer, 1974b, Experiment 1) than in a group trained with food reward (Kubena & Barry, 1969b).

The two groups trained to discriminate chlordiazepoxide from saline show some unique attributes. A group trained to discriminate chlordiazepoxide (10 mg/kg) from saline by pressing alternative levers for shock escape (Krimmer, 1974b, Experiment 4) was the only one with a higher ED_{50} for chlordiazepoxide than for pentobarbital. Incomplete generalization of the chlordiazepoxide response to the other drugs is suggested by the unusually high ED_{50} for alcohol in this group and by the unusually high ED_{50} for pentobarbital and the failure of generalization to alcohol in a group trained with a higher dose of chlordiazepoxide (20 mg/kg) to press alternative levers for food pellets (Krimmer & Barry, unpublished). Greater differentiation from the other drugs in animals trained with the higher chlordiazepoxide dose is further suggested by a report by Overton (1966) that the drug response trained with an even higher dose of chlordiazepoxide (30 mg/kg) was not dependably elicited by pentobarbital.

The i.p. route of administration was used in all the experiments with rats summarized in Table 1.3. In the study reported by Krimmer and Barry (1976), the same animals were trained at a prior stage at 15 sec after i.v. infusion of pentobarbital (5 mg/kg) or saline. In tests at 15 sec after i.v. infusion, the ED_{50} for this drug was 3.0 mg/kg, and the drug response generalized to low doses of chlordiazepoxide (ED_{50} = 1.6 mg/kg) and alcohol (ED_{50} = 248 mg/kg). In the unpublished study by Krimmer and Barry of rats trained to discriminate chlordiazepoxide (20 mg/kg) from the saline condition, another group was trained at the same time interval (60 min) after oral administration. Tests also with oral administration showed a lower ED_{50} for chlordiazepoxide (4.2 instead of 6.1 mg/kg) but a higher ED_{50} for pentobarbital (20 instead of 12.9 mg/kg).

Incomplete generalization from one sedative agent to another is suggested by other studies that used food reward for alternative, equivalent responses. Bueno, Carlini, Finkelfarb, and Suzuki (1976) found that in rats trained to discriminate alcohol (1200 mg/kg) from saline in a T-maze, the ED_{50} for alcohol was between 400 and 600 mg/kg, whereas the ED_{50} for pentobarbital (between 10 and 20 mg/kg) was relatively high. Colpaert *et al.* (1976b) trained rats to discriminate chlordiazepoxide (5 mg/kg) from saline, administered orally, in a lever-pressing response. The ED_{50} for the drug response was 2.76 mg/kg in tests with chlordiazepoxide (orally), lower than the ED_{50} of 2.97 mg/kg for a measure of ataxia (at the same 30-min interval after i.p. injection). In contrast, the ED_{50} was 11 mg/kg in tests with pentobarbital (orally), higher than the ED_{50} of 6.38 mg/kg for ataxia after i.p. injection.

In rats trained to discriminate pentobarbital from saline, the role of the CNS depressant effects in the discriminative drug stimulus has been studied by tests in which pentobarbital was combined with stimulants that might be expected to antagonize the depressant drug effect. The pentobarbital response has been shown to be effectively antagonized by sufficiently high doses of bemegride (Krimmer, 1974a; Overton, 1966) and by pentylenetetrazol (Krimmer, 1974a) but not by amphetamine, picrotoxin, or doxapram (Krimmer, 1974a). Similarly, in gerbils, Johansson and Jarbe (1975) found that the pentobarbital signal was antagonized by bemegride and pentylenetetrazol but not by cocaine. These findings suggest that the discriminative pentobarbital signal is based on a fairly specific component of the central sedative effect. Greater specificity of the alcohol stimulus is suggested by a further finding that the discriminative alcohol response was antagonized by bemegride but not by the other stimulant agents tested (Krimmer, 1974a).

In studies of some stimulants, a discriminative response has been trained with amphetamine or its isomers (Bueno *et al.*, 1976; Harris & Balster, 1968; Huang & Ho, 1974; Jones, Hill, & Harris, 1974; Kuhn, Appel, & Greenberg, 1974; Schechter & Rosecrans, 1973; Waters *et al.*, 1972). The amphetamine response was elicited by methylphenidate (Ritalin), cocaine, ephedrine, and norephedrine, while the saline response was generally elicited by mescaline, fenfluramine, LSD, atropine, nicotine, THC, and caffeine. There have also been reports, however, that the amphetamine response was elicited by THC (Bueno *et al.*, 1976) and by an extremely high dose (2 mg/kg) of nicotine (Huang & Ho, 1974). The discriminative stimulus depends on a change in the functioning of the CNS because *para*-hydroxyamphetamine, which has the peripheral but not central effects of *d*-amphetamine, was not discriminable from saline in animals trained to discriminate *d*-amphetamine or *l*-amphetamine from saline (Jones *et al.*, 1974).

Figure 1.3 portrays a possible two-dimensional structure for the relationship among various depressant and stimulant drugs, based on evidence reviewed by Barry (1974) and on further findings reported by Krimmer (1974b). The drugs are located on a dimension labeled relaxation–tension and on an independent dimension labeled disinhibition–passivity. Thus, the pentobarbital stimulus is shown as a combination of relaxation and disinhibition. The long line away from the nondrug point indicates a strong discriminative effect with moderate doses. In comparison with pentobarbital, alcohol is similar with respect to relaxation but dissimilar because it produces passivity instead of disinhibition. Chlordiazepoxide is similar to pentobarbital but is a weaker discriminative stimulus.

The drugs shown in Figure 1.3 are located on the basis of tests with animals trained to discriminate pentobarbital, alcohol, or chlordiazepoxide from saline. The stimulant antagonists are located on the basis of tests in which they are administered together with pentobarbital or alcohol. Therefore, the diagram is based on incomplete information, especially with regard to the stimulants, and it is undoubtedly oversimplified. The labels for the two dimensions refer to

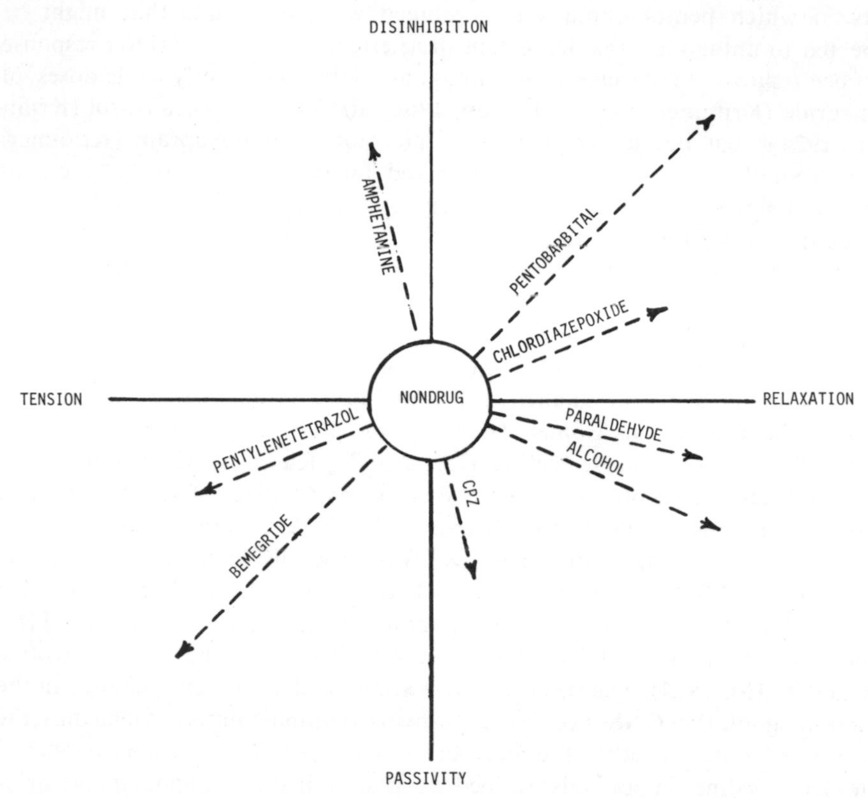

Figure 1.3. Diagram of differential positions of several depressant and stimulant drugs in two hypothetical dimensions of effect, one identified as a continuum from relaxation to tension, the other from passivity to disinhibition. Strength of discriminative stimulus effect of a moderate dose is indicated by distance from neutral nondrug point.

global concepts rather than to specific conditions or events in the CNS. Future research is likely to indicate additional dimensions of important stimulus characteristics.

In spite of these shortcomings, however, Figure 1.3 may help to summarize the positions of several prototypical drugs and to predict the positions of other drugs with respect to their discriminative signals. One implication of the model is that training with higher doses of one sedative drug should enhance differentiation from other drugs, as indicated by greater separation of the lines as they diverge from the common neutral point. Training with different dosages of chlordiazepoxide supports this hypothesis. The discriminative signals might converge, however, at doses high enough to cause the uniform condition of anesthesia.

Experimental manipulation of neurotransmitters may provide evidence about the central mechanisms for discriminative drug stimuli, leading to more specific classifications than the general concepts in Figure 1.3. Pretreatment with *para*-chlorophenylalanine (PCPA) to reduce brain 5-hydroxytryptamine (5-HT) levels or with *alpha*-methyltyrosine (AMT) to reduce brain catecholamine (CA) levels did not alter the discriminative response elicited by pentobarbital, thereby suggesting that these neurotransmitters are not involved in the pentobarbital discriminative action (Rosecrans, Goodloe, Bennett, & Hirschhorn, 1973). An absence of CA involvement is consistent with the failure of amphetamine to antagonize the pentobarbital response since amphetamine is a structural analog of the catecholamines and mimics many of their actions. In comparison with the pentobarbital stimulus, a differential feature of the alcohol stimulus is indicated by a report (Schechter, 1973a) that in animals trained to discriminate alcohol from saline, pretreatment with PCPA disrupted the drug response, while the saline response remained intact.

A finding that pretreatment with AMT disrupted the discriminative amphetamine response suggests involvement of the catecholamines (Kuhn *et al.*, 1974). Further evidence is that *d*-amphetamine is much more potent than *l*-amphetamine as a discriminative stimulus (Huang & Ho, 1974; Jones *et al.*, 1974; Schechter & Rosecrans, 1973) because these isomers are about equally potent as CNS stimulants but show a similar difference in potency measured by gnawing behavior, which is thought to be mediated by a central dopaminergic system (Jones *et al.*, 1974; Schechter & Rosecrans, 1973).

Morphine and Other Depressants

The discriminative-stimulus characteristics of morphine have been studied only recently (Hill, Jones & Bell, 1971; Hirschhorn & Rosecrans, 1974a,b; Rosecrans *et al.*, 1973). Morphine is discriminable from saline, but rather high doses (4.5 to 36 mg/kg) have been used. These doses are sufficient to produce a high degree of analgesia, but successful use of shock escape as an incentive and the reported faster reaction times for shock escape observed on morphine training days suggest that the analgesic property of morphine is not the discriminative signal (Hill *et al.*, 1971). The characteristically rapid development of tolerance to morphine may have accounted for the animals' failure to learn a discriminative response when training occurred in blocks of 18 consecutive daily sessions with one of the conditions (morphine, 9 mg/kg or saline). Discriminative responding was achieved when the conditions were alternated on consecutive training days or when the block-type training procedure was used for pentobarbital (20 mg/kg) and saline (Hill *et al.*, 1971). Another study indicates only limited reduction of the discriminative response learned with morphine (10 mg/kg) and saline following supplemental morphine administration to rats, but training continued during the chronic treatment, and it is possible the animals were able to learn a progressively more difficult discrimination (Hirschhorn & Rosecrans, 1974a).

The discriminative response for morphine is disrupted by naloxone, a narcotic antagonist, but not by pretreatment with AMT, a catecholamine depletor (Rosecrans *et al.*, 1973), nor by two 5-HT antagonists (cyproheptadine and methysergide) or atropine (Hirschhorn & Rosecrans, 1974b). An earlier study by Rosecrans *et al.* (1973) showed that pretreatment with PCPA, a serotonin depletor, did disrupt morphine discrimination. This finding suggests that some effects of PCPA other than 5-HT depletion might have disrupted the morphine discrimination, but the difference in incentive (shock escape in the early study and food reward in the later study) may also account for the different results.

The phenothiazines, another class of drugs with depressant actions, have produced ambiguous results and seem to have weak stimulus characteristics. Generally, animals trained to discriminate chlorpromazine (CPZ, 4 mg/kg, [i.p.] from saline made the CPZ response in tests with the phenothiazines (acepromazine and perphenazine) but not with prochlorperazine of the same class nor with imipramine (a dibenzazepine compound with a structure similar to that of CPZ). Animals trained with imipramine (20 mg/kg, i.p.) and saline did not make the drug response when tested with CPZ or acepromazine (Stewart, 1962).

Overton (1966) reported that CPZ (5 mg/kg, i.p.) was indiscriminable from no injection. Animals trained to discriminate the same dose of CPZ from pentobarbital (20 mg/kg, i.p.) made the CPZ choice when tested with no drug, indicating that no strong perceived effects were produced by CPZ in his T-maze shock-escape situation.

Barry, Steenberg, Manian, and Buckley (1974) used much lower doses of CPZ (initially 2 mg/kg and later reduced to 1 mg/kg, i.p.) in the single-lever shock-avoidance or food-approach procedure. Discrimination of the drug from saline developed slowly during prolonged training, indicating a weak discriminative stimulus property of CPZ. The drug response was elicited by a CPZ metabolite (7-hydroxy-CPZ), whereas the saline response was elicited by two other CPZ metabolites (7,8-dihydroxy-CPZ and 3,7-dihydroxy-CPZ) and also by quaternary CPZ, which produces the peripheral but not central effects of CPZ.

Cholinergic Agonists and Antagonists

Several drug discrimination studies have been on a nicotinic agonist (nicotine) and a muscarinic antagonist (atropine). Animals trained to discriminate nicotine from saline made the drug response in a test with lobeline, a nicotine agonist used as a smoking deterrent (Overton, 1969), but the nondrug response was made in tests with caffeine, pentobarbital, chlordiazepoxide, epinephrine, apomorphine, physostigmine, amphetamine (Morrison & Stephenson, 1969), and arecoline (Schechter & Rosecrans, 1972f). Therefore, a high degree of specificity is indicated for the discriminative nicotine stimulus. In animals trained to discriminate .4 mg/kg nicotine from saline, tolerance was indicated by diminution of the drug response after 20 doses of .4 mg/kg in 5 days (Schechter & Rosecrans, 1972c).

Nicotine stimulates peripheral and central ganglia, but several studies provide evidence for a central site of action for the discriminative stimulus actions of the drug. The discriminative nicotine response is disrupted by mecamylamine, a centrally active ganglionic blocker (Hirschhorn & Rosecrans, 1974c; Morrison & Stephenson, 1969; Overton, 1969; Schechter & Rosecrans, 1971a), but is not blocked by the peripherally acting ganglionic blocker chlorisondamine. Morrison and Stephenson (1969) cautioned that the rats may have been responding directly to some peripheral change caused by the central action of nicotine, which was not adequately mimicked by the other drugs tested.

Additional studies suggest the existence of differential nicotinic and muscarinic receptor sites within the CNS for discriminative learning. Pretreatment with atropine (an antimuscarinic) had no effect on the discriminable properties of nicotine (Hirschhorn & Rosecrans, 1974c; Schechter & Rosecrans, 1971b) but did inhibit the drug response trained with arecoline, a muscarinic agonist (Schechter & Rosecrans, 1972b). Pretreatment with mecamylamine disrupted the nicotine but not the arecoline response in animals trained to discriminate nicotine from arecoline (Schechter & Rosecrans, 1972a). These authors have also shown the involvement of brain biogenic amines with discrimination of nicotine from saline. The discrimination remained intact following 5-HT depletion by PCPA pretreatment. However, the nicotine response was significantly decreased following NE depletion induced by AMT pretreatment, suggesting that NE and not 5-HT is involved in the discriminative stimulus effects of nicotine (Schechter & Rosecrans, 1972d). Contrary to this conclusion, Hirschhorn and Rosecrans (1974c) found that AMT and also other adrenergic antagonists (propranolol and dibenamine) failed to affect the nicotine response.

Further evidence for a central site of the discriminative effect of nicotine is provided by Schechter (1973b). Shock escape in a three-compartment chamber was used to train rats to discriminate nicotine from saline. Nicotine (.4 mg/kg) or saline was administered s.c. to some animals, while others were trained with nicotine (.011 mg per rat) or saline administered intraventricularly. Both groups of animals maintained the discriminative responding when nicotine was administered by the alternative administration route and dose.

Atropine and other cholinergic antagonists have also been used as training drugs (Overton, 1966, 1969; Kubena & Barry, 1969b). Overton (1966) trained animals to discriminate atropine (150 mg/kg, i.p.) from saline. The high dose of atropine strongly suggests CNS involvement. This is further substantiated in tests that show the drug response was elicited by other antimuscarinic agents (scopolamine, homatropine HBr, and cyclopentolate) that cross the blood–brain barrier but not by quaternary compounds (atropine methyl bromide and atropine methyl nitrate) that do not readily enter the CNS. High doses of atropine have a depressant action (Innes & Nickerson, 1975), but attempts to mimic this effect with pentobarbital (5–20 mg/kg) failed to elicit the atropine response, and pentobarbital did not antagonize the response to atropine (Overton, 1966).

Kubena and Barry (1969b) used a much lower dose of atropine (10 mg/kg, i.p.) and saline to train a discriminative response with an approach–avoidance procedure. The drug response generalized to scopolamine but not to the quaternary atropine methyl bromide.

Marihuana and Hallucinogens

The drugs in this category produce mood changes and disturbances in thinking. The discriminative stimulus attributes indicate a clear-cut distinction between marihuana and such other drugs, as LSD, mescaline, and psilocybin, which give rise to predominantly visual hallucinations at low doses (Barry, 1974; Winter, 1974b).

The discriminative marihuana stimulus in laboratory animals has generally been tested by using THC, considered to be the major psychoactive constituent of marihuana. The initial demonstrations were by Kubena and Barry (1972), Henriksson and Jarbe (1972), and Bueno and Carlini (1972). A highly specific marihuana stimulus was indicated by Kubena and Barry (1972), who found that in rats trained to discriminate THC (4 mg/kg) from its vehicle, the drug response was elicited by marihuana-extract distillate and by crude marihuana extract but not by various other pharmacological agents, including several hallucinogens (mescaline, LSD, psilocybin). More detailed information was given by Barry and Kubena (1972). Jarbe and Henriksson (1974) likewise found a high degree of specificity of the THC stimulus. Bueno *et al.* (1976) found that the THC response was not elicited by amphetamine, although in another group of rats the amphetamine response was elicited by THC.

Barry and Krimmer (1976) trained rats with a lower dose of THC (2 mg/kg, i.p.) and elicited the drug response in tests with oral (p.o.) and i.v. routes of administration. Ferraro *et al.* (1974) trained rhesus macaque monkeys with an oral dose of THC and thereby expanded these studies of drug discrimination beyond the species (laboratory rats) on which most of the studies are based.

There is evidence that tolerance develops less to the discriminative THC stimulus than to other effects of this drug. Jarbe and Henriksson (1973) showed that in rats trained in a T-shaped water maze sensitivity to the discriminative effect of THC was only slightly decreased by tolerance to THC. Bueno and Carlini (1972) demonstrated tolerance to repeated THC injections by measuring rats' performance of a rope-climbing test and then training these tolerant rats to discriminate THC (1 mg/kg) from its vehicle with food reward for pressing the correct lever. These animals achieved 70–80% correct choices, indicating that while tolerance had developed to one behavioral task, some THC effect could still be perceived. Hirschhorn and Rosecrans (1974b) reported limited tolerance to the discriminative THC stimulus (4 mg/kg) in a two-lever food-rewarded task. They established tolerance by using higher doses, for longer chronic treatment, than in previous studies during discrimination training.

In common with THC, hallucinogens produce mood changes and disturbances in thinking; in addition, they cause delusions and predominantly visual hallucinations at low doses. Greenberg, Kuhn, and Appel (1975) trained rats to discriminate the extremely low LSD dose of .01 mg/kg from saline, following prior training to discriminate .08 mg/kg from saline. Overton (1969) demonstrated discriminative stimulus effects of Ditran, a potent anticholinergic with hallucinogenic properties. The number of trials necessary to train the response increased with doses below 15 mg/kg but likewise increased with higher doses, suggesting that toxic drug effects (seizures, convulsions) interfered with learning.

Different groups of animals have been trained to discriminate LSD or mescaline from saline, whereas another group failed to discriminate between dosages of these two drugs estimated to be equivalent from the data on animals trained to discriminate one of the drugs from saline (Hirschhorn & Winter, 1971). In another study, the LSD response was elicited in tests with mescaline and psilocybin but not with amphetamine (Schechter & Rosecrans, 1972e). These results suggest that qualitatively similar characteristics can be produced by equivalently potent doses of LSD, mescaline, and psilocybin. Cameron and Appel (1973) reported that animals trained to discriminate LSD from saline made a partial drug response in a test with THC.

Winter (1973, 1974a) has compared the stimulus properties of mescaline with 2,3,4-trimethoxyphenylethylamine (2,3,4-TMPEA), a structural isomer of mescaline, and has presented a cautious case for using the discriminative properties of hallucinogens as a means of identifying such drugs preclinically. Mescaline and 2,3,4-TMPEA were discriminated from saline but not from each other, showing that the rats were using similar characteristics of both drugs as discriminative stimuli. Hallucinogenic effects in humans are caused by mescaline but not by 2,3,4-TMPEA. If the same differential effects apply to rats, the discriminable stimulus for mescaline and 2,3,4-TMPEA is probably a drug effect other than hallucinogenic activity. DMPEA, another nonhallucinogenic structural isomer of mescaline about which much more is known, is discriminable from mescaline (Winter, 1974a,b).

SUMMARY

Drugs have distinctive effects that enable them to become signals for differential motor responses. Such unconditional stimuli as food pellets or escape from painful shock become associated with the motor responses by instrumental learning and with the drug condition by classical conditioning.

The discriminative drug signal is probably not based primarily on a strong unconditional rewarding or punishing effect because the conditional drug stimulus is subordinated to the different unconditional stimulus. The slow change in the drug effect precludes the sudden onset that characterizes most sensory signals of effective conditional stimuli. The ease with which drug effects can be established as discriminative signals suggests, therefore, that the stability and pervasiveness of the drug effect provide important advantages.

An effect on the CNS seems to be a necessary attribute of the discriminative drug stimulus. Amnesic dissociation between drug and nondrug states seems to be limited to situations in which the drug state is novel and therefore does not explain the discrimination learning.

The usual unconditional stimulus for drug discrimination is food reward or shock escape. The drug and nondrug conditions are most often associated with differential but equivalent responses, such as pressing different levers in a chamber or turning in opposite directions in a T-maze. In some experiments, the alternative responses have been qualitatively different, such as approach and avoidance, or active and passive avoidance.

Tests at different time intervals indicate a more rapid onset and longer duration of action of the drug signal than is seen for behavioral effects. A moderately high dosage seems necessary for the discriminative signal, but in most studies the training has been with rather high dosages, and there is evidence that for some drugs the training dosage partly determines the sensitivity to effects of lower doses.

When discrimination has been established between the drug and nondrug condition, tests with other drugs establish the degree to which the drug response is specific to that drug or generalizes to other drugs. A substantial degree of specificity is indicated by the existence of several categories of drugs that are differentiated from each other. These categories include (*1*) sedatives and "minor tranquilizers" (barbiturates, benzodiazepines, meprobamate, chloral hydrate, alcohol), generally counteracted by the stimulant bemegride; (*2*) narcotic analgesic (morphine), antagonized by naloxone or nalorphine; (*3*) "major tranquilizer" (chlorpromazine); (*4*) muscarinic agonist (arecoline), blocked by atropine; (*5*) nicotinic agonist (nicotine), blocked by mecamylamine; (*6*) hallucinogens (LSD, mescaline); (*7*) marihuana (THC and some of its metabolites and analogs), (*8*) stimulants (amphetamine, methylphenidate, cocaine).

Greater specificity of the drug stimulus might be demonstrated by training a discrimination between two drugs in the same category, as has been done with pentobarbital and alcohol. Two dimensions of variation from the nondrug condition (disinhibition and relaxation) have been proposed to account for differences among the sedative agents. These dimensions constitute a speculative inference about generalized emotional conditions that may be associated with the pervasive, generalized effects of these drugs on the CNS.

REFERENCES

Appel, J. B. Association of aversive and reinforcing stimuli during intermittent punishment. *Psychol. Rep.*, 1968, *22*, 267–271.

Barry, H., III. Prolonged measurements of discrimination between alcohol and nondrug states. *J. Comp. Physiol. Psychol.*, 1968, *65*, 349–352.

Barry, H., III. Classification of drugs according to their discriminable effects in rats. *Federation Proc.*, 1974, *33*, 1814–1824.

Barry, H., III, Koepfer, E., & Lutch, J. Learning to discriminate between alcohol and nondrug condition. *Psychol. Rep.*, 1965, *16*, 1072.

Barry, H., III, & Krimmer, E. C. Pentobarbital effects perceived by rats as resembling several other depressants but not alcohol. *Proc. 80th Annual Convention*, Amer. Psychol. Assoc., 1972, *7*, 849–850.

Barry, H., III, & Krimmer, E. C. Discriminative Δ^9-tetrahydrocannabinol stimulus tested with several doses, routes, intervals, and a marihuana extract. In M. C. Braude & S. Szara (Eds.), *The pharmacology of marihuana*. Vol. 2. New York: Raven, 1976. Pp. 535–538.

Barry, H., III, & Kubena, R. K. Discriminative stimulus characteristics of alcohol, marihuana and atropine. In J. M. Singh, L. H. Miller, & H. Lal (Eds.), *Drug addiction*. Vol. 1. *Experimental pharmacology*. Mt. Kisko, New York: Futura, 1972. Pp. 3–16.

Barry, H., III, Steenberg, M. L., Manian, A. A., & Buckley, J. P. Effects of chlorpromazine and three metabolites on behavioral responses in rats. *Psychopharmacologia*, 1974, *34*, 351–360.

Berger, B. D. Conditioning of food aversions by injections of psychoactive drugs. *J. Comp. Physiol. Psychol.*, 1972, *81*, 21–26.

Bindra, D., & Reichert, H. The nature of dissociation: Effects of transitions between normal and barbiturate-induced states on reversal learning and habituation. *Psychopharmacologia*, 1967, *10*, 330–344.

Bliss, D. K. Theoretical explanations of drug-associated behaviors. *Federation Proc.* 1974, *33*, 1787–1796.

Brown, A., Feldman, R. S., & Moore, J. W. Conditional discrimination learning based upon chlordiazepoxide. *J. Comp. Physiol. Psychol.*, 1968, *66*, 211–215.

Bueno, O. F. A., & Carlini, E. A. Dissociation of learning in marihuana tolerant rats. *Psychopharmacologia*, 1972, *25*, 49–56.

Bueno, O. F. A., Carlini, E. A., Finkelfarb, E., & Suzuki, J. S. Δ^9-Tetrahydrocannabinol, ethanol, and amphetamine as discriminative stimuli—generalization tests with other drugs. *Psychopharmacologia*, 1976, *46*, 235–243.

Cameron, O. G., & Appel, J. B. A behavioral and pharmacological analysis of some discriminative properties of d-LSD in rats. *Psychopharmacologia*, 1973, *33*, 117–134.

Cappell, H., & LeBlanc, A. E. Conditioned aversion by amphetamine: Rates of acquisition and loss of the attenuating effects of prior exposure. *Psychopharmacologia*, 1975, *43*, 157–162.

Chute, D. L., & Wright, D. C. Retrograde state dependent learning. *Science*, 1973, *180*, 878–880.

Colpaert, F. C., Desmedt, L. K. C., & Janssen, P. A. J. Discriminative stimulus properties of benzodiazepines, barbiturates and pharmacologically related drugs; relation to some intrinsic and anticonvulsant effects. *European J. Pharmacol.*, 1976, *37*, 113–124. (b)

Colpaert, F. C., Niemegeers, C. J. E., & Janssen, P. A. J. Theoretical and methodological considerations on drug discrimination learning. *Psychopharmacologia*, 1976, *46*, 169–177. (a)

Conger, J. J. The effects of alcohol on conflict behavior in the albino rat. *Quart. J. Studies Alc.*, 1951, *12*, 1–29.

Dollard, J., & Miller, N. E. *Personality and psychotherapy*. New York: McGraw-Hill, 1950.

Ferraro, D. P., Gluck, J. P., & Morrow, C. W. Temporally-related stimulus properties of Δ^9-tetrahydrocannabinol in monkeys. *Psychopharmacologia*, 1974, *35*, 305–316.

Frumkin, K. Differential potency of taste and audiovisual stimuli in the conditioning of morphine withdrawal in rats. *Psychopharmacologia*, 1976, *46*, 245–248.

Garcia, J., Hankins, W. G., Robinson, J. H., & Vogt, J. L. Bait shyness: Tests of CS–US mediation. *Physiol. Behav.*, 1972, *8*, 807–810.

Garcia, J., & Koelling, R. A. Relation of cue to consequence in avoidance learning. *Psychon. Sci.*, 1966, *4*, 123–124.

Goodwin, D. W. Alcoholic blackout and state-dependent learning. *Federation Proc.*, 1974, *33*, 1833–1835.

Greenberg, I., Kuhn, D. M., & Appel, J. B. Behaviorally induced sensitivity to the discriminable properties of LSD. *Psychopharmacologia*, 1975, *43*, 229–232.

Harris, R. T., & Balster, R. L. Discriminative control by dl-amphetamine and saline of lever choice and response patterning. *Psychon. Sci.*, 1968, *10*, 105–106.

Harris, R. T., & Balster, R. L. An analysis of the function of drugs in the stimulus control of operant behavior. In T. Thompson & R. Pickens (Eds.), *Stimulus properties of drugs*. New York: Appleton, 1971. Pp. 111–132.

Henriksson, B. G., & Jarbe, T. U. C. Δ^9-Tetrahydrocannabinol used as discriminative stimulus for rats in position learning in a T-shaped water maze. *Psychon. Sci.*, 1972, *27*, 25–26.

Hill, H. E., Jones, E., & Bell, E. C. State dependent control of discrimination by morphine and pentobarbital. *Psychopharmacologia*, 1971, *221*, 305–313.

Hirschhorn, I. D., & Rosecrans, J. A. Morphine and Δ^9-tetrahydrocannabinol: Tolerance to the stimulus effects. *Psychopharmacologia*, 1974, *36*, 243–253. (a)

Hirschhorn, I. D., & Rosecrans, J. A. A comparison of the stimulus effects of morphine and lysergic acid diethylamide (LSD). *Pharmacol. Biochem. Behav.*, 1974, *2*, 361–366. (b)

Hirschhorn, I. D., & Rosecrans, J. A. Studies on the time course and the effect of cholinergic and adrenergic receptor blockers on the stimulus effect of nicotine. *Psychopharmacologia*, 1974, *40*, 109–120. (c)

Hirschhorn, I. D., & Winter, J. C. Mescaline and lysergic acid diethylamide (LSD) as discriminative stimuli. *Psychopharmacologia*, 1971, *22*, 64–71.

Huang, J. T., & Ho, B. T. Discriminative stimulus properties of d-amphetamine and related compounds in rats. *Pharmacol. Biochem. Behav.*, 1974, *2*, 669–673.

Innes, I. R., & Nickerson, M. Atropine, scopolamine, and related antimuscarinic drugs. In L. Goodman & A. Gilman (Eds.), *The pharmacological basis of therapeutics*. (5th ed.) New York: Macmillan, 1975, Pp. 514–532.

Jarbe, T. U. C., & Henriksson, B. G. Open-field behavior and acquisition of discriminative response control in Δ^9-THC tolerant rats. *Experientia*, 1973, *29*, 1251–1253.

Jarbe, T. U. C., & Henriksson, B. G. Discriminative response control produced with hashish, tetrahydrocannabinols (Δ^8-THC and Δ^9-THC), and other drugs. *Psychopharmacologia*, 1974, *40*, 1–16.

Johansson, J. O., & Jarbe, T. U. C. Antagonism of pentobarbital induced discrimination in the gerbil. *Psychopharmacologia*, 1975, *41*, 225–228.

Jones, C. N., Hill, H. F., & Harris, R. T. Discriminative response control by d-amphetamine and related compounds in the rat. *Psychopharmacologia*, 1974, *36*, 347–356.

Krimmer, E. C., Selective antagonism of discriminable properties of pentobarbital by several stimulants. *Federation Proc.*, 1974, *33*, 550. (a)

Krimmer, E. C. Drugs as discriminative stimuli. *Dissertation Abstr. Internat.*, 1974, *35*, 4572-B.(b)

Krimmer, E. C., and Barry, H., III. Differential stimulus characteristics of alcohol and pentobarbital in rats. Proc. 81st Ann. Convention Amer. Psychol. Assoc., 1973, *8*, 1005–1006.

Krimmer, E. C., & Barry, H., III. Discriminative pentobarbital stimulus in rats immediately after intravenous administration. *European J. Pharmacol.*, 1976, *38*, 321–327.

Kubena, R. K., & Barry, H., III. Two procedures for training differential responses in alcohol and nondrug conditions. *J. Pharm. Sci.*, 1969, *58*, 99–101. (a)

Kubena, R. K., & Barry, H., III. Generalization by rats of alcohol and atropine stimulus characteristics to other drugs. *Psychopharmacologia*, 1969, *15*, 196–206. (b)

Kubena, R. K., & Barry, H., III. Stimulus characteristics of marihuana components. *Nature* 1972, *235*, 397–398.

Kuhn, D. M., Appel, J. B., & Greenberg, I. An analysis of some discriminative properties of d-amphetamine. *Psychopharmacologia*, 1974, *39*, 57–66.

Lynch, J. J., Fertziger, A. P., Teitelbaum, H. A., Cullen, J. W., & Gantt, W. H. Pavlovian conditioning of drug relations: Some implications for problems of drug addiction. *Conditional Reflex*, 1973, *8*, 211–233.

Mackintosh, N. J. A theory of attention: Variations in the associability of stimuli with reinforcement. *Psychol. Rev.*, 1975, *82*, 276–298.

Morrison, C. F., & Stephenson, J. A. Nicotine injection as the conditioned stimulus in discrimination learning. *Psychopharmacologia*, 1969, *15*, 351–360.

Overton, D. A. Discriminative behavior based on the presence or absence of drug effects. *Amer. Psychologist*, 1961, *16*, 453.

Overton, D. A. State-dependent or "dissociated" learning produced with pentobarbital. *J. Comp. Physiol. Psychol.*, 1964, *57*, 3–12.

Overton, D. A. State-dependent learning produced by depressant and atropine-like drugs. *Psychopharmacologia*, 1966, *10*, 6–31.

Overton, D. A. Dissociated learning in drug states (state-dependent learning). In D. H. Efron, J. O. Cole, J. Levine, & J. R. Wittenborn (Eds.), *Psychopharmacology: A review of progress, 1957–1967*. Public Health Service Publ. no. 1836. Washington, D.C.: U.S. Govt. Printing Office, 1968. Pp. 918–930.

Overton, D. A. Control of T-maze choice by nicotine, antinicotine, and antimuscarinic drugs. *Proc. 77th Ann. Convention, Amer. Psychol. Assoc.*, 1969, *4*, 869–870.

Overton, D. A. State-dependent learning produced by addicting drugs. In S. Fisher & A. M. Freedman (Eds.), *Opiate addiction: Origin and treatment*. Washington, D.C.: V. H. Winston, 1973. Pp. 61–75.

Overton, D. A. Experimental methods for the study of state-dependent learning. *Federation Proc.*, 1974, *33*, 1800–1813.

Pavlov, I. P. *Conditioned reflexes* (C. V. Anrep, trans.). London: Oxford Univ. Press, 1927.

Roffman, M., & Lal, H. Stimulus control of hexobarbital narcosis and metabolism in mice. *J. Pharmacol. Exp. Therap.*, 1974, *191*, 358–369.

Roffman, M., Reddy, C., & Lal, H. Control of morphine-withdrawal hypothermia by conditional stimuli. *Psychopharmacologia*, 1973, *29*, 197–201.

Rosecrans, J. A., Goodloe, M. H., Jr., Bennett, G. J., & Hirschhorn, I. D. Morphine as a discriminative cue: Effects of amine depletors and naloxone. *European J. Pharmacol.*, 1973, *21*, 252–256.

Schechter, M. D. Ethanol as a discriminative cue: Reduction following depletion of brain serotonin. *European J. Pharmacol.*, 1973, *24*, 278–281. (a)

Schechter, M. D. Transfer of state-dependent control of discriminative behaviour between subcutaneously and intraventricularly administered nicotine and saline. *Psychopharmacologia*, 1973, *32*, 327–335. (b)

Schechter, M. D., & Rosecrans, J. A. C.N.S. effect of nicotine as the discriminative stimulus for the rat in a T-maze. *Life Sci.*, 1971, *10*, Part I, 821–832. (a)

Schechter, M. D., & Rosecrans, J. A. Behavioral evidence for two types of cholinergic receptors in the CNS. *European J. Pharmacol.*, 1971, *15*, 375–378. (b)

Schechter, M. D., & Rosecrans, J. A. Effects of mecamylamine on discrimination between nicotine- and arecoline-produced cues. *European J. Pharmacol.*, 1972, *17*, 179–182. (a)

Schechter, M. D., & Rosecrans, J. A. Atropine antagonism of arecoline-cued behavior in the rat. *Life Sci.*, 1972, *11*, Part I, 517–523. (b)

Schechter, M. D., & Rosecrans, J. A. Behavioral tolerance to an effect of nicotine in the rat. *Arch. Intern. Pharmacodyn.*, 1972, *195*, 52–56. (c)

Schechter, M. D., & Rosecrans, J. A. Nicotine as a discriminative stimulus in rats depleted of norepinephrine or 5-hydroxytryptamine. *Psychopharmacologia,* 1972, *24,* 417–429. (d)

Schechter, M. D., & Rosecrans, J. A. Lysergic acid diethylamide (LSD) as a discriminative cue: Drugs with similar stimulus properties. *Psychopharmacologia,* 1972, *26,* 313–316. (e)

Schechter, M. D., & Rosecrans, J. A. Nicotine as a discriminative cue in rats: Inability of related drugs to produce a nicotine-like cueing effect. *Psychopharmacologia,* 1972, *27,* 379–387. (f)

Schechter, M. D., & Rosecrans, J. A. D-Amphetamine as a discriminative cue: Drugs with similar properties. *European J. Pharmacol.,* 1973, *21,* 212–216.

Siegel, S. Evidence from rats that morphine tolerance is a learned response. *J. Comp. Physiol. Psychol.,* 1975, *89,* 498–506.

Steiner, S. S., Beer, B., & Shaffer, M. M. Escape from self-produced rates of brain stimulation. *Science* 1969, *163,* 90–91.

Stewart, J. Differential responses based on the physiological consequences of pharmacological agents. *Psychopharmacologia,* 1962, *3,* 132–138.

Trost, J. G., & Ferraro, D. P. Discrimination and generalization of drug stimuli in monkeys. In J. M. Singh & H. Lal (Eds.), Drug addiction. Vol. 3. *Neurobiology and influences on behavior.* New York: Stratton Intercontinental Medical Book Corp., 1974. Pp. 223–239.

Wallgren, H., & Barry, H., III. Actions of alcohol. Vol. 1, *Biochemical, Physiological, and Psychological Aspects.* Amsterdam, Elsevier, 1970.

Waters, W. H., Richards, D. W., III, & Harris, R. T. Discriminative control and generalization of the stimulus properties of d,l-amphetamine in the rat. In J. M. Singh, L. Miller, & H. Lal (Eds.), *Drug addiction.* Vol. 1. *Experimental pharmacology.* Mount Kisco, New York: Futura, 1972, Pp. 87–89.

Wikler, A., & Pescor, F. T. Classical conditioning of a morphine abstinence phenomenon, reinforcement of opioid-drinking behavior and "relapse" in morphine-addicted rats. *Psychopharmacologia,* 1967, *10,* 255–284.

Williams, D. R., & Barry, H., III. Counter conditioning in an operant conflict situation. *J. Comp. Physiol. Psychol.,* 1966, *61,* 154–156.

Winter, J. C. A comparison of the stimulus properties of mescaline and 2,3,4-trimethoxyphenylethylamine. *J. Pharmacol. Exp. Therap.,* 1973, *185,* 101–107.

Winter, J. C. The effects of 3,4-dimethoxyphenylethylamine in rats trained with mescaline as a discriminative stimulus. *J. Pharmacol. Exp. Therap.,* 1974, *189,* 741–747. (a)

Winter, J. C. Hallucinogens as discriminative stimuli. *Federation Proc.,* 1974, *33,* 1825–1832. (b)

Wise, R. A., Yokel, R. A., & deWit, H. Both positive reinforcement and conditioned aversion from amphetamine and from apomorphine in rats. *Science,* 1976, *191,* 1273–1275.

Wright, D. C. Differentiating stimulus and storage hypotheses of state-dependent learning. *Federation Proc.,* 1974, *33,* 1797–1799.

Wright, D. C., & Chute, D. L. State dependent learning produced by post trial intrathoracic administration of sodium pentobarbital. *Psychopharmacologia,* 1973, *31,* 91–94.

I

Mechanism of Drugs as Discriminative Stimuli

2

Current Trends in the Study of Drugs as Discriminative Stimuli

LeRoy G. Frey and J. C. Winter
Department of Pharmacology and Therapeutics
School of Medicine
State University of New York at Buffalo

Throughout its brief history, pharmacology has been a uniquely hybrid discipline. In the search for an understanding of drug action, it has proven useful to combine the elements of many branches of science. One of the more recent combinations is the union of pharmacology with experimental psychology to yield what is now called behavioral pharmacology. Although its roots are in the work of Lashley and Skinner, behavioral pharmacology did not bloom fully until the discovery, less than 25 years ago, of drugs useful in the treatment of disturbed human behavior.

As might be expected, the way of behavioral pharmacology has not always been smooth. Since the first reports by Sidman (1955) and Dews (1955) of the use of operant behavioral techniques in the study of drug action, there have been numerous calls for more truly interdisciplinary research. Too often the interaction of drugs and behavior has been studied by pharmacologists who had little appreciation or understanding of the well-established principles of the analysis of behavior or by psychologists who were essentially untrained in the principles of drug action. Even in those instances in which the techniques and principles of psychology and pharmacology have been skillfully combined, the results have not always been entirely satisfying. This, we believe, is due not to any fundamental error in research strategy but to the fact that the merger of pharmacology and psychology has opened an area of investigation in which the variables are more adequately described by the product of the parent variables than by their sum.

The observation, 25 years ago, by Conger (1951) that rats could be trained to respond differentially following the injection of ethanol and saline had little immediate impact on the development of behavioral pharmacology. Only quite recently, following the pioneering work of Barry and of Overton, have a significant number of pharmacologists and psychologists sensed the potential importance of this line of investigation. This delayed recognition has in it a virtue; we can benefit from lessons already learned: the need to examine a range of doses, the importance of ongoing behavior in determining the observed actions of a drug, the hazards, as Dews (1974) put it, of "simplistic stereotype notions of behavioral phenomena," and so forth. In this brief prefatory chapter, we shall attempt to identify certain questions regarding the investigation of drugs as discriminative stimuli, questions that we believe must be addressed if orderly and efficient advances are to be made.

A drug whose primary effect in man is to produce euphoria, or depression of mood, or hallucinations presents a considerable barrier to relevant investigation in animals. The difficulty lies not in the fact that these phenomena lack reality but that, to a large extent, we depend on verbal communication to establish their presence. This fact notwithstanding, thousands of investigations of hallucinogens and euphoriants and mood depressors have been carried out using, as dependent variables, effects that it can only be hoped are related to the actions of these drugs in man.

In the past, nearly all studies of the interaction of psychoactive drugs and operant behavior have focused on disruption or, less commonly, facilitation of ongoing behavior. As an alternative approach, the study of psychoactive drugs as controlling or discriminative stimuli may offer distinct advantages in terms of both specificity of action and relevance to man. This is obviously not a novel idea, but pharmacologists have seldom pursued it. For example, morphine has been the subject of thousands of investigations, yet prior to 1973 only two studies reported its use as a discriminative stimulus (Belleville, 1964; Hill, Jones, & Bell, 1971). While too few data have been collected to permit definitive conclusions to be drawn with respect to the range of drugs that may function as discriminative stimuli or to the relationship between discriminative properties and other pharmacologic effects, and although differences between drugs in terms of the ease with which they assume discriminative control are peculiar and poorly understood, it is already known that several classes of drugs can function as discriminative stimuli. It seems well established that infrahumans can distinguish between drugs of distinct pharmacological classes; furthermore, there is substantial evidence that the stimulus properties of some drugs are quite specific. For example, the stimuli produced by morphine appear, on the basis of tests of generalization, stereospecificity, and antagonism by naloxone, to be typical opiate effects (Hirschhorn, this volume; Winter, 1975a). Because a variety of mechanisms may account for the stimulus properties of drugs, however, we should not expect all classes of pharmacological agents to exhibit the same degree of specificity.

Those drugs that are of interest because of their actual or potential effects on human behavior have most often been studied in animals for one of two reasons: to establish their mechanisms of action, a goal with several distinct meanings that depend on the orientation of the investigator, or to predict the actions in man of drugs that have not yet been administered to human subjects. The latter activity has been widely practiced in the pharmaceutical industry and is commonly referred to as "screening." The probability that a study of the stimulus effects of drugs will be useful in establishing mechanisms of action or in preclinical identification of drugs for human use cannot be expected to be the same for all classes of drugs. For example, in normal human subjects, amphetamine enhances physical performance, an effect that has a quantifiable counterpart in animals and seems to be related to subjective reports of a sense of well-being and euphoria. Thus, it is not surprising that in work reported earlier (e.g., Huang & Ho, 1974) and in the data presented by Wilson and Buelke in the present volume, classification of amphetamine and related drugs according to their stimulus properties closely parallels that based on subjective effects in man or on reinforcing properties in animals.

In contrast with amphetamine, a drug like chlorpromazine has rather unremarkable effects in normal subjects, and its importance stems from its ability to ameliorate the subjective effects as well as the behavioral consequences of schizophrenia. In the absence of any evidence that schizophrenia afflicts species other than *Homo sapiens,* we are at a considerable disadvantage in devising plausible experiments with animals. The stimulus properties of chlorpromazine as well as those properties assessed by standard "neuroleptic screens" may be unrelated to the pharmacologic actions essential for the alleviation of schizophrenia.

Nonetheless, the stimulus properties of antipsychotic drugs may yet prove useful. For example, a sizable body of information suggests that phenothiazines and therapeutically related agents exert their beneficial as well as certain of their adverse effects by blockade of dopaminergic receptors in the central nervous system (CNS). Such blockade would be expected to occur in normal and in schizophrenic humans and in animals, although in each group the behavioral consequences may be quite different. To the extent that such speculation on the origin and treatment of schizophrenia is valid, the search for an animal model of schizophrenia may be unnecessary; we need seek only methods for determining the antagonism of dopamine in the CNS. Instead of schizophrenic rats, we need dopaminergically hyperstimulated rats. In this regard, it is of interest that at least one drug house has already examined the effects of antipsychotic agents on the stimulus properties of apomorphine (Colpaert, Niemegeers, Kuyps, & Janssen, 1975), a drug assumed to act by direct stimulation of dopaminergic receptors. Schechter and Cook (1975) reported the antagonism by haloperidol of the stimulus effects of amphetamine, an agent that seems to act indirectly as a dopaminergic agonist and that has long been known to produce a syndrome resembling paranoid schizophrenia in man.

If one assumes, as a working hypothesis, that the discriminative stimulus properties of psychoactive drugs in animals are related in a meaningful manner to the subjective effects of these drugs in man, we would suggest that any unique value of drug stimuli as a tool for investigating drug mechanisms resides in their complexity. We should resist temptations to oversimplify our views of both drug stimulus phenomena and the mechanisms proposed to account for them. For example, Schechter (p. 115) outlines some of the pharmacological caveats appropriate when proposing a functional relationship between brain substances and behavior. In addition, as Kelleher and Morse (1968) pointed out, uncertainty in interpretations of drug-behavior interactions arises not only because of our incomplete understanding of drug effects but also because of our incomplete understanding of the behaviors upon which drug effects are measured. Twenty years ago Dews (1955) demonstrated that both the direction and the magnitude of drug effects on behavior can depend on the specific behavior under investigation. The degree to which drug stimuli, as measured through operants, are relative to the discrimination paradigm used to generate them has not been established. Discriminative stimuli are usually assessed through response rates or response choices. Is a differential response rate task more specific for stimulus generalization testing, or are rate effects a confounding factor absent from a response task? Put another way, how are rate effects per se related to the stimulus properties of drugs? In addition, because the time course of multiple drug effects are often not parallel, to what extent can pretreatment times and training doses influence the outcome of drug stimulus generalization tests? Can the route of administration influence responding controlled by drug stimuli? Krimmer and Barry (pp. 187–189) present evidence suggesting that the stimulus effects of pentobarbital are qualitatively similar following intraperitoneal and intravenous administration, an observation of considerable practical as well as theoretical significance.

The majority of experiments designed to distinguish the role of central and peripheral sites of action have favored the view that drug-induced stimuli arise in the CNS. For example, stimulus control exerted by nicotine is antagonized by mecamylamine, a ganglionic blocking agent that readily crosses the blood-brain barrier, but it is undiminished by chlorisondamine (Morrison & Stephenson, 1969) or hexamethonium (Schechter & Rosecrans, 1971), quaternary ammonium ganglionic blockers that enter the brain slowly following peripheral administration. In related experiments, Overton (1971) found that, in contrast with atropine, stimulus control was not achieved by atropine methylnitrate over the course of 50 training sessions. Taking a more direct approach Schechter (1973) observed nicotine-appropriate responding following intraventricular injection of nicotine in rats trained via the subcutaneous route. Browne, Harris, and Ho (1974) reported similar results with mescaline in rats trained with intraperitoneal injections and tested intraventricularly. Further progress is expected in localizing the sites at which drug-induced stimuli arise. For example, Hirschhorn, Hayes, and Rosecrans (1975) found that rats trained with electrical stimulation of the dorsal raphe nucleus as a discriminative stimulus emit stimulation-appropriate responses when treated with lysergic acid diethylamide (LSD). Further evidence for an interaction between the midbrain raphe nuclei and hallucinogens of the mescaline/LSD type is provided by Browne (this volume).

Despite the considerable body of evidence that suggests that the stimulus properties of many drugs originate in the CNS, we remain confident that much is to be learned in the investigation of drug-induced peripheral stimuli. For example, we are unaware of any attempts to extend the observation made some 15 years ago by Cook, Davidson, Davis, and Kelleher (1960) that emission of a conditioned avoidance response can be controlled by intravenous injection of epinephrine, norepinephrine, or acetylcholine, drugs generally assumed not to enter the CNS when administered peripherally. We believe it prudent to assume, until proved otherwise, that peripheral effects may represent a significant component of the stimulus properties of drugs.

One interpretation of drug-mediated responding is that "when rats learn a drug versus nondrug discrimination they [may] actually learn a rather generalized 'normal versus abnormal' discrimination . . ." (Overton, 1974). Although this is a plausible hypothesis, substantial experimental evidence favors the idea that at least in certain circumstances the stimuli produced by drugs are quite specific. For example, after examining a range of doses of drug Y in subjects trained with drug X versus saline and finding no dose of Y that mimics X, we could conclude that Y is without stimulus properties. In experiments recently reported (Winter, 1975a), however, subjects trained with morphine versus saline continued to respond in a fashion appropriate for the saline condition when tested with doses of ethanol that equal and exceed a dose of ethanol sufficient to establish stimulus control when paired with saline. Similar results were obtained in subjects trained with ethanol and tested with morphine (Winter, 1975a) or trained with mescaline and tested with *d*-amphetamine and cocaine (Winter, 1975d).

The observation that saline-appropriate responding may occur despite the presence of a drug in a concentration adequate, under other circumstances, to exert stimulus control leads us to a speculation that we shall call the third-state hypothesis: An animal trained with drug X versus saline will respond in a fashion appropriate for the saline condition when presented with stimuli that resemble neither those of drug X nor those of saline. This hypothesis explains adequately the results cited in the preceding paragraph (Winter, 1975a,d), but, it must be added preliminary data from experiments in which combinations of drugs were given have provided no support for the hypothesis (Winter, unpublished). For example, rats trained with mescaline versus saline and tested with mescaline plus one of a range of doses of morphine continue to respond as if only mescaline had been administered. Similar experiments with mescaline and morphine in amphetamine-trained rats have likewise yielded negative results.

Despite our present inability to provide direct experimental support for the third-state hypothesis, we believe that it is relevant to the interpretation of pharmacological antagonism of the stimulus properties of drugs in subjects trained with drug versus saline. Indeed, Johansson and Järbe (1975) tested the possibility of a third state produced by the combination of pentobarbital and bemegride (*vide infra*). In a typical test of antagonism, subjects are trained with drug X versus saline. Combinations of X and a range of doses of drug Z, a purported antagonist of drug X, are then administered. Saline-appropriate responses following X plus Z are interpreted as a blockade by Z of the stimulus effects of X. The third-state hypothesis offers an alternative explanation: Saline-appropriate responding results not from the absence of the effects of X but from the production by the combination of X and Z of a third state unlike either of the training conditions.

Because of the tentative nature of the third-state hypothesis, our major concern is to examine the pitfalls that may be encountered when defining classical drug antagonism in terms of drug-induced stimuli. This may be done in several ways. We may wish to demonstrate specificity of action, for example, the combination of morphine plus naloxone results in saline-appropriate responding, but naloxone has no effect on the stimulus properties of ethanol (Winter, 1975c) or pentobarbital (Rosecrans, Goodloe, Bennet, & Hirschhorn, 1973). Johansson and Jarbe (1975) have taken a more elegant approach. After having observed saline-appropriate responding in pentobarbital-trained gerbils tested with pentobarbital plus bemegride, they attempted to train with the combination versus saline. Discriminated responding was not observed, and the authors concluded that bemegride plus pentobarbital yielded a state indistinguishable from that following saline. (It should be noted that the antagonism of pentobarbital by bemegride observed by Johansson and Jarbe is most likely to be of a type that has been called physiological or functional antagonism, i.e., arising from a balance between opposite effects at different sites rather than pharmacological antagonism caused by interaction at a common receptor.)

There has been a tendency among authors to regard the results of generalization tests (cross-tests) in an all-or-none fashion. For example, subjects trained to discriminate the effects of drug X and saline may then be tested with a range of doses of drug Y. If the data fall on a smooth curve joining saline-appropriate responding and drug X-appropriate responding, interpretation is unambiguous: Y mimics X. Likewise, if, over a range of doses of Y, responding never deviates from that appropriate for the saline condition, we conclude simply that Y is devoid of stimulus effects or, invoking the third-state hypothesis, that the stimulus properties of Y differ from those of X. Data reported in the literature have not always been so straightforward, however. The general situation, which we shall henceforth refer to as "intermediate

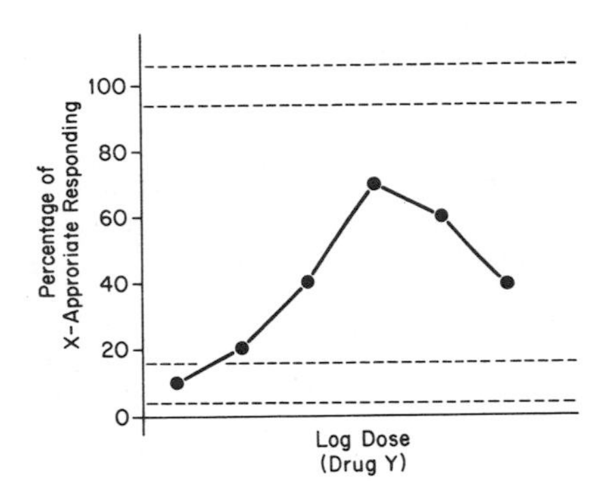

Figure 2.1. Hypothetical data that illustrate intermediate results. Ordinate: Responses appropriate for the drug-X training condition expressed as a percentage of total responses. Abscissa: Dose of drug Y plotted on a log scale. Dotted lines indicate the range of results following saline (10 ± 5) and drug X (100 ± 5).

results," is shown in Figure 2.1, in which the data indicate that, following certain doses of drug Y, responding approaches that appropriate for the X-condition but then falls back at higher doses. In several instances, intermediate results closely resembling the hypothetical data of Figure 2.1 have been interpreted as a failure of Y to mimic X. On narrow grounds, this conclusion is justified. We believe, however, that there is an important difference between the conclusion that responding under drug Y was not appropriate for the drug X training condition and the conclusion that responding under drug Y was appropriate for neither training condition.

The first step in attacking the problem of intermediate results is adequate statistical analysis of cross-test data. It is our practice to apply a repeated measures test to the differences between cross-tests and the two training conditions. If such tests indicate that we are not pursuing an illusion born of random variation, we may proceed to the important task of interpretation. In the past, such interpretation has seldom extended beyond a statement that "partial mimicry" was observed or the speculation that complete mimicry would have been observed had not behaviorally suppressant ("toxic") effects interfered. If we are to progress beyond speculation, experimental analysis is clearly required. For example, if drug Z yields intermediate results in chlorpromazine-trained subjects and later is found to be interchangeable with atropine, we might reasonably attribute the intermediate results to anticholinergic effects of drug Z. Likewise, any intermediate results obtained in morphine-trained animals could be challenged with a specific narcotic antagonist like naloxone. Alternatively, subjects may be trained with a morphinelike drug versus saline and the effects of naloxone then examined, as Appel and his colleagues (this volume) have done with pentazocine. In any event, we shall be better served by testable hypotheses than by those that merely account for past observations.

The role federal funding plays in the orientation of pharmacological research is no more evident than in behavioral pharmacology. Hence, it is not surprising that the majority of investigations reported in this volume center on agents that carry the sociopharmacological label "drug of abuse." Indeed, it has been suggested that there is a link between the "abuse liability" of a drug and its efficacy as a discriminative stimulus. It is fortunate for those of us who may find this suggestion useful in preparing grant applications that stimulus properties are characteristic of such a wide variety of drugs, and abuse liability is so ill-defined as to make the suggestion virtually immune to refutation.

The point we wish to make is that nearly all current investigations are implicitly anthropocentric, and the species differences of greatest concern are those that may occur between human subjects and either rodents or infrahuman primates. Although the results of clinical studies have rarely been expressed in terms of the stimulus properties of drugs, it is apparent that any investigation in which the subjective effects of drugs are observed may readily be interpreted in terms of drug stimuli. In actual practice, a one-to-one correlation of data from human subjects with data from animals is seldom possible because of the absence of a comparable experimental design. With the recent increase in interest in the stimulus properties of drugs and the accumulation of a significant body of data from animals, it is now essential that a limited number of crucial experiments be conducted in man.

The interpretation of existing clinical data must be undertaken with considerable caution, for the data are readily found compatible with a range of sometimes contradictory laboratory findings. A recent example is provided by cyclazocine, a drug with narcotic agonist and antagonist properties that is also hallucinogenic (Haerzten, 1970). In terms of the stimulus properties of cyclazocine in rats, obtained data indicate that it mimics both LSD (Hirschhorn & Rosecrans, 1975) and morphine (Winter, unpublished). These findings seem in agreement with the range of effects of cyclazocine in man. However, closer examination of the clinical data reveals that cyclazocine is readily distinguished from both LSD and morphine (Haertzen, 1970). Thus, the results in rats may plausibly be interpreted as being at variance with the results in humans. Once again, answers will not be found in re-examination and reinterpretation of existing data but must be sought in additional experimentation.

Although it is our opinion that studies of the discriminative stimulus properties of drugs will yield data with a high probability of relevance to man, the question remains of whether truly novel pharmacological effects arise in man as a consequence of his unique phyletic level. The extent to which the stimulus properties of drugs in animals reflect the subjective effects of the same drugs in man must be tested by clinical verification of predictions made by studies in animals and vice versa. For example, in view of the demonstration that the stimulus properties of mescaline (Browne & Ho, 1974; Winter, 1975b; Browne, this volume), LSD, and DOM (Winter, 1975c) can be attenuated by serotonin antagonists, a clinical assessment of the efficacy of such agents in diminishing the effects of hallucinogenic drugs in man seems warranted.

A parsimonious assumption is that the stimulus properties of drugs represent classical effects in new clothing. General support for this assumption comes from the fact that classifications of drugs according to their stimulus properties (e.g., Barry, 1974) have yielded categories titled depressant, anticholinergic, hallucinogenic, and so on; that is, much the same categories as would be arrived at without consideration of stimulus effects. The pattern is clearer still when we examine a specific category like the narcotic analgesics. Subjects trained with morphine versus saline yield morphine-appropriate responding when tested with a variety of narcotic analgesics but not after the administration of depressants of the ethanol-barbiturate type. Furthermore, the stimulus effects, like analgesic activity and reinforcing properties, are stereospecific; levorphanol mimics morphine, dextrophan does not (Shannon and Holtzman, 1975; Winter, 1975c). More variable results, perhaps intermediate results in the sense used earlier, have been obtained with drugs such as cyclazocine and pentazocine, but this again is as would be expected from existing clinical and animal data.

In light of the foregoing remarks, we must confront the question of whether a study of the stimulus properties of drugs is likely to provide new insights regarding drug action. At the present time, most efforts are aimed at validation, a process of determining whether stimulus effects make sense when placed in the context of the rest of pharmacology. This validation phase will never be complete, but we believe that it has progressed to the point that we may confidently consider the study of the stimulus effects of drugs as a tool in seeking new therapeutic agents and as a new approach in determining the mechanism of action of established drugs. With regard to seeking new psychotherapeutic agents, we must be wary of the trap laid for so many other "screens": a possible absence of any necessary relationship between the desired therapeutic activity and that aspect of the drug's action to which our screen is sensitive (*vide supra*). Nonetheless, the day may soon be here when a chemical fresh from the organic chemist will be administered to groups of rats trained with imipramine, amphetamine, morphine, phenobarbital, ethanol, meprobamate, diazepam, chlorpromazine, and LSD, respectively, and questions asked in much the same fashion as when Martin and his colleagues in days past asked of their addict–patients, "Is that good dope?"

REFERENCES

Barry, H., III. Classification of drugs according to their discriminable effects in rats. *Federation Proc.*, 1974, *33*, 1814–1824.

Belleville, R. E. Control of behavior by drug-produced internal stimuli. *Psychopharmacologia*, *5*, 95–105, 1964.

Browne, R. G., Harris, R. T., & Ho, B. T. Stimulus properties of mescaline and N-methylated derivatives: Difference in peripheral and direct central administration. *Psychopharmacologia*, 1974, *39*, 43–56.

Browne, R. G., & Ho, B. T. The role of serotonin (5-HT) in the discriminative stimulus properties of mescaline. *Proc. of the Society for Neurosciences* 1974, *4*, 155.

Colpaert, F. C., Niemegeers, C. J. E., Kuyps, J. J. M. D., & Janssen, P. A. J. Apomorphine as a discriminative stimulus, and its antagonism by haloperidol. *European J. Pharmacol.*, 1975, *32*, 383–386.

Conger, J. J. The effects of alcohol on conflict behavior in the albino rat. *Quart. J. Studies Alc.*, 1951, *12*, 1–29.

Cook, L., Davidson, A., Davis, D. J., & Kelleher, R. T. Epinephrine, norepinephrine, and acetylcholine as conditioned stimuli for avoidance behavior. *Science*, 1960, *131*, 990–991.

Dews, P. B. Studies on behavior. I. Differential sensitivity to pentobarbital of pecking performance in pigeons depending on the schedule of reward. *J. Pharmacol. Exp. Therap.*, 1955, *113*, 393–401.

Dews, P. B. What is analgesia? In M. C. Braude, L. S. Harris, E. L. May, J. P. Smith, & J. E. Villarreal (Eds.), *Narcotic antagonists. Advan. Biochem. Psychopharmacol.*, 1974, *8*, 235–243.

Haertzen, C. A. Subjective effects of narcotic antagonists cyclazocine and nalorphine on the Addiction Research Center Inventory (ARCI). *Psychopharmacologia*, 1970, *18*, 366–377.

Hill, H. E., Jones, B. E., & Bell, E. C. State dependent control of discrimination by morphine and pentobarbital. *Psychopharmacologia*, 1971, *22*, 305–313.

Hirschhorn, I. D., Hayes, R. L., & Rosecrans, J. A. Discriminative control of behavior by electrical stimulation of the dorsal raphe nuclei: Generalization to LSD. *Brain Res.*, 1975, *86*, 134–138.

Hirschhorn, I. D., & Rosecrans, J. A. Generalization of morphine and LSD stimulus properties to narcotic analgesics. *Federation Proc.*, 1975, *34*, 787.

Huang, J.-T., & Ho, B. T. Discriminative stimulus properties of *d*-amphetamine and related compounds in rats. *Pharmacol. Biochem. Behav.*, 1974, *2*, 669–673.

Johansson, J. O., & Järbe, T. U. C. Antagonism of pentobarbital induced discrimination in the gerbil. *Psychopharmacologia*, 1975, *41*, 225–228.

Kelleher, R. T., & Morse, W. H. Determinents of the specificity of behavioral effects of drugs. In Reviews of physiology, biochemistry and experimental pharmacology. Vol. 60. New York: Springer-Verlag, 1968. Pp. 2–56.

Morrison, C. F., & Stephenson, J. A. Nicotine injections as the conditioned stimulus in discrimination learning. *Psychopharmacologia*, 1969, *15*, 351–360.

Overton, D. A. Discriminative control of behavior by drug states. In T. Thompson & R. Pickens (Eds.), *Stimulus properties of drugs*. New York: Appleton, 1971. Pp. 87–110.

Overton, D. A. Experimental methods for the study of state-dependent learning. *Federation Proc.*, 1974, *33*, 1800–1813.

Rosecrans, J. A., Goodloe, M. H., Jr., Bennett, G. J., & Hirschhorn, I. D. Morphine as a discriminative cue: Effects of amine depletors and naloxone. *European J. Pharmacol.*, 1973, *21*, 252.

Schechter, M. D. Transfer of state-dependent control of discriminative behavior between subcutaneously and intraventricularly administered nicotine and saline. *Psychopharmacologia*, 1973, *32*, 327–335.

Schechter, M. D., & Cook, P. G. Dopaminergic mediation of the interoceptive cue produced by *d*-amphetamine in rats. *Psychopharmacologia*, 1975, *42*, 185–193.

Schechter, M. D., & Rosecrans, J. A. CNS effect of nicotine as the discriminative stimulus for the rat in a T-maze. *Life Sci.*, 1971, *10*, 821.

Shannon, H. E., & Holtzman, S. G. Evaluation of the discriminative effects of morphine in the rat. *Federation Proc.*, 1975, *34*, 786.

Sidman, M. Technique for assessing the effects of drugs on timing behavior. *Science*, 1955, *122*, 925–926.

Winter, J. C. The stimulus properties of morphine and ethanol. *Psychopharmacologia*, 1975, *44*, 209–214. (a)

Winter, J. C. Blockade of the stimulus properties of mescaline by a serotonin antagonist. *Arch. Intern. Pharmacodyn.*, 1975, *214*, 250–252. (b)

Winter, J. C. Stimulus properties of indoleamine and phenethylamine hallucinogens: Antagonism by cinanserin and pizotyline. *Federation Proc.*, 1975, *34*, 767. (c)

Winter, J. C. The effects of 2,5-dimethoxy-4-methylamphetamine (DOM), 2,5-dimethoxy-4-ethylamphetamine (DOET), *d*-amphetamine, and cocaine in rats trained with mescaline as a discriminative stimulus. *Psychopharmacologia*, 1975, *44*, 29–32. (d)

3

Discriminative Properties of *l*-Amphetamine: Stimulus Generalization[1]

Marvin C. Wilson and Judy Buelke
Department of Pharmacology and Research
School of Pharmacy
University of Mississippi

INTRODUCTION

Differential responding in rats has been demonstrated successfully when *d*-, *l*-, or *dl*-amphetamine was used as a discriminative stimulus (Harris & Balster, 1968, 1971; Huang & Ho, 1974a,b; Jones, Hill, & Harris, 1974; Schechter & Rosecrans, 1973; Waters, Richards, & Harris, 1972). Acquisition of several discriminated tasks, however, requires a longer training period when *l*-amphetamine is employed as the stimulus cue than when *d*-amphetamine is used (Harris & Balster, 1971; Jones *et al.*, 1974). It is generally acknowledged that *d*- and *l*-amphetamine are of similar potency peripherally (Innes & Nickerson, 1975) but that *d*-amphetamine is more efficacious in engendering a variety of central effects. Jones *et al.* (1974) reported that when 4-hydroxyamphetamine was administered to rats trained on a discriminated schedule for food reinforcement, using either *d*- or *l*-amphetamine versus saline as cue, responding on the drug-paired lever was not significantly different from nontest saline days. These data suggest that centrally active drugs produce an internal environment more readily discriminable than do compounds that act primarily in the periphery.

[1] This research was supported, in part, by the Research Institute of Pharmaceutical Sciences, University of Mississippi, and by NIMH Grant No. 21618.

The experiments presented here dealt with two aspects of discriminated operant performance. The stimulus cues were *l*-amphetamine and saline. First, a range of doses of *l*-amphetamine, and various other tests compounds whose central and/or peripheral actions are similar to those of *l*-amphetamine, were administered to rats after their stable operant performance was established. In this manner, dose–response effects concerning CNS stimulation, as measured by avoidance responding and/or efficiency and disruption of food-motivated behavior, may be estimated for each drug administered. Second, the efficacy of *l*-amphetamine as a discriminative stimulus, at a dosage that minimally disrupts established operant performance, and stimulus generalization from this to other dosages of *l*-amphetamine and to other test compounds were investigated. In combination, these procedures resulted in some information concerning the discriminative cue sufficient to produce stimulus generalization from 1 mg/kg of *l*-amphetamine.

MATERIALS AND METHODS

Subjects and Apparatus

Twelve male Wistar rats (National Laboratories, O'Fallon, Missouri), weighing 200 to 225 gm at the onset of the study, were housed individually. Four animals used in the avoidance procedure were maintained on a total of 15 gm/day of rat chow; 8 animals used in the food-reinforcement procedures received a total of 15 gm/day of food, with supplemental chow fed after each session if the animals did not receive that amount as reinforcers. All rats had free access to water except during the experimental sessions.

Two-lever operant chambers (BRS SC-004), modified as shown in Figure 3.1 and enclosed in sound-attenuated cubicles (BRS SCH-001) were used in all experiments. The chambers were equipped with a grid flooring and an airflow, noise-attenuation blower. Three stimulus lights were located above each lever, and two food cups were present. Ambient illumination was provided by a 7-W house light. During the avoidance procedure, electric shocks were delivered through the grid floor by a BRS-LVE shock generator (SG902) and scrambler (SG903).

Contingencies were programmed and data recorded on electromechanical programming equipment.

Avoidance Procedure

Four rats were trained on a discriminated, free-operant avoidance schedule. The procedure allowed the subject to avoid delivery of a shock stimulus by making the appropriate response, but it did not permit the rat to escape, that is, terminate the shock once it was delivered. The stimulus for foot shock was 1.5 mA dc passed through the grid floor of the chamber for .3 sec. Shocks were programmed every 5 sec as long as no responding on the correct lever

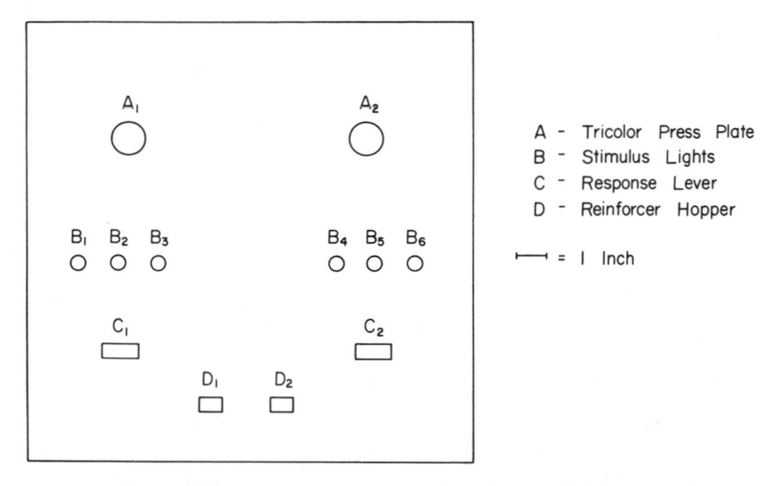

Figure 3.1. Scaled schematic of the rat intelligence panel.

occurred (shock–shock interval, S–S). Each response on the correct lever postponed shock delivery for 35 sec (response–shock interval, R–S). Responses on the other lever had no consequence but were recorded as incorrect responses.

During the first 30 sec of the R–S interval, the middle stimulus light (B_5) located over the correct lever was illuminated. During the last 5 sec of the R–S interval, the light was turned off. Thus, light offset was used as the warning signal that a shock delivery was approaching. The light was not illuminated again until an avoidance response was made, and thus remained off during the S–S interval as well. Light B_4 flashed after each correct response, and light B_6 was illuminated for the duration of each shock delivery.

Animals were placed in the experimental chamber for 50-min sessions, Monday through Friday. No training sessions were given. The first drug-test day occurred 2 weeks after stabilization of responding at an avoidance efficiency of 90%, that is, when 90% of possible shocks were avoided. All drugs were administered intraperitoneally (i.p.) 10 min before the session began, and drug-test days were scheduled no more frequently than once per week. Monday was never a drug-test day. Each dosage tested was administered to every subject in a nonordered sequence. The mean number of correct responses and the mean number of shocks received per nondrug session over the entire study were defined as the control values (100%). Mean responses and mean shocks received following pretreatment with each dose of a test compound were expressed as the percentage of the control value.

Reinforcement Procedure

Four subjects were trained to respond on a fixed-ratio 10 (FR-10) schedule of reinforcement for 45-mg food pellets (P. J. Noyes, Lancaster, New Hampshire). Presentation of a food pellet occurred after the tenth response following delivery of the previous pellet. Only responses on the correct lever were reinforced. Responses on the other lever had no consequence but were recorded. Subjects were shaped to lever-press by successive approximation. After several reinforcements were presented on a continuous reinforcement schedule, the response requirement was gradually increased from one until all subjects were responding on a FR-10 schedule of reinforcement. Stimulus lights were not used in this study, and only the illumination of the house light signaled food availability. One-hour sessions were conducted Monday through Friday.

The first drug pretreatment session did not occur until food intake per session varied less than 10% during five consecutive sessions. All drug treatments were administered intraperitoneally 10 min before the start of the session, and drug-test days were scheduled no more frequently than once a week, and never on Mondays. Each dose of each compound tested was administered in a nonordered sequence, at least once to all four animals. The mean number of food pellets received per nondrug session over the entire study was defined as baseline control (100%). The mean number of pellets received following pretreatment with each dose of a compound was expressed as the percentage of this value.

Discrimination Procedure

Four rats were trained on a discriminated FR-10 schedule for 45-mg food pellets. Experimental sessions were conducted Monday through Friday. Initially, animals were injected intraperitoneally (i.p.) with saline, 10 min before the sessions. Shaping procedures were the same as described for food reinforcement. Only responses on the right lever were reinforced, and no stimulus lights were used. Illumination of the house light signaled the onset of the session. After establishing stable FR-10 performance on this lever, the animals were treated i.p. with 1 mg/kg of *l*-amphetamine 10 min before each session. When *l*-amphetamine was administered, only responses on the left lever were reinforced. During both stimulus conditions, responding on the incorrect lever had no consequence but was recorded.

After 2 weeks of treatment with *l*-amphetamine, the dosage schedule was altered so that two successive *l*-amphetamine pretreatment sessions were followed by two consecutive saline sessions in a double-alternation sequence. The schedule was chosen to equalize the probability of a given stimulus condition following the same or other stimulus condition. This stimulus schedule remained in effect for 6 weeks, after which extinction test sessions were included on Fridays. The alternating stimulus schedule was then in effect from Monday through Thursday. During extinction test sessions, animals were injected i.p. 10 min before the session with saline, 10 mg/kg of *l*-amphetamine, or varying doses of test compounds. During these sessions, the food-delivery apparatus was disconnected, and the location and sequence of the first 100 responses were recorded. These sessions were terminated following 100 responses or after 10 min, whichever occurred first. The animals were then returned to their home cages and fed.

Each dose of the test compounds was administered at least once to the four subjects. In most instances, a given dose of a test drug was administered to some subjects before the session that followed a saline session and to others before the session that followed a *l*-amphetamine session. If any animal failed to respond at least 50 times during the allotted 10-min extinction period, the data were not used for statistical comparisons. The percentage of initial and total responses made on the left (*l*-amphetamine) lever were determined for both stimulus conditions during the study. On extinction-test days, these percentages were determined and then compared to baseline discrimination performance values using χ^2, $p < .01$.

Test Compounds

All drugs were dissolved in .9% saline to a concentration such that the appropriate dose of each compound was delivered at a volume of .25 ml/100 gm of body weight. Doses represent mg/kg of the HCl salt form of each drug. Solutions were prepared within 7 days before use and were refrigerated until that time. All subjects received injections in transfer cages and were not placed into the experimental chamber until the sessions began.

The drugs used in this study were supplied by the following companies: *l*-amphetamine HCl and *d*-amphetamine HCl (Smith, Kline and French Laboratories), diethylpropion HCl (Merrell National Laboratories), fenfluramine HCl (A. H. Robins Co.), methylphenidate HCl and phenmetrazine HCl (Ciba-Geigy, Inc.), chlorphentermine HCl (Warner-Lambert), methoxamine HCl (Burroughs Wellcome), amantadine HCl (E. I. Dupont Co.).

RESULTS

Baseline Performance

Baseline performance levels in both the discriminated avoidance and food-reinforcement procedures remained stable over the course of the study. Mean avoidance efficiency, that is, percentage of shocks avoided, during nondrug sessions for the entire period was 94%. The mean number of responses was 328, and the mean number of shocks received was 37.8 (Table 3.1). Response rate during nontreatment sessions of the food-reinforcement paradigm averaged 59.4 per min, and the mean number of reinforcers received per session was 297. Mean values calculated on a monthly basis did not vary from these levels by more than 10%.

Stimulus control of behavior in the discrimination procedure was evident after 15 sessions of double-alternation stimulus conditions. No tendency

Table 3.1
Correlations between the Effects of Drug Treatment on Discriminated Avoidance and Food-Reinforced Behaviors and Generalization of the Stimulus Properties of These Treatments to Those of *l*-Amphetamine

	Discriminated avoidance			Food reinforcement		Generalization test	
Drug and dose	Responses (% of control $\bar{X}$)	*N*	Shocks (% of control $\bar{X}$)	*N*	Pellets (% of control $\bar{X}$)	*N*	Percentage of responses on l-amphetamine lever
Baseline behavior							
Nondrug days	100 ± 10[a]	348	100 ± 10[a]	329	100 ± 8[a]	—	—
Absolute	(328 ± 33)	—	(37 ± 4)	—	(297 ± 24)	—	—
Amphetamine days	—	—	—	—	—	172	87
Saline days	—	—	—	—	—	183	11
Saline						12	33[b,c,d]
l-amphetamine							
.25	—	—	—	—	—	4	25[b,c]
.5	—	—	—	—	—	7	43[b,c]
1	99	4	72	4	84	14	50[b,c,e]
2.5	102	4	76	4	70	4	100[c]
5	143[g]	4	58	4	55	—	nc[f]
10	144[g]	4	79	4	36	—	—
20	122[g]	4	139	4	5	—	—
d-Amphetamine							
.125	—	—	—	—	—	5	0[b]
.25	197[g]	4	82	—	—	6	33[b,c]
.5	273[g]	4	82	4	92	4	100[c]
1	253[g]	4	63	4	69	5	100[c]
2.5	210[g]	6	84	4	6	—	nc[f]
5	233[g]	6	113	4	0		

Methylphenidate							
.5	113	4	24	—	—	—	—
1	69	4	26	4	102	8	63[b,c]
2.5	87	4	18	4	92	7	58[b,c]
5	178	4	21	4	86	5	60[b,c]
10	189	4	26	4	41	—	nc[f]
Phenmetrazine							
.5	—	—	—	—	—	4	75[c]
1	98	4	92	4	89	4	75[c]
2.5	118	4	76	4	88	—	—
5	184	4	113	4	86	4	100[c]
10	370[g]	4	116	4	58	—	nc[f]
20	352[g]	4	121	4	7	—	—
Amantadine							
25	113	4	142	4	108	4	100[c]
50	134	4	150	4	47	7	87[c]
100	161	4	171	4	62	—	nc[f]
Diethylpropion							
.5	108	4	82			5	20[b]
1	98	4	53	4	103	8	38[b,c]
2.5	150	4	63	4	93	—	—
5	206[g]	4	108	4	31	4	75[c]
10	372[g]	4	50	4	31	—	nc[f]
20	327[g]	4	58	4	0	—	—
Chlorphentermine							
1	84	4	92	4	101	4	0[b]
2.5	64	4	74	4	110	9	66[b,c]
5	83	4	74	4	105	—	—
10	80[g]	4	100	4	78	6	66[b,c]
20	87[g]	4	153	4	17	4	50[b,c]
40	84[g]	4	187	—	—	—	nc[f]

Table 3.1 (Continued)

Drug and dose	Discriminated avoidance			Food reinforcement		Generalization test	
	Responses (% of control $\bar{X}$)	*N*	Shocks (% of control $\bar{X}$)	*N*	Pellets (% of control $\bar{X}$)	*N*	Percentage of responses on l-amphetamine lever
Fenfluramine							
.5	—	—	—	—	—	5	20[b]
1	71	4	74	4	88	5	20[b]
1.5	78	4	89	4	57	5	60[b,c]
2.5	75	4	100	4	—	—	nc[f]
5	72	4	258	4	—	—	—
Methoxamine							
2	86	4	132	4	115	4	75[c]
4	95	4	145	4	112	4	75[c]
8	78	4	191	4	60	—	nc[f]

[a] Mean equals 100% ± two standard errors.
[b] Significantly less than baseline amphetamine days, $p \leq .01$.
[c] Significantly greater than baseline saline days, $p \leq .01$.
[d] Significantly different from amphetamine test days, $p \leq .05$.
[e] Significantly different from saline days, $p \leq .05$.
[f] Fifty response criterion not met by two or more subjects injected with this dose.
[g] Responding on *incorrect* lever significantly increased above that occurring on control days.

toward improvement of the discrimination occurred during the next 8 months of the study. Percentages of correct initial responses (IRs) in *l*-amphetamine and saline sessions was 87% and 89%, respectively. Correct lever responses during the first 10 min in all *l*-amphetamine and saline-cued sessions were greater than 99% of total responding. During the training period and throughout the study, 1 mg/kg of *l*-amphetamine did not appear to produce anorexia, since the number of reinforcers received during *l*-amphetamine-cued sessions versus the number obtained in saline-cued sessions was not statistically different when compared by Student's *t*-test.

Avoidance Performance

Mean changes in avoidance responding after pretreatment with various doses of the test compounds are presented in Table 3.1. Phenmetrazine, diethylpropion, *l*-amphetamine, *d*-amphetamine, and amantadine administration increased the rate of avoidance responding. Except in trials with diethylpropion and amandatine, a biphasic relationship was evident between responding and the number of shocks received. Initially, as responding increased, the number of shocks avoided increased above control values. Further increments in dosage, however, decreased the number of shocks avoided even though responding was also elevated. In the case of amantadine, changes in response rate and the number of shocks received were directly proportional to dosage. Over the range of amantadine dosages tested, no enhancement of shock-avoidance performance was noted. Treatment with diethylpropion resulted in a dose-related enhancement of responding and, in general, a fairly constant decrease in the number of shocks received over the range of dosages tested.

Pretreatment with methylphenidate resulted in a marked decrease in the number of shocks received by all subjects, while a biphasic change in responding was observed, that is, a decrement in responding at moderate doses, followed by rates greater than baseline levels as the dosage was increased. Even when the response rate was significantly decreased, the number of shocks received was reduced from control values by 75%. A decrease in both responses emitted and shocks received was noted following low doses of chlorphentermine. As the dosage of this drug was increased (20–40 mg/kg), a dose-related increment in shocks received was found, with only minor concurrent change in the amount of responding. The same general dose–response pattern was observed after pretreatment with fenfluramine and methoxamine.

Responses on the incorrect lever were not altered by administration of fenfluramine, methoxamine, or amantadine and remained, as during nontreatment sessions, at near-zero levels. Moderate to high doses of all other compounds tested resulted in noticeable enhancement of such responding, for example, fewer than 500 but more than 25 responses per session. One animal substantially increased its nonreinforced behavior, for example, more than 50 responses per session, after all doses of *l*-amphetamine, *d*-amphetamine, and methylphenidate. Larger doses of the psychomotor stimulants were not administered because of a potentially lethal interaction with shock (data obtained in this laboratory).

Fixed-Ratio Food Performance

Over the range of dosages administered, all drugs produced a dose-related decrease in the mean number of reinforcers obtained during test sessions (Table 3.1). In no case was enhancement of responding on the nonreinforced lever observed. Larger dosages of these compounds were not tested because response suppression at the highest dosage tested was essentially complete during the early part of the session. In some cases, for example, after methylphenidate and amantadine administration, behavioral recovery occurred during the latter part of the session.

Generalization Testing

Initial and total response data obtained during extinction test sessions are presented in Table 3.1 and Figure 3.2. It should be noted that the percentage of initial responses (IRs) on test days for both saline and 1 mg/kg *l*-amphetamine, differs significantly from baseline performance levels. Test-day IR distributions for these stimulus doses, however, differ significantly from each other. During generalization testing, an attempt was made to keep constant all external variables, for example, temperature of solutions, site of injection, and so on, which might signal an upcoming extinction session, and to administer each test dose on days following both stimulus conditions. Only in the case of 1 mg/kg of *l*-amphetamine did a correlation appear between the stimulus condition on the previous day and the IR choice during the extinction test session. This dosage was tested 14 times; 8 of the tests being run on days following baseline sessions with saline as the cue. On these days, only once did a subject make the correct IR on the *l*-amphetamine lever. There were no discernible differences between test and baseline days until 10 responses on the correct lever had occurred that resulted in absence of reinforcement. This absence might have influenced the location of additional responses but not of the initial response. Percentages of mean total responses on the *l*-amphetamine lever during extinction tests for saline and 1 mg/kg of *l*-amphetamine were 32% and 67%, respectively, with no relationship noted between percentage of total responding and the previous day's cue.

For the test drugs, response location during extinction tests after at least one dose of *l*-amphetamine, *d*-amphetamine, phenmetrazine, diethylpropion, methoxamine, and amantadine was quantitatively similar to the 87%-correct responses found on *l*-amphetamine baseline days. In general, this relationship was dose-related, the greater percentage of correct IRs occurring with the

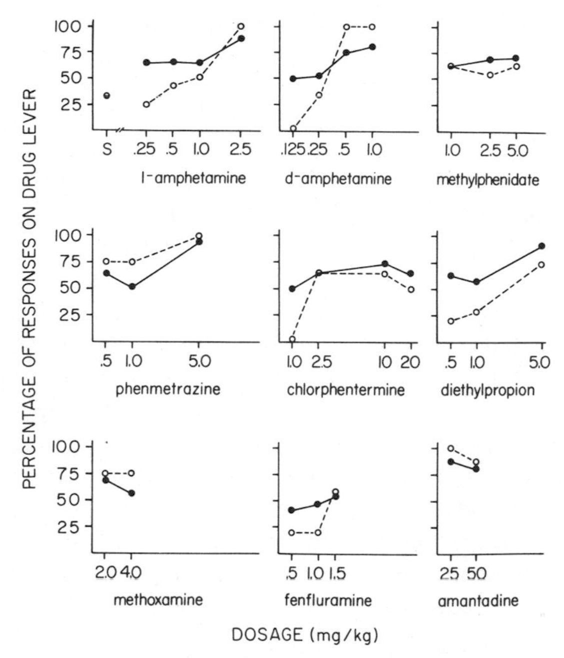

Figure 3.2. Percentages of intelligence panel responses during extinction test sessions for all stimulus conditions during which at least 2 subjects met response criterion. These responses were made on the lever that resulted in food reinforcement under the 1 mg/kg *l*-amphetamine stimulus condition. Open circles represent percentage of initial responses on the drug lever; closed circles represent the percentage of mean total responses on the drug lever of the first 100 responses.

higher dosages. Low dosages of fenfluramine and chlorphentermine resulted in IR percentages not different from those of saline baseline behavior. All dosages of methylphenidate and high dosages of chlorphentermine and fenfluramine, however, resulted in IR percentages significantly different from both saline and amphetamine baseline values. With further dosage increments of these three agents, the animals failed to meet test criterion.

Individual subjects showed some consistent choice patterns not evident in the grouped data. Means presented in Figure 3.2 suggest that of all the drugs tested, fenfluramine is the least similar to 1 mg/kg of *l*-amphetamine. One subject, however, consistently made its initial response on the *l*-amphetamine lever; its percentage of total responding on this lever was also at or above the mean for the *l*-amphetamine stimulus dose ($\geq 67\%$). A second animal never responded initially on the *l*-amphetamine lever after fenfluramine administration but always chose the *l*-amphetamine lever for IR after phenmetrazine or diethylpropion injection. Another animal always chose the *l*-amphetamine lever for the IR after methylphenidate administration, but was the only subject not to respond to methoxamine as to the stimulus dose of *l*-amphetamine.

For all drugs tested, total responding during extinction sessions tended to show a similar, though less obvious, dose–response relationship as did the corresponding IR data (Figure 3.2).

DISCUSSION

Although *l*-amphetamine, *d*-amphetamine, methylphenidate, phenmetrazine, diethylpropion, and chlorphentermine are commonly classified as psychomotor stimulants, it seems that the stimulus properties of these compounds vary at different dosages. Fenfluramine, although it is not a psychomotor stimulant, was included in this study because of its anorexic action, the mechanism of which is believed to differ from that of the psychomotor stimulants. Methoxamine is reported to be an indirectly acting, purely alpha-sympathomimetic compound, which exerts little central effect (Innes & Nickerson, 1975). The compound would be expected to approximate closely the peripheral sympathomimetic effects of the psychomotor stimulants and thus have peripheral stimulus properties in common with them. Most sympathomimetics exert some beta-adrenergic action and therefore increase the rate and force of myocardial activity. Methoxamine, however, tends to reflexly decrease heart rate.

Amantadine exerts central and peripheral dopaminergic action (Von Voigtlander & Moore, 1971) and would be expected, at some dosage, to mimic those stimulus properties of *l*-amphetamine that are the result of increased dopaminergic activity. Scheel-Kruger (1972), however, has reported that *l*-amphetamine does not produce effects thought to be the result of increased central dopaminergic activity until doses of 10–20 mg/kg are administered i.p. It is likely, therefore, that 1 mg/kg *l*-amphetamine exerts most of its actions either via direct or sympathomimetic rather than dopaminergic mechanisms.

The effects of the various drugs on FR food-reinforced behavior found in the present study agree with the results of other studies. Several investigators (Appel & Freedman, 1968; Owen, 1960; Sparber & Peterson, 1974; Tilson & Sparber, 1973) have demonstrated that the parenteral administration to rats of *d*-amphetamine (>1 mg/kg) depresses FR food-reinforced behavior. Owen (1960) reported that *l*-amphetamine administered subcutaneously (>5 mg/kg) produces a similar decrement in behavior. Sparber and Peterson (1974) reported that administration of 1 mg/kg of *l*-amphetamine i.p. produced only a slight decrement in this behavior, but that 4 mg/kg significantly decreased responding equivalent to the decrement observed following 1 mg/kg of *d*-amphetamine. In the present study, 1 mg/kg of *d*-amphetamine and 2.5 mg/kg of *l*-amphetamine produced equivalent suppression of food-reinforced responding.

Cox and Maickel (1972), using rats, reported similar data for *d*-amphetamine in a 2-hr, free-feeding paradigm. Food intake was decreased 50% by i.p. administration of 1.5–2 mg/kg of *d*-amphetamine. A similar reduction in food consumption occurred after i.p. administration of 2 mg/kg of fenfluramine, 6 mg/kg of chlorphentermine, 8.5 mg/kg of diethylpropion, and 13 mg/kg of phenmetrazine. These results are quite analogous to those presented in this study. The lowest dosages tested that produced a comparable degree of response suppression were 1.5 mg/kg of fenfluramine, about 15 mg/kg of chlorphentermine, 5 mg/kg of diethylpropion, and 10 mg/kg of phenmetrazine. Comparable effects were also produced by 10 mg/kg of methylphenidate, 50 mg/kg of amantadine, and 8 mg/kg of methoxamine. For all drugs tested, the degree of response suppression in the FR food-reinforcement procedure was directly related to the dosage.

We found the effects of drug treatment on avoidance behavior also similar to data previously reported. Cox and Maickel (1972) reported that i.p. administration to rats of .8 mg/kg of *d*-amphetamine, 4.3 mg/kg of phenmetrazine, and 2.8 mg/kg of diethylpropion increased responding by 50% in an unsignaled avoidance paradigm. Souskova, Benesova, and Roth (1964) demonstrated that 2 mg/kg of phenmetrazine administered subcutaneously to rats reduced avoidance and escape response latencies and increased intertrial-interval responding in a conditioned-avoidance paradigm. These data agree with ours in that 2.5 mg/kg of diethylpropion increased response rate by 50% and an 84% increase resulted from treatment with 5 mg/kg of phenmetrazine. Several studies (Carlton & Didamo, 1961; Goldberg & Ciofalo, 1969; Rech & Stolk, 1970; Weissman, Koe, & Tenen, 1966) demonstrated that i.p. administration of .5–2 mg/kg of *d*-amphetamine enhanced avoidance responding in a continuous-avoidance procedure. The present results also support these data.

Cox and Maickel (1972) also reported that 4 mg/kg of fenfluramine administered i.p. decreased avoidance responding by 50%. A lesser reduction in responding was seen with a similar dose in the present study. Cox and Maickel (1972) further reported that 16 mg/kg of chlorphentermine did not increase avoidance responding by 50%. In the present study, chlorphentermine, at dosages up to 40 mg/kg, did not increase avoidance responding. It must be assumed, therefore, that fenfluramine and chlorphentermine differ from *l*-amphetamine, *d*-amphetamine, phenmetrazine, and diethylpropion with respect to their effects on avoidance responding.

Bernstein and Latimer (1968) reported that 7.5 mg/kg of methylphenidate administered i.p. to rats, increased the response rate in an unsignaled avoidance paradigm by 100%. In our study, 5 mg/kg of this drug increased the rate by 78%. Stretch and Skinner (1967) observed increased avoidance responding after injecting 4–16 mg/kg of methylphenidate but not after 2 mg/kg. The number of shocks received by rats at this lowest dosage, however, was less than the number received by controls. At the highest dose, shock rate again approached baseline levels. This is very similar to the data we obtained with this drug, in which 1 and 2.5 mg/kg decreased responding but also greatly reduced shock frequency. Shock frequency decreased concurrently with an increased response rate following treatment with 5 and 10 mg/kg of methylphenidate. A similar relationship between response rate and shock frequency also was observed after administration of 2.5 and 5 mg/kg of chlorphentermine and 1 mg/kg of fenfluramine. These data suggest that these drugs, in low dosages, enhance the efficiency of avoidance responding. In contrast, the relationship between response rate and shock frequency was not found after treatment with *d*-amphetamine, *l*-amphetamine, and phenmetrazine. With these drugs, lower dosages resulted in increased response rates, which correlated with a decrease in shock frequency. As the dosage increased, however, shock frequency increased toward, and often exceeded, control values. At the higher dosages, much of the responding occurred after shock onset. Rech (1964) reported that very large doses (78 mg/kg) of *d*-amphetamine i.p. blocked escape responding in a conditioned-avoidance paradigm. Furthermore, doses larger than 4 mg/kg of *d*-amphetamine decreased avoidance responding. This decrease perhaps could be attributed either to toxic results of a drug-shock interaction or to the introduction of a behavior, for example, stereotypy, which competed with emission of the avoidance response. In the present study, these mechanisms may have also resulted in decreased avoidance responding. However the increased stimulus conditions associated with escape responding as compared to avoidance responding, may have been sufficient, to break the stereotypic chaining and thereby allow the expression of escape responding.

Perhaps the competing set of behaviors was not elicited by diethylpropion. At high dosages, this drug increased responding as much as any other drug tested; however, shock frequency remained below control values. This effect resembles that seen after treatment with high dosages of methylphenidate. Diethylpropion, therefore, also differs from *l*-amphetamine, *d*-amphetamine, and phenmetrazine in affecting this behavior. As such, it appears there is no correlation between maximal enhancement of avoidance responding and effects on shock frequency. Diethylpropion (20 mg/kg) increased responding 327% and decreased shock frequency to 58% of control levels. Treatment with such other drugs as 20 mg/kg of *l*-amphetamine and 5 mg/kg of *d*-amphetamine, which produced less of an increase in responding, or with 20 mg/kg of phenmetrazine, which produced an equal increase in response frequency, resulted in increased shock frequency.

Increases in shock frequency occurred following treatment with fenfluramine and methoxamine. The increase produced by the administration of 5 mg/kg of fenfluramine was associated with grossly observable sedation, which apparently contributed to the decrease in avoidance responding. The sedative properties of fenfluramine have been well documented (Innes & Nickerson, 1975). However, methoxamine treatment increased shock frequency at dosages that did not affect response rate. The subjects did not appear to be sedated. Nonspecific central depression is not suggested as the mechanism of methoxamine's effect on shock frequency, as the drug did not disrupt food-reinforced behavior.

An analysis of incorrect-lever responding demonstrated that increases in avoidance responding were also associated with increases in incorrect-lever responding. This effect followed treatment with *l*-amphetamine, *d*-amphetamine, phenmetrazine, chlorphentermine, and diethylpropion. Incorrect-lever responding did not increase concurrently with appropriate avoidance responding after treatment with methylphenidate or amantadine. It is of interest that, even though avoidance responding decreased after treatment with chlorphentermine, incorrect-lever responding increased. These data again support the contention that methylphenidate and chlorphentermine differ from the amphetamines, phenmetrazine, and diethylpropion in their effects on avoidance behavior. In both the food-reinforced and avoidance paradigms, incorrect-lever responding did not occur during control sessions. Incorrect responding, likewise, never occurred during sessions following drug pretreatment in the food-reinforcement paradigm. If the increase in incorrect-lever responding resulted from an increase in unconditioned locomotor activity, one could have expected a similar increase in responding on both the correct and incorrect levers in the food-reinforcement paradigm, especially since the drug treatments and experimental cubicles used in the food-reinforcement procedure were identical to those used in the avoidance study.

The effects of amantadine resemble those of high doses of *l*- and *d*-amphetamine, phenmetrazine, and diethylpropion on both food-reinforced and shock-avoidance behavior. In general, with these five agents and with methylphenidate, the threshold dosage for increasing avoidance responding was less than the threshold dosage for disrupting food-reinforced behavior. Chlorphentermine did not disrupt food-reinforced behavior until dosages were used that increased shock frequency. Even at these dosages, however, avoidance responding was still below control values. These data do not seem to imply complete anorexic selectivity for chlorphentermine, but they do indicate that although anorexic selectivity may not be complete, chlorphentermine is the least stimulating psychomotor stimulant tested at potential "anorexic" dosages.

It has been demonstrated repeatedly that various centrally acting drugs induce a "state" that can serve as an effective cue for animals in discrimination tasks. In such studies, in which all environmental stimuli remain constant except the drug administered before testing, the subjects are trained so that response choice is appropriate to the drug state. The subjects' behavior is said to be controlled by the stimulus properties of the drug. Overton (1971) has suggested that chemicals possessing similar pharmacological actions produce similar interoceptive cues, that is, similar stimulus properties or discriminable states.

Several research groups have reported the ability of *d*-amphetamine (Huang & Ho, 1974a,b; Jones *et al.*, 1974; Schechter & Rosecrans, 1973) and *dl*-amphetamine (Harris & Balster, 1968, 1971; Waters *et al.*, 1972) to serve as discriminative stimuli in rats. Harris and Balster (1971) employed operant techniques and found that the stimulus properties of 1 mg/kg *dl*-amphetamine generalized to 1 mg/kg of *d*-amphetamine and to 4.5 mg/kg of methylphenidate. Waters *et al.* (1972), using a two-lever food-reinforcement procedure, found that animals could be trained to discriminate between saline and 2.5 mg/kg of *dl*-amphetamine (group 1), between .3 and 2.5 mg/kg of *dl*-amphetamine (group 2), but not between saline and .3 mg/kg of *dl*-amphetamine (Group 3). Treatment with 1 mg/kg of *dl*-amphetamine generalized equally to the two stimulus conditions when tested in subjects from Group 1, but almost completely to the 2.5 mg/kg condition in the second group. These authors suggested that the different groups were using different interoceptive cues produced by the 2.5 mg/kg training dose in discriminating the two training states.

Schechter and Rosecrans (1973), using a three-chambered maze and a shock-escape paradigm, reported that subjects could be trained to discriminate between saline and 4 mg/kg of *d*-amphetamine. Treatment with either 4 or 8 mg/kg of *l*-amphetamine generalized to the stimulus conditions produced by *d*-amphetamine. Treatment with either 4 or 8 mg/kg of fenfluramine, .4 mg/kg of nicotine, .048 mg/kg of LSD, or 7.4 to 29.7 mg/kg of mescaline did not generalize to the training dose of *d*-amphetamine. More recently, Schechter and Cook (1975) reported that the interoceptive cue produced by 2 mg/kg of *d*-amphetamine generalized to 1 mg/kg of *d*-amphetamine and to 1.25, 2.5, and 5 mg/kg of apomorphine but not to .5 mg/kg of *d*-amphetamine. In addition, these authors pretreated the subjects with a variety of central and peripheral monamine-depleting agents and the dopamine-blocking agent, haloperidol. Results suggested that the stimulus set produced by 2 mg/kg of *d*-amphetamine was mediated by dopaminergic systems.

Huang and Ho (1974a, b) used a two-lever food-reinforcement procedure and found that rats could distinguish between saline and .8 mg/kg of *d*-amphetamine states. Little generalization to the training dosage (.8 mg/kg) of *d*-amphetamine was observed after treatment with .2 or .4 mg/kg of *d*-amphetamine. The stimulus properties of *l*-amphetamine (1.6 but not .8 mg/kg), methamphetamine (.8 mg/kg), ephedrine (8 but not 2 or 4 mg/kg), cocaine (7.5 mg/kg), and methylphenidate (2.5 but not .5 or 1 mg/kg) were reported as similar to the cues produced by *d*-amphetamine. Mescaline, nikethamide, picrotoxin, and strychnine possessed stimulus properties that differed from those of the training dosage of *d*-amphetamine.

Jones *et al.* (1974), using the same training dose of *d*-amphetamine and the same procedure, reported that the stimulus properties of this dosage generalized to 1.6 and 2.4 mg/kg of *l*-amphetamine. However, generalization was not found with lower doses of *l*-amphetamine or with *para*-hydroxyamphetamine, a peripheral sympathomimetic agent. Subjects trained to discriminate between .8 mg/kg of *l*-amphetamine and saline responded on the drug lever when treated with .8 mg/kg of *d*-amphetamine or 2.4 mg/kg of *l*-amphetamine but not when treated with *para*-hydroxyamphetamine. These results suggest that central rather than peripheral interoceptive cues functioned in establishing the initial discrimination between drug and saline. Our results are similar to those reported by Jones and colleagues. In our study, doses of 2.5 mg/kg of *l*-amphetamine and .5 mg/kg of *d*-amphetamine generalized to the 1 mg/kg training dose employed. Larger doses of both *d*- and *l*-amphetamine sufficiently disrupted behavior so that generalization-test criteria were not met.

A study by Schechter and Rosecrans (1973) and our data failed to show generalization between fenfluramine and *d*- or *l*-amphetamine, respectively. These results suggest that anorexia, per se, is not the primary interoceptive cue during development of the drug–saline discrimination, as even with higher anorexic doses generalization to the amphetamine state was not complete. Further evidence on this point is provided by our generalization testing with phenmetrazine. At dosages that did not disrupt food-reinforced behavior, phenmetrazine was perceived as *l*-amphetamine. This also occurred with .5 mg/kg of *d*-amphetamine, 25 mg/kg of amantadine, and 2 and 4 mg/kg of methoxamine.

Our study did not find significant generalization between methylphenidate and *l*-amphetamine and disagrees, therefore, with the work of Huang and Ho (1974b). The disparity may have resulted from different amphetamine "states" or cues being engendered by 1 mg/kg of *l*-amphetamine and .8 mg/kg of *d*-amphetamine in establishing appropriate differential response patterns.

A positive correlation between enhancement of avoidance responding and generalization to *l*-amphetamine was not complete. Doses of *l*-amphetamine (2.5 mg/kg), phenmetrazine (1 mg/kg), and methoxamine (2 and 4 mg/kg) did not increase avoidance responding but resulted in significant differential responding on the *l*-amphetamine lever during generalization testing. Furthermore, drug dosages that enhanced avoidance responding sometimes were not perceived as *l*-amphetamine in the generalization tests, that is, .25 mg/kg of *d*-amphetamine and 5 mg/kg of methylphenidate. The absence of total correlation between disruption of food-reinforced responding or enhanced avoidance responding and generalization to *l*-amphetamine is not surprising, as the training dose of *l*-amphetamine did not markedly affect either of these behaviors.

The results from generalization testing with amantadine suggest that the stimulus cues produced by 1 mg/kg of *l*-amphetamine may involve dopaminergic mechanisms. This hypothesis is supported by the report of Schechter and Cook (1975), who found generalization between *d*-amphetamine and apomorphine, a central dopaminergic stimulant. However, dosages of *d*-amphetamine and phenmetrazine much lower than those reported to cause stereotypy (Randrup & Munkvad, 1970), which is thought to be mediated by dopaminergic systems, still generalized well to the training dosage of *l*-amphetamine. In addition, Scheel-Kruger (1972) demonstrated that *l*-amphetamine administered i.p. failed to produce signs of stereotypy in rats even at a dose of 5 mg/kg. Therefore, the discriminative properties of a larger dose of *d*- or *l*-amphetamine may be based on dopaminergic activity, but the stimulus properties of 1 mg/kg of *l*-amphetamine possibly are not.

Positive generalization between methoxamine and *l*-amphetamine, however, does suggest that at least the peripheral alpha-adrenergic system may be involved in generating interoceptive cues associated with 1 mg/kg of *l*-amphetamine. This hypothesis is not supported by the results of Jones *et al.* (1974), who found no generalization from *d*- or *l*-amphetamine to *para*-hydroxyamphetamine. This apparent discrepancy may have resulted from methoxamine's exerting some additional behavioral or central action since both shock frequency and food-reinforced responding were increased to some extent by doses of this drug. Perhaps a central noradrenergic cue was produced by methoxamine that generalized to the *l*-amphetamine state.

In summary, the stimulus cues produced by 1 mg/kg of *l*-amphetamine *versus* those produced by saline are adequately distinctive to maintain differential responding. It is likely that these stimuli are of a compound nature, and that individual animals may cue on individual aspects of the stimulus complex. Individual subjects consistently reacted to some of the test compounds as being either like *l*-amphetamine or like saline, in contrast to the rest of the subjects. Results from tests with methoxamine and fenfluramine suggest this, in particular. The methoxamine data also suggest that the peripheral components of the *l*-amphetamine cue may be of substantial, if not of primary, importance. That the 1 mg/kg training dose did not markedly increase avoidance responding or disrupt food-reinforced behavior supports this hypothesis. From these studies, however, it seems improbable that purely peripheral drug actions produced a stimulus complex capable of establishing and maintaining the observed discriminated performance. The generalization from *l*-amphetamine to amantadine suggests dopaminergic mediation of at least some of the components of the *l*-amphetamine stimulus complex. The contribution of dopaminergic systems to the stimulus properties of the training dose of *l*-amphetamine used here, however, is not consistently obvious. Therefore, the neurochemical and neuropharmacological nature of the *l*-amphetamine stimulus cues awaits further analysis.

SUMMARY

Groups of four rats each were trained on either a discriminated free-operant avoidance schedule, a fixed-ratio 10 (FR-10) food-reinforcement schedule, or a discriminated FR-10 schedule for food reinforcement. Intraperitoneal injections of saline or 1 mg/kg of *l*-amphetamine were used as stimulus cues in the discriminated reinforcement procedure. Alterations in avoidance and food-reinforced behavior were measured following administration of various dosages of *l*-amphetamine, methylphenidate, phenmetrazine, diethylpropion, chlorphentermine, fenfluramine, methoxamine, and amantadine. From these results, dosages were selected for use in generalization testing during extinction sessions of the discrimination procedure. Of the drugs administered, behavior observed following at least one dose of *l*-amphetamine, *d*-amphetamine, phenmetrazine, diethylpropion, methoxamine, and amantadine was not significantly different from 1 mg/kg of *l*-amphetamine baseline performance on the basis of initial-response choice during the extinction-test sessions. All dosages tested of fenfluramine, methylphenidate, and chlorphentermine were found to be either not different from saline baseline behavior or significantly different from both saline and *l*-amphetamine values. These results suggest that the interoceptive cue that establishes the *l*-amphetamine-saline discrimination may be based on peripheral or central noradrenergic or dopaminergic activation or a combination of these factors. The "cue" was not based on anorexic activity or on increased activity as measured by enhanced avoidance responding.

ACKNOWLEDGMENTS

The authors express their sincere appreciation to Mr. Sammy Burrell, Mrs. Kathy Fisackerly, Mrs. Judy Flowers, and Dr. John Holbrook for their technical assistance in carrying out the reported research; to Mrs. Hind Hatoum for her statistical assistance in the preparation of this manuscript; and to Dr. John Bedford for his constructive criticism and review of the manuscript.

REFERENCES

Appel, J. B., & Freedman, D. X. Tolerance and cross-tolerance among psychotomimetic drugs. *Psychopharmacologia*, 1968, *13*, 267–274.

Bernstein, B., & Latimer, C. Behavioral facilitation: The interaction of imipramine and desipramine with amphetamine, alpha-pipradrol, methylphenidate and thozalinone. *Psychopharmacologia*, 1968, *12*, 338–345.

Carlton, P., & Didamo, P. Augmentation of the behavioral effects of amphetamine by atropine. *J. Pharmacol. Exp. Therap.*, 1961, *132*, 91–96.

Cox, R., & Maickel, R. Comparison of anorexigenic and behavioral potency of phenethylamines. *J. Pharmacol. Exp. Therap.*, 1972, *181*, 1–9.

Goldberg, M. E., & Ciofalo, V. B. Alteration of the behavioral effects of amphetamine by agents which modify cholinergic function. *Psychopharmacologia*, 1969, *14*, 142–149.

Harris, R. T., & Balster, R. L. Discriminative control by *dl*-amphetamine and saline of lever choice and response patterning. *Psychon. Sci.*, 1968, *10*, 105–106.

Harris, R. T., & Balster, R. L. An analysis of the function of drugs in the stimulus control of operant behavior. In T. Thompson & R. Pickens (Eds.), *Stimulus properties of drugs*. New York: Appleton, 1971. Pp. 111–132.

Huang, J., & Ho, B. T. Effects of nikethamide, picrotoxin and strychnine on amphetamine state. *European J. of Pharmacol.*, 1974, *29*, 175–178. (a)

Huang, J., & Ho, B. T. Discriminative stimulus properties of d-amphetamine and related compounds in rats. *Pharmacol. Biochem. Behav.*, 1974, *2*, 669–673. (b)

Innes, I., & Nickerson, M. Norepinephrine, epinephrine and the sympathomimetic amines. In L. Goodman & A. Gilman (Eds.), *The pharmacological basis of therapeutics*. (5th ed.) New York: Macmillan, 1975. Pp. 477–513.

Jones, C., Hill, H., & Harris, R. Discriminative response control by d-amphetamine and related compounds in the rat. *Psychopharmacologia*, 1974, *36*, 347–356.

Overton, D. A. Discriminative control of behavior by drug state. In T. Thompson & R. Pickens (Eds.), *Stimulus properties of drugs*. New York: Appleton, 1971. Pp. 87–110.

Owen, J. The influence of *dl, d,* and *l*-amphetamine and *d*-methamphetamine on a fixed-ratio schedule. *J. Exp. Anal. Behav.,* 1960, *3,* 293–310.

Randrup, A., & Munkvad, I. Biochemical, anatomical and psychological investigations of stereotyped behavior induced by amphetamines. In E. Costa & S. Garattini (Eds.), *Amphetamines and related compounds.* New York: Raven Press, 1970. Pp. 695–713.

Rech, R. Antagonism of reserpine behavioral depression by *d*-amphetamine. *J. Pharmacol. Exp. Therap.,* 1964, *146,* 369–376.

Rech, R., & Stolk, J. Amphetamine drug interactions that relate brain catecholamines to behavior. In E. Costa & S. Garattini, (Eds.), *Amphetamines and related compounds.* New York: Raven Press, 1970. Pp. 385–414.

Schechter, M., & Cook, P. Dopaminergic mediation of the interoceptive cue produced by d-amphetamine in rats. *Psychopharmacologia,* 1975, *42,* 185–193.

Schechter, M., & Rosecrans, J. D-amphetamine as a discriminative cue: Drugs with similar stimulus properties. *European J. Pharmacol.,* 1973, *21,* 212–216.

Scheel-Kruger, J. Behavioral and biochemical comparison of amphetamine derivatives, cocaine, benztropine and tricyclic antidepressant drugs. *European J. Pharmacol.,* 1972, *18,* 63–73.

Schuster, C. R., Dockens, W. J., & Woods, J. H. Behavioral variables affecting the development of amphetamine tolerance. *Psychopharmacologia,* 1966, *9,* 170–182.

Souskova, M., Benesova, O., & Roth, Z. The effect of chlorpromazine, phenmetrazine, imipramine and physostigmine on the exploratory and conditioned avoidance reaction in rats with different excitability of the central nervous system. *Psychopharmacologia,* 1964, *5,* 447–456.

Sparber, S. B., & Peterson, D. W. Operant behavioral demonstration of qualitative difference between the *d* and *l* isomers of amphetamine. In E. Usdin & S. Snyder (Eds.), *Frontiers in catecholamine research.* New York: Pergamon Press, 1974. Pp. 969–972.

Stretch, R., & Skinner, N. Methylphenidate and stimulus control of avoidance behavior. *J. Exp. Anal. Behav.,* 1967, *10,* 485–493.

Tilson, H. A., & Sparber, S. B. The effects of *d*- and *l*-amphetamine on fixed interval and fixed ratio behavior in tolerant and nontolerant rats. *J. Pharmacol. Exp. Therap.,* 1973, *187,* 72–79.

Von Voigtlander, P. F., & Moore, K. E. Dopamine: Release from the brain in vivo by amantadine. *Science* 1971, *174,* 408–410.

Waters, W. H., III, Richards, D. W., & Harris, R. T. Discriminative control and generalization of the stimulus properties of dl amphetamine in the rat. In J. M. Singh, L. Miller, & H. Lal (Eds.), *Drug addiction.* Vol. 1. *Experimental pharmacology.* Mt. Kisco, New York: Futura, 1972. Pp. 87–98.

Weissman, A., Koe, B. K., & Tenen, S. Antiamphetamine effects following inhibition of tyrosine hydroxylase. *J. Pharmacol. Exp. Therap.,* 1966, *151,* 339–352.

4

Discriminative Stimulus Properties of Central Stimulants

Beng T. Ho and Mary McKenna
Texas Research Institute of Mental Sciences
and
The University of Texas Health Sciences Center at Houston

The three central stimulants, *d*-amphetamine, cocaine, and methylphenidate, share certain common neurochemical mechanisms in eliciting excitatory behavior. The purpose of our study was to compare the discriminative stimulus properties of the three stimulants and to determine the underlying neurobiochemical basis of their control of behavior.

Amphetamine-induced stimulation of motor activity has been correlated with a noradrenergic system; stereotyped behavior has been associated with a dopaminergic system (Taylor & Snyder, 1971; Snyder, Taylor, Coyle, & Meyeroff, 1970b). More recently, Thornberg and Moore (1973) showed that amphetamine exerts its locomotor stimulant effect via a dopaminergic mechanism.

Amphetamine is a potent central nervous system (CNS) stimulant and inhibitor of catecholamine uptake (Glowinski, Axelrod, & Iverson, 1966; Snyder, Kuhan, Green, Coyle, & Shaskan, 1970a). Its action is related to inhibition of reuptake following release at the synaptic level (Raiteri, Levi, & Federico, 1974; Snyder *et al.*, 1970a; Von Voightlander & Moore, 1973). Amphetamine inhibits catecholamine synthesis uniquely in the corpus striatum as measured by tyrosine hydroxylase activity (Harris & Baldessarini, 1973). This effect may be a result of a neurophysiological feedback mechanism involving stimulation of dopamine receptors or gamma-aminobutyric acid neurons in the nigrostriatal pathway (Harris & Baldessarini, 1975).

The influence of amphetamine on the serotonergic system has not been extensively studied. Levels of serotonin (5-HT) in brain are unaffected by *d*-

amphetamine (Pletscher, Bartholini, Bruderer, Burkard, & Gey, 1964). Amphetamines block reuptake and facilitate release of 5-HT in nerve endings (Wong, Horng, & Fuller, 1973). Knapp, Mandell, and Geyer (1974) showed that amphetamines decrease 5-HT biosynthesis by reducing the conversion of tryptophan to 5-HT.

Methylphenidate is similar to amphetamine in its behavioral effects, that is, it increases locomotor activity and produces stereotyped behavior. This central-stimulant action has been correlated with a catecholaminergic neuronal system, specifically release of dopamine (DA; Costall & Naylor, 1974). Chiueh and Moore (1975) showed that methylphenidate preferentially releases dopamine from a reserpine-sensitive storage pool. Direct action by methylphenidate on DA receptors has been reported by Thornburg and Moore (1973).

Breese, Cooper, and Hollister (1975) suggested involvement of 5-HT in the action of methylphenidate, reporting that *p*-chlorophenylalanine potentiated methylphenidate activity while 5-HTP reduced it. Destruction of 5-HT neurons with 5,7-dihydroxytryptamine also potentiated the methylphenidate-induced activity. Thus, drugs that alter 5-HT function affect the central-stimulant properties of methylphenidate. Shih, Khachaturian, Barry, and Hanin (1976) suggest that the involvement of a cholinergic mechanism is responsible for methylphenidate's therapeutic effect on hyperactive children.

The neurochemical actions of cocaine resemble those of amphetamine and methylphenidate. Scheel-Kruger (1972), using whole rat brain, reported that cocaine reduces endogenous norepinephrine (NE) levels while increasing normetanephrine levels. DA levels remained unchanged. In contrast, tricyclic antidepressants, which inhibit catecholamine reuptake, altered neither DA nor NE levels. Scheel-Kruger thus concluded that cocaine releases NE. Increased plasma levels of NE were found in cats after cocaine administration (Whitby, Hertting, & Axelrod, 1960); examination of adrenal, heart, and spleen tissue indicated the elevation to result from cocaine-induced NE reuptake inhibition.

Cocaine interacts also with the serotonergic system. Friedman, Gershon, and Rotrosen (1975) reported decreased 5-HT turnover in rats. Segawa, Kuruma, Takatsuka, and Takagi, (1968) found that cocaine inhibits 5-HT uptake, and Knapp and Mandell (1972) showed that cocaine inhibits tryptophan uptake.

Dopaminergic mechanisms play a role in cocaine activity. Cocaine blocks the uptake of tritiated DA by brain tissues (Ross & Renyi, 1967) and induces DA release in stimulated brain slice studies (Farnebo & Hamberger, 1971; Starke & Montel, 1973). Thus, cocaine affects noradrenergic, dopaminergic, and serotonergic systems by blocking catecholamine reuptake and inhibiting 5-HT turnover.

Amphetamine and related compounds have been studied extensively for their discriminative-stimulus properties (Ho & Huang, 1975; Huang & Ho, 1974; Jones, Hill, & Harris, 1974; Overton, 1971; Schechter & Rosecrans,

1973; Tilson, Baker, & Gylys, 1975). Discrimination studies have been useful for classification of drugs (Barry, 1974; Silverman & Ho, 1976), and some reports have shown that stimulants do not generalize to hallucinogens or other drug classes and vice versa. Rats trained to discriminate amphetamine versus saline states do not generalize to ethanol (Ando, 1975), DOM (STP), DOET (Ho & Huang, 1975), LSD, or mescaline (Ho & Huang, 1975; Schechter & Rosecrans, 1973). Conversely, amphetamine does not generalize to the drug condition in rats trained to discriminate ethanol versus saline (Schechter, 1974), mescaline versus saline (Winter, 1975), or LSD versus saline (Schechter & Rosecrans, 1972).

Stimulants, however, generalize to other stimulants, and hallucinogens generalize to other hallucinogens. Thus, drugs with similar stimulus properties may have a similar mechanism of action. Generalization to amphetamine responding in animals trained to discriminate amphetamine from saline occurs with such other stimulants as cocaine (Ando, 1975; Huang & Ho, 1974), methamphetamine, methylphenidate (Huang & Ho, 1974) and *l*-amphetamine (Schechter & Rosecrans, 1973). Generalization among the hallucinogens is also known. For example, DOM (STP) and DOET generalize to mescaline stimulus properties (Winter, 1975), as do mescaline and psilocybin to LSD stimulus properties (Schechter & Rosecrans, 1972). The consistency of generalization within drug classes, when drug discrimination methods are used, is useful for classifying and comparing drugs. In the present study, reliability of this test measure would be further substantiated if cocaine-discriminating animals could recognize amphetamine and other stimulants as being similar to cocaine. Furthermore, generalization of stimulants in animals trained to discriminate cocaine versus saline and amphetamine versus saline would suggest that the mechanism by which the central stimulants exert their cueing effects is similar.

To determine the involvement of any monoaminergic system in drug-discrimination behavior, pretreatment is generally done with an inhibitor of rate-limiting enzyme in monoamine biosynthesis or with a releasor known to deplete that monoamine. If drug-induced responding is affected by monoamine deficiency resulting either from blocked synthesis or depletion, then involvement of the monoamine is indicated. Drugs used in this approach include α-methyl-*p*-tyrosine (α-MT) (an inhibitor of tyrosine hydroxylase, the rate-limiting enzyme in CA biosynthesis), *p*-chlorophenylalanine (an inhibitor of tryptophan hydroxylase, the rate-limiting enzyme in 5-HT biosynthesis), and reserpine, a depletor of 5-HT and catecholamines from storage vesicles. Amphetamine discrimination was shown to be antagonized by α-MT (Kuhn, Appel, & Greenberg, 1974; Schechter & Cook, 1975; Ho & Huang, 1975); the inhibitor, however, did not affect cocaine (McKenna & Ho, unpublished data) or methylphenidate discrimination (Silverman & Ho, 1976). Contrarily, our group has found that reserpine selectively antagonizes responding produced by cocaine and methylphenidate.

In some cases, when the action of a drug is not mediated indirectly through a monoamine, as described above, the drug acts directly on the receptor. We chose antagonists or receptor-blocking agents to test the neurochemical mechanism of drug discriminative behavior because, regardless of whether the drug action is direct or indirect, monoamine participation would be ruled out if the action is not blocked by pretreatment with a receptor blocker. Blocking agents commonly used are atropine (cholinergic), cinanserin or methysergide (serotonergic), phenoxybenzamine or phentolamine (α-adrenergic), propranolol (β-adrenergic), haloperidol or pimozide (dopaminergic). While the other blocking agents fail to block the discriminative properties of amphetamine, pimozide acts as an antagonist (Ho & Huang, 1975). These results, together with the finding of Schechter and Cook (1975) with haloperidol, suggest a role for dopamine in amphetamine discriminative behavior. Further substantiation was obtained from the studies of Ho and Huang (1975) in which L-DOPA, a precursor of DA, was found capable of generating the amphetamine-cueing effect. The dopaminergic response was also confirmed by the use of apomorphine, a dopaminergic receptor stimulant (Schechter & Cook, 1975). The present study was done to determine whether a similar dopaminergic mechanism is responsible for producing cocaine discriminative responding.

METHODS

Subjects

Forty male Sprague-Dawley rats, initially weighing 200–250 gm and obtained from Horton Laboratories, Los Gatos, California, served as subjects. They were individually housed. Water was freely accessible throughout the experiment, and animals were maintained at 80–85% of expected free-feeding weight with Purina Rat Chow. The daily schedule was 12 hours light/12 hours dark.

Apparatus

Five two-lever operant chambers (Scientific Prototype, Model A-100) were enclosed in sound-attenuating chambers (Scientific Prototype, Model SPC-300) and equipped with fans to circulate air. A brass food tray centered below the levers was connected to a pellet dispenser (BRS/Foringer Model PDC) located behind the lever panel. Noyes 45-mg food pellets served as reinforcement. Lighting in the chamber was provided by 7-W house lights. Behavioral scheduling and data collection were controlled by Grason Stradler programming equipment.

Training

Twenty rats were trained to discriminate .8 mg/kg of *d*-amphetamine versus 1 ml/kg of saline. A second group of 20 rats was trained to discriminate 10 mg/kg of cocaine versus 1 ml/kg of saline. All injections were given intraperitoneally (i.p.) 15 min before training sessions. Sessions were conducted 5 days a week. Duration of each training session was 30 min. Before injections, all animals were shaped to respond with both operant levers so that only responses made on the appropriate lever at intervals longer than 15 sec were reinforced (DRL 15-sec schedule). Following schedule training, animals were injected daily with saline or the appropriate drug. Reinforcement during sessions was contingent on administration of saline or drug; that is, in sessions following drug injections, only responses made on the drug lever were reinforced; in sessions following saline injections, only responses made on the saline lever were reinforced.

Extinction Testing

Every fourth day during acquisition training, animals were tested for accuracy. Extinction testing involved the absence of reinforcement. Animals were considered accurate when lever responding reached 80% or more on the appropriate lever. After saline administration, at least 80% of the responses were on the saline lever; after drug administration, at least 80% of the responses were on the drug lever. When animals attained 80% accuracy under both conditions, generalization testing of novel drugs began. Training sessions continued for 4 days a week, with generalization testing on the fifth day. Stimulants for generalization testing were administered intraperitoneally 15 min before testing. Phentolamine was administered 45 min, propranolol 60 min, and pimozide 4 hr prior to testing.

Drugs

Dosages of all drugs are expressed as salt form except for pimozide, which is a free base. Solutions were prepared with isotonic saline containing .9% benzyl alcohol as a preservative; pimozide was moistened with a few drops of glacial acetic acid dissolved in 30% propylene glycol. Methylphenidate hydrochloride and phentolamine hydrochloride were kindly supplied by CIBA Pharmaceutical Company, Summit, New Jersey, pimozide by Professor Paul Janssen of Janssen Pharmaceutical Research Laboratories, Belgium. Cocaine hydrochloride, *d*-amphetamine sulfate, and propranolol hydrochloride were purchased from commercial sources.

RESULTS

The two central stimulants, cocaine and methylphenidate, generalized to the amphetamine discriminative stimulus properties (Table 4.1). A dose–response study indicated that three times the amphetamine training dose (2.5 mg/kg) is required for methylphenidate to produce generalization to the amphetamine stimulus. At least six times the amphetamine training dose (5 mg/kg) is

Table 4.1
Generalization by Cocaine and Methylphenidate to *d*-Amphetamine Discriminative Effect

	Dose (mg/kg)	Amphetamine lever choice (%)	
		5 min	10 min
Saline	(1 ml/kg)	10.6 ± 4.7	16.1 ± 1.1
d-Amphetamine	.4	44.2 ± 8[a]	48.3 ± 9.2[b]
	.8	81.1 ± 2	82.5 ± 2.8
Methylphenidate	1	40 ± 7.7[c]	45.6 ± 6.3[c]
	2.5	89.9 ± 7.5	88.1 ± 6.8
Cocaine	2.5	49.2 ± 18.3	30.4 ± 13.1
	5	77.4 ± 10.7	77.4 ± 9.6
	7.5	92.7 ± 4.2	90 ± 5.1
Methylphenidate + *d*-amphetamine	1 + .4	98 ± 2	92.7 ± 4.7
Cocaine + *d*-amphetamine	2.5 + .4	91.6 ± 5.2	91.6 ± 5.2

Note: Each value represents the mean (± S.E.M.) of five animals. All animals were used more than once but not for the same generalization test.

Significantly different from *d*-amphetamine (.8 mg/kg) control group: $p < .003$ ([a]); $p < .05$ ([b]); $p < .08$ ([c]) (Mann–Whitney U-Test).

required before cocaine generates the amphetamine cue. While 1 mg/kg of methylphenidate, 2.5 mg/kg of cocaine, and .4 mg/kg of amphetamine resulted in about 50% of the amphetamine-correct responding, combining these subthreshold doses of methylphenidate or cocaine with amphetamine produced the amphetamine response.

Methylphenidate, 2.5 mg/kg, and *d*-amphetamine, .8 mg/kg, generated the cocaine response (Table 4.2). Relative potency of stimulants tested in

Table 4.2
Generalization by *d*-Amphetamine and Methylphenidate to Cocaine Discriminative Effect

	Dose (mg/kg)	Cocaine lever choices (%)	
		5 min	10 min
Saline	(1 ml/kg)	13.7 ± 4.3	15.9 ± 4.3
Cocaine	2.5	19.6 ± 10[a]	23.2 ± 10.2[a]
	5	86.2 ± 4.5	83.7 ± 3.3
	10	83.4 ± 7.1	86.6 ± 5.3
Methylphenidate	2.5	75.5 ± 9.8	82.5 ± 6.7
d-Amphetamine	.25	29.1 ± 9.9[b]	45.6 ± 12.4[c]
	.8	79.4 ± 4.3	86.2 ± 3.5
d-Amphetamine + cocaine	.25 + 2.5	73.3 ± 11.5	76.6 ± 12

Note: Each value represents the mean (± S.E.M.) of five animals.

Significantly different from cocaine (5 mg/kg) control group: $p < .02$ ([a]); $p < .006$ ([b]); $p < .07$ ([c]) (Mann–Whitney U-Test).

this group of animals corresponds to data obtained with amphetamine-discriminating animals. In the animals trained to discriminate cocaine from saline, *d*-amphetamine, .25 mg/kg, and cocaine, 2.5 mg/kg, produced subthreshold cocaine-appropriate responses. Combination of these subthreshold doses generated the cocaine cue.

Use of catecholamine-receptor blockers provided further characterization of cocaine discriminative stimulus properties. Phentolamine, an α-adrenergic receptor blocker, propranolol, a β-adrenergic receptor blocker, and pimozide, a dopaminergic receptor blocker, were tested for their effects on discriminative response control by cocaine. Phentolamine neither blocked nor generated cocaine response. In combination with a subthreshold dose (2.5 mg/kg) of cocaine, however, it produced cocaine responding (Table 4.3). The potentiation of cocaine responding was also observed with propranolol. In fact, propranolol administration not only potentiated cocaine responding but resulted in generalization of propranolol to cocaine. Pimozide did not generate or potentiate cocaine responding; instead, it blocked the cocaine discrimination (Table 4.4).

In summary, dopaminergic receptor blockade produced by pimozide abolished cocaine discriminative stimulus properties. The α-adrenergic receptor blocker, phentolamine, potentiated cocaine cueing, while the β-

Table 4.3
Effects of Adrenergic Receptor-Blocking Agents on Cocaine-Induced Discriminative Responding

	Dose (mg/kg)	Cocaine lever choice (%)	
		5 min	10 min
Cocaine	5	86.2 ± 4.5	85.7 ± 3.3
Phentolamine	5	12.4 ± 6	26.9 ± 7.5
	10	16.4 ± 4.1	23.1 ± 4.2
Phentolamine + cocaine	10 + 2.5	70.7 ± 14.9[b]	86.5 ± 5.7[a]
Cocaine	2.5	12.6 ± 3.4	17.1 ± 3.3
Propranolol	10	18.2 ± 4	18.6 ± 4.2
	15	87.9 ± 3.1	86.7 ± 4.2
Propranolol + cocaine	10 + 2.5	90.6 ± 3.8[c]	93.2 ± 2.8[d]
Cocaine	2.5	49.3 ± 14.4	48 ± 15.2
Propranolol + phentolamine + cocaine	5 + 5 + 2.5	92.7 ± 3.6[b]	90.8 ± 2.5[c]
Cocaine	2.5	44.4 ± 12.7	49.6 ± 11.6
Propranolol + phentolamine	10 + 5	35.7 ± 10.9	41.8 ± 10.3

$p < .002$ ([a]); $p < .02$ ([b]); $p < .05$ ([c]); $p < .06$ ([d]) compared to individual cocaine (2.5 mg/kg) control (Mann–Whitney U-Test).

Table 4.4
The Effect of Dopamine Receptor-Blocking Agent on Cocaine-Induced Discriminative Responding

	Dose (mg/kg)	Cocaine lever choice (%)	
		5 min	10 min
Pimozide + cocaine	1 + 5	39.7 ± 1.6[a]	47.8 ± 1.2[a]
Cocaine	5	96.5 + 1.3	98.5 + .9
Pimozide + saline	1	15.7 ± 6.8	19.6 ± 9

$p < .001$ ([a]) compared to cocaine control (Mann–Whitney U-Test).

noradrenergic receptor blocker, propranolol, not only potentiated cocaine responding but actually generated a cocaine response.

DISCUSSION

Like *d*-amphetamine, cocaine affects dopaminergic, noradrenergic, and serotonergic systems. It inhibits 5-HT turnover and DA and NE release and blockade of reuptake. Our unpublished results with cinanserin, a serotonergic receptor blocker, indicate serotonin does not play a role in cocaine discriminative stimulus properties. Cocaine discrimination seems to be mediated by the catecholamines. DA receptor blockade abolished cocaine responding, suggesting that DA receptors are essential for the cocaine cue. Since our preliminary study showed the antagonism of cocaine discrimination by reserpine, the action of cocaine is therefore mediated by release of DA. The finding that noradrenergic receptor blockade produces potentiation of cocaine discriminative responding suggests a noradrenergic inhibitory role. It is perhaps feasible to visualize the cocaine discriminative behavior as governed by two opposite neuronal systems, with the dopaminergic neuronal system as excitatory, while an inhibitory or regulatory function is provided through the noradrenergic neuronal system. The blockade of noradrenergic receptors would then suppress the inhibition, allowing activation of the dopaminergic system. This might account for the potentiation of cocaine responding in the presence of phentolamine and propranolol.

In addition to possessing many common neurochemical properties, amphetamine and cocaine share such similar pharmacological activities as increasing motor activity, producing stereotyped behavior, exerting antifatigue activity, reducing appetite and thirst, as well as inducing paranoid psychosis (Bejerot, 1970; Simon, 1973). In this study, we have shown further that the discriminative stimulus properties of these drugs are similar. This finding substantiates the reliability of drug discrimination for use in drug classification. Dose–response curves for amphetamine, cocaine, and methylphenidate in cocaine and amphetamine-discriminating animals are similar. The relative potency obtained from these dose-response curves is amphetamine > methylphenidate > cocaine.

Previously, the generalization of cocaine and methylphenidate to the amphetamine discrimination was attributed to involvement of a common mechanism for this class of pharmacological agents (Huang & Ho, 1974). Results of the present study support the hypothesis by showing the generalization of cocaine-appropriate responding by *d*-amphetamine or methylphenidate alone and the production of amphetamine and cocaine responding by a combination of subthreshold doses of two stimulants.

Intraventricular administration of amphetamine and use of parahydroxyamphetamine, which is active peripherally but devoid of central stimulant properties, show amphetamine stimulus properties to be centrally, not peripherally, mediated (Jones *et al.*, 1974; Richards, Harris, & Ho, 1973). A central site of action for cocaine discrimination has been suggested by a preliminary study (McKenna & Ho, unpublished data), showing N-methylcocaine, a quaternary compound presumably being excluded from the CNS, does not generalize to cocaine.

REFERENCES

Ando, K. The discriminative control of operant behavior by intravenous administration of drugs in rats. *Psychopharmacologia,* 1975, *45,* 47–50.

Barry, H., III. Classification of drugs according to their discriminable effects in rats. *Federation Proc.,* 1974, *33,* 1814–1824.

Bejerot, N. A comparison of the effects of cocaine and synthetic central stimulants. *Brit. J. Addict.,* 1970, *65,* 35–37.

Breese, G. R., Cooper, B. R., & Hollister, A. S. Involvement of brain monoamines in the stimulant and paradoxical inhibitory effects of methylphenidate. *Psychopharmacologia,* 1975, *44,* 5–10.

Chiueh, C. C., & Moore, K. E. The relative importance of dopaminergic and noradrenergic neuronal systems for the stimulation of locomotor activity induced by amphetamine and other drugs. *Neuropharmacol.,* 1975, *12,* 853–866.

Costall, B., & Naylor, R. J. The involvement of dopaminergic systems with the stereotyped behavior patterns induced by methylphenidate. *J. Pharm. Pharmacol.,* 1974, *26,* 30–33.

Farnebo, L. O., & Hamberger, B. Drug-induced changes in the release of ^{3}H-monoamines from field stimulated rat brain slices. *Acta Physiol. Scand.,* 1971, *371,* 35–44.

Friedman, E., Gershon, S., & Rotrosen, J. Effects of acute cocaine treatment on the turnover of 5-hydroxytryptamine in the rat brain. *Brit. J. Pharmacol.,* 1975, *54,* 61–64.

Glowinski, J., Axelrod, J., and Iverson, L. L. Regional studies of catecholamines in the rat brain. Effects of drugs on the disposition and metabolism of ^{3}H-norepinephrine and ^{3}H-dopamine. *J. Pharmacol. Exp. Therap.,* 1966, *153,* 30–41.

Harris, J. E., & Baldessarini, R. J. Amphetamine-induced inhibition of tyrosine hydroxylation in homogenates of rat corpus striatum. *J. Pharm. Pharmacol.,* 1973, *25,* 755–757.

Harris, J. E., & Baldessarini, R. J. Amphetamine-induced inhibition of tyrosine hydroxylation in homogenates of rat corpus striatum. *Neuropharmacol.,* 1975, *14,* 457–471.

Ho, B. T., & Huang, J. T. Role of dopamine in *d*-amphetamine-induced discriminative responding. *Pharmacol. Biochem. Behav.,* 1975, *3,* 1085–1092.

Huang, J. T., & Ho, B. T. Discriminative stimulus properties of *d*-amphetamine and related compounds in rats. *Pharmacol. Biochem. Behav.,* 1974, *2,* 669–673.

Jones, C. N., Hill, H. F., & Harris, R. T. Discriminative response control by *d*-amphetamine and related compounds in the rat. *Psychopharmacologia,* 1974, *36,* 347–356.

Knapp, S., & Mandell, A. J. Narcotic drugs: Effects on the serotonin biosynthetic systems of the brain. *Science,* 1972, *177,* 1209–1211.

Knapp, S., Mandell, A. J., & Geyer, M. A. Effects of amphetamine on regional tryptophan hydroxylase activity and synaptosomal conversion of tryptophan to 5-hydroxytryptamine in rat brain. *J. Pharmacol. Exp. Therap.,* 1974, *189,* 676–689.

Kuhn, D. M., Appel, J. B., & Greenberg, I. An analysis of some discriminative properties of *d*-amphetamine. *Psychopharmacologia,* 1974, *38,* 57–66.

Overton, D. A. Discriminative control of behavior by drug state. In T. Thompson & P. Pickens (Eds.), *Stimulus properties of drugs.* New York: Appleton, 1971. Pp. 87–110.

Pletscher, A., Bartholini, G., Bruderer, H., Burkard, W. P., & Gey, K. F. Chlorinated arylalkylamines affecting the cerebral metabolism of 5-hydroxytryptamine. *J. Pharmacol. Exp. Therap.,* 1964, *145,* 344–350.

Raiteri, M., Levi, G., & Federico, R. *d*-Amphetamine and the release of ^{3}H-norepinephrine from synaptosomes. *European J. Pharmacol.,* 1974, *28,* 237–240.

Richards, D. W., III, Harris, R. T., & Ho, B. T. Central control of *d*-amphetamine-induced discriminative stimuli. Abstract 3rd annual meeting, Society for Neuroscience, San Diego, California, 1973.

Ross, S. B., & Renyi, A. L. Inhibition of the uptake of tritiated catecholamines by antidepressant and related agents. *European J. Pharmacol.,* 1967, *2,* 181–186.

Schechter, M. D. Effect of propranolol, *d*-amphetamine and caffeine on ethanol as a discriminative cue. *European J. Pharmacol.,* 1974, *29,* 52–57.

Schechter, M. D., & Cook, P. G. Dopaminergic mediation of the interoceptive cue produced by *d*-amphetamine in rats. *Psychopharmacologia,* 1975, *42,* 185–193.

Schechter, M. D., & Rosecrans, J. A. Lysergic acid diethylamide (LSD) as a discriminative cue: Drugs with similar stimulus properties. *Psychopharmacologia,* 1972, *26,* 313–316.

Schechter, M. D., & Rosecrans, J. A. *d*-Amphetamine as a discriminative cue: Drugs with similar stimulus properties. *Eur. J. Pharmacol.,* 1973, *21,* 212–216.

Scheel-Kruger, J. Behavioral and biochemical comparison of amphetamine derivatives, cocaine, benztropine and tricyclic anti-depressant drugs. *European J. Pharmacol.,* 1972, *18,* 63–73.

Segawa, T., Kuruma, I., Takatsuka, K., & Takagi, H. The influences of drugs on the uptake of 5-hydroxytryptamine by synaptic vesicles of rabbit brain stem. *J. Pharm. Pharmacol.,* 1968, *20,* 800–801.

Shih, T. M., Khachaturian, Z. S., Barry, H., & Hanin, I. Cholinergic mediation of the inhibitory effect of methylphenidate on neuronal activity in reticular formation. *Neuropharmacol.,* 1976, *15,* 55–60.

Silverman, P. B., & Ho, B. T. Discriminative response control by psychomotor stimulants. In H. Lal (Ed.), Discriminative stimulus properties of drugs. New York: Raven Press, 1976.

Simon, P. Psychopharmacological profile of cocaine. In E. Usdin & S. H. Snyder (Eds.), Frontier in catecholamine research. New York: Pergamon, 1973. Pp. 1043–1044.

Snyder, S. H., Kuhan, M. J., Green, A. I., Coyle, J. T., & Shaskan, E. G. Uptake and subcellular localization of neurotransmitters in the brain. *Intern. Rev. Neurobiol.,* 1970, *13,* 127. (a)

Snyder, S. H., Taylor, K. M., Coyle, J. T., and Meyeroff, J. L. The role of brain dopamine in behavioral regulation and the actions of psychotropic drugs. *Am. J. Psychiat.,* 1970, *127,* 199–207. (b)

Starke, K., & Montel, H. Alpha-receptor-mediated modulation of transmitter release from central noradrenergic neurons. *Naunyn-Schmiedebergs Arch. Exp. Pathol. Pharmakol.,* 1973, *279,* 53–60.

Taylor, K. M., & Snyder, S. Differential effects of *d*- and *l*-amphetamine on behavior and on catecholamine disposition in dopamine and norepinephrine containing neurons of the rat brain. *Brain Res.,* 1971, *28,* 295–309.

Thornburg, J. E., & Moore, K. E. The relative importance of dopaminergic and noradrenergic neuronal systems for the stimulation of locomotor activity induced by amphetamine and other drugs. *Neuropharmacology,* 1973, *12,* 853–866.

Tilson, H. A., Baker, T. G., & Gylys, J. A. A comparison of the discriminative stimulus properties of R-2, 5-dimethoxy-4-methylamphetamine (R-DOM) and S-amphetamine in the rat. *Psychopharmacologia,* 1975, *44,* 225–228.

Von Voightlander, P. F., & Moore, K. E. Involvement of nigrostriatal neuron in the *in vivo* release of dopamine by amphetamine, amantadine and tyramine. *J. Pharmacol. Exp. Therap.,* 1973, *184,* 542.

Whitby, L. G., Hertting, G., & Axelrod, J. Effect of cocaine on the disposition of noradrenaline labeled tritium. *Nature,* 1960, *187,* 604–605.

Winter, J. C. The effects of 2,5-dimethoxy-4-methylamphetamine (DOM), 2,5-dimethoxy-4-ethylamphetamine (DOET), *d*-amphetamine and cocaine in rats trained with mescaline as a discriminative stimulus. *Psychopharmacologia,* 1975, *44,* 29–32.

Wong, D. T., Horng, J. S., & Fuller, R. W. Kinetics of serotonin accumulation into synaptosomes of rat brain—effects of amphetamines and chloroamphetamines. *Biochem. Pharmacol.,* 1973, *22,* 311–322.

5

The Role of Serotonin in the Discriminative Stimulus Properties of Mescaline

Ronald G. Browne
Psychiatry Department
School of Medicine
University of California, San Diego

INTRODUCTION

Toxic models of functional psychoses have played a prominent role in the development of biological psychiatry as a scientific discipline. The history may be traced at least to Thudichum (1884), who felt that many forms of insanity were the result of intoxications. A considerable number of substances have been measured or sought in the body fluids of schizophrenics. Most biochemical hypotheses relating to schizophrenia revolved around the idea that schizophrenia is caused by toxic material formed in brain cells from faulty metabolism or coming to the brain from other parts of the body or from sources outside the body proper, for example, intestinal bacteria. Theories about the origin of the toxin implicated disorders of energy metabolism or electron transport, functions or metabolism of nervous transmitter amines, or autoimmunization reactions. A currently popular idea is based on the observation that natural substances like mescaline produce abnormal mental states. The basic assumption—that the mental states produced by drugs are comparable to naturally occurring psychosis—is open to much speculation. Nevertheless, several points suggest that the study of psychotomimetic drugs may contribute to an understanding of psychoses.

The use of drugs to ameliorate naturally occurring psychoses has suggested that if chemical treatment is effective, the disorder must have a chemical cause. Many psychotomimetic drugs (e.g., mescaline, *N,N*-dimethyltryptamine, or bufotenin) are methylated derivatives of neurotransmitter amines. The hypothesis that abnormal neurotransmitter methylation is involved in psychosis is supported by investigations of an enzyme in rabbit lung (Axelrod, 1962), normal human brain (Mandell, Buckingham, & Segal, 1971), and rat brain that could *N*-methylate indoleethylamines to form the hallucinogenic substances *N,N*-dimethyltryptamine (DMT) and bufotenin both *in vitro* (Mandell *et al.*, 1971; Saavadra & Axelrod, 1972) and *in vivo* (Saavadra & Axelrod, 1972). In addition, Friedhoff, Schweitzer, and Miller (1972) suggested that mammalian liver may be capable of synthesizing mescaline.

Biochemical, behavioral, and physiological studies have attempted to elucidate the mechanism of action of psychotomimetic drugs. These studies, however, have often been either inconsistent or lacking in specificity with nonhallucinogenic drugs that produce similar effects. The recent use of psychoactive drugs as discriminative stimuli has opened a new vista to explore the mechanism of action of such drugs. While the nature of the discriminative cue produced by drugs is unknown, drugs with similar pharmacological properties tend to produce similar cues. When animals were trained to discriminate a particular drug state and then tested with other drugs, response control generalized between drugs of the same pharmacological classification (Barry & Kubena, 1972; Barry, Steenberg, Manian, & Buckley, 1974; Kubena & Barry, 1969; Overton, 1966, 1967, 1968). For example, Schechter and Rosecrans (1972a) trained rats to discriminate lysergic acid diethylamide (LSD) from saline. Subjects responded similarly to the cueing effect of LSD when tested with mescaline but not with amphetamine (Schechter & Rosecrans, 1972a, 1973). Furthermore, when equivalent doses of LSD and mescaline were used as the alternating drug state, no discrimination was established (Hirschhorn & Winter, 1971), which suggests that LSD and mescaline produce similar interoceptive cues. Control of dosage and injection-testing intervals is important in such studies since generalization decrements have occurred when subjects were tested with doses significantly different from training doses (Overton, 1972; Waters, Richards, & Harris, 1972; Huang & Ho, 1974; Kuhn, Appel, & Greenberg, 1974) or at different time intervals (Hirschhorn & Rosecrans, 1974a; Huang & Ho, 1974; Kuhn *et al.*, 1974; Overton, 1972; Schechter & Rosecrans, 1971).

While the mechanism by which drugs produce interoceptive cues remains unclear, several studies have attempted to dissociate peripheral from central nervous system (CNS) effects of drug cues. Drugs that exhibit central activity like *dl*-amphetamine or sodium pentobarbital acquire discriminative response control much more readily than drugs that do not cross the blood–brain barrier (e.g., atropine methyl nitrate, gallamine). Further evidence for CNS locus of discriminative control is based on the work of Morrison and Stephenson (1969), who used nicotine discrimination in a two-lever operant task for water reinforcement. After rats had acquired a saline versus nicotine discrimination, they were tested with drugs (amphetamine, adrenaline, pentobarbital, physostigmine, caffeine, or galline) designed to mimic nicotine's peripheral effects. The results indicated that rats did not generalize to the cue produced by nicotine. More conclusive evidence for a CNS locus of discriminative control was the observation that stimulus generalization occurred following direct administration of drugs into the brain. Browne, Harris, and Ho (1974) demonstrated that rats trained to discriminate systemically administered mescaline from saline exhibited a dose-dependent generalization to the mescaline state following intraventricular injection of mescaline. Similar results have been observed for amphetamine (Richards, Harris, & Ho, 1973) and nicotine (Schechter, 1973b) administered intraventricularly. These studies on the use of drug discrimination paradigms indicate that such paradigms offer a highly specific and quantitative method for evaluating the behavioral effects of centrally acting drugs.

A variety of pharmacological manipulations have been employed in an attempt to elucidate the neurochemical mechanisms by which drugs act as discriminative stimuli. First, it is essential to determine if the discriminative stimulus properties of a drug are mediated by the parent compound or by a metabolite of the drug. This was especially important in the behavioral activity of mescaline in which evidence implicated a metabolite (Friedhoff & Goldstein, 1962; Harley-Mason, Laird, & Smythies, 1958; Smythies & Sykes, 1964). Generalization studies using all known metabolites of mescaline do not indicate a role of a metabolite in the interoceptive cueing properties of mescaline (Browne & Ho, 1975). Other studies that examined stimulus generalization produced by metabolites of the training drug also yielded negative results. For example, the *para*-hydroxylated metabolite of amphetamine did not generalize to the amphetamine state (Huang & Ho, 1974; Jones, Hill, & Harris, 1974; Richards *et al.*, 1973).

Other approaches have examined generalization, antagonism, or potentiation of the cue produced by the training drug following administration of drugs known to enhance or diminish the effects of putative neurotransmitters. Perhaps the most extensive work using the pharmacological pretreatment approach utilized nicotine as a cue. In rats trained to discriminate nicotine from saline, the nicotinic-cholinergic receptor antagonist mecamylamine completely blocked the stimulus properties of nicotine (Hirschhorn & Rosecrans, 1974a; Schechter & Rosecrans, 1972b,c). Quaternary analogs of nicotine, which do not cross the blood–brain barrier, produced saline responses (Schechter and Rosecrans, 1972c). Similarly, nicotinic-cholinergic receptor antagonists, which do not cross the blood–brain barrier (chlorisondamine, hexamethonium), had no effect on the nicotine cue (Schechter & Rosecrans, 1972c). Arecoline (a muscarinic-cholinergic receptor stimulant) was devoid of nicotinelike cueing properties (Schechter & Rosecrans, 1972c) and could not be antagonized by mecamylamine (Schechter & Rosecrans, 1972b). The cue produced by arecoline, however, was antagonized by atropine, a muscarinic-cholinergic receptor antagonist (Schechter & Rosecrans, 1972d). These results strongly suggest that the cue produced by nicotine is mediated by central nicotinic receptor stimulation.

Other studies examined the effect of depleting brain neurotransmitters on the discriminative properties of various drugs. For example, reductions in brain catecholamines with the tyrosine hydroxylase inhibitor alpha-methyl-*para*-tyrosine (α-MPT) has been shown to antagonize the stimulus properties of amphetamine (Kuhn *et al.*, 1974). Depletion of brain serotonin (5-HT) with the tryptophan hydroxylase inhibitor *para*-chlorophenylalanine (PCPA) completely inhibited the interoceptive cue produced by ethanol (Schechter, 1973a). Reduction of brain 5-HT content but not catecholamine levels reportedly abolished the cue produced by morphine (Rosecrans, Goodloe, Bennett, & Hirschhorn, 1973), which was also antagonized by the narcotic antagonist naloxone (Hirschhorn & Rosecrans, 1974b; Rosecrans *et al.*, 1973).

Since the early demonstration by Gaddum (1953) that very low concentrations of LSD antagonized certain actions of 5-HT, many attempts to implicate 5-HT in the mechanism of action of psychomimetic drugs have been reported (Anden, Corrodi, Fuxe, & Hokfelt, 1968; Appel, Lovell, & Freedman, 1970; Boakes, Bradley, Briggs, & Dray, 1970; Freedman, Gottlieb, & Lovell, 1970; Tilson & Sparber, 1972; Cheng, Long, Barfknecht, & Nichols, 1973; Dyer & Gant, 1973; Haigler & Aghajanian, 1973; Anden, Corrodi, Fuxe, & Meek, 1974; Dyer, 1974). While increasing evidence suggests that these drugs directly stimulate 5-HT receptors both peripherally and centrally (Anden *et al.*, 1968, 1974; Cheng *et al.*, 1973; Dyer, 1974; Dyer & Gant, 1973; Haigler & Aghajanian, 1973), there is as yet very little behavioral data to support this contention. Indeed, the only consistent finding is that depletion of brain 5-HT potentiates the behavioral effects of LSD (Appel *et al.*, 1970; Cameron & Appel, 1973; Tonge & Leonard, 1972). Attempts to examine the influence of serotonergic receptor blockers on the behaviorally disrupting effects of psychotomimetic agents have produced conflicting data showing either potentiation (Tonge & Leonard, 1972), antagonism (Corne & Pickering, 1967; Winter, 1969), or no effect (Hirschhorn & Rosecrans, 1974b).

The use of mescaline as a discriminative stimulus in the control of operant behavior by interoceptive cueing has been well demonstrated in several behavioral paradigms (Browne *et al.*, 1974; Hirschhorn & Winter, 1971; Schechter & Rosecrans, 1972a; Winter, 1973). The present studies were undertaken to evaluate the role of 5-HT in the discriminative stimulus properties of mescaline. It was hypothesized that if the cue produced by mescaline is related to 5-HT antagonism, then centrally acting 5-HT receptor blockers should produce stimulus properties similar to mescaline. Alternatively, if the mescaline cue is a concomitant of 5-HT receptor stimulation, then these same 5-HT antagonists should reduce the discriminability of mescaline. Furthermore, it was anticipated that depletion of brain 5-HT with either PCPA or midbrain raphe lesions should potentiate the discriminability of mescaline, as has been reported for LSD (Cameron & Appel, 1973).

METHODS

Subjects

Male Sprague-Dawley rats (225–250 gm) obtained from Horton Laboratories (Oakland, California) served as subjects. Throughout the study, the rats were housed individually in home cages with water freely available. Purina Rat Chow was fed after daily experimental sessions and on weekends in quantities to maintain the rats between 80 and 85% of their expected free-feeding weight based on the supplier's growth chart.

Behavioral Apparatus

Five two-lever operant chambers (Scientific Prototype, Model A-100) enclosed in sound-attenuating chambers (Scientific Prototype, Model SPC-300) and equipped with fans to circulate fresh air were used. The two operant levers (Scientific Prototype, Model PCS-100) separated by 8 cm were mounted on the manipulandum approximately 3 cm above the grid floor of the operant chamber. A brass food tray located on the panel between the levers was connected to a pellet dispenser (BRS/Foringer Model PDC) situated behind the panel. Reinforcement consisted of single 45-mg Noyes pellets (Standard Formula). Illumination was provided by a 7-W house light mounted in the ceiling of the sound-attenuated chamber. Behavioral contingencies and data collection were controlled by programming equipment (Grason-Stadler 1200 series) located in the same room.

Discrimination Training

Pretraining

On the first day of pretraining, the rats were placed in the operant chambers for 30 min with noncontingent delivery of food pellets on a variable interval schedule (VI 60 sec) as well as on a continuous reinforcement (CRF) schedule programmed on both levers. Two additional daily 30-min sessions of only CRF on either lever was given before introducing a differential reinforcement of low response rate (DRL) schedule. When a DRL schedule is in effect, a specified amount of time must elapse between the delivery of one reinforcement and the availability of the next reward. Premature responses reset the interval timer. The first response after the specified interval has elapsed will result in reinforcement. Responses on the lever designated "incorrect" also reset the interval timer to prevent chaining of responses between the two levers. Two days on one lever under DRL-5 were followed by 2 days of training on the other level under DRL-5. Subsequently, 2 days on each lever on a DRL-10 followed by 2 days on each lever on DRL-15 were given. DRL-15 (unlimited hold) served as the schedule of reinforcement throughout the remainder of the experiment.

Training and extinction testing

Twenty-five rats were run in five daily sessions per week. Intraperitoneal injections of either 15 mg/kg of mescaline (as the hydrochloride salt) in isotonic saline or saline alone were given 15 min before the 30-min sessions. Injection volumes were 1 ml/kg. The sequence of injections was a counterbalanced order of all the possible permutations of drug (D) and saline (S) with the limitation that not more than two consecutive sessions followed either drug or saline. On session days following drug injections, only responses made on the right lever were reinforced; on session days following saline injections, only responses made on the left lever were rewarded. Responses of the inappropriate lever resulted in the resetting of the DRL interval timer.

On the first day and every fourth session thereafter, no responses during the first 5 min of the 30-min session were reinforced (extinction). The degree of discrimination between mescaline and saline is defined as the percentage of total responses made on the appropriate lever in the absence of reinforcement. The semirandom lever sequence used in this design resulted in an equal number of extinction tests in which the drug state was the same or different from the one imposed on the previous day. Each subject received five extinction tests following mescaline and five tests following saline.

Generalization Testing

Time-course and dose-response

In order to test the degree of generalization after mescaline at various doses or at various times after injection, the rats were given a series of four daily training sessions in the order SDDS and tested during a 10-min extinction on a fifth session. No training occurred after these testing sessions nor for the following two days. On these test days, the subjects were injected with mescaline 15 mg/kg i.p. and allowed to remain in their home cages for 15 to 360 min before the initiation of the 10-min extinction test. For dose–response determinations, the rats were injected with mescaline (5–25 mg/kg) 15 min before the test. The order in which doses and times after injection of mescaline were tested was randomized between subjects and across weekly tests so that no rat ever experienced the same time course or dose of drug more than once.

Serotonin antagonists

The rats were run in the SDDS sequence of four daily sessions and one 10-min extinction test session per week. On these test days the subjects were pretreated with cinanserin, methysergide, cyproheptadine, or xylamidine tosylate as specified below; 15 min before the initiation of testing, subjects received an i.p. injection of either mescaline (15 mg/kg) or saline. The order in which pretreatment drugs were tested was randomized between subjects and across weekly tests so that the rats did not receive the same pretreatment drug twice.

Serotonin depletion with PCPA

The rats were not run for 2 weeks to allow any residual effects of the earlier experiments to dissipate. To ensure maintenance of the discrimination, the subjects were then given 4 days of training in the order SDDS and one extinction test, with half the rats tested with saline and the other half with mescaline. The subjects were then injected i.p. with the methyl ester of PCPA (100 mg/kg) for 3 days; 48 hr after the last injection of PCPA the rats were tested with either saline or mescaline (7.5 mg/kg) as described above.

Drugs

Dosages of all drugs are expressed as salts. Mescaline (Sigma Chemical Co.) was administered as the hydrochloride salt in isotonic saline. Cinanserin ([2′-(3,3-dimethylaminopropyl)thiocinnamanilide]; Nutritional Biochemicals) was administered in saline. Methysergide (Sansert®, Sandoz Pharmaceuticals, New Jersey) was administered as the maleate in saline. Cyproheptadine hydrochloride (Merck, Sharp and Dohme Research Laboratories, New Jersey) was dissolved in 15% aqueous propylene glycol. Xylamidine tosylate (The Burroughs Wellcome Research Laboratories, England) was dissolved in saline. All injection volumes were 1 ml/kg. Time intervals between administering the pretreatment compounds and the test drugs are given in Table 5.1.

Pentobarbital Discrimination

In order to evaluate whether the above pharmacological manipulations are indicative of mescaline's mode of action or, alternatively, whether drug pretreatments are generally nonspecific and render the rats incapable of making accurate drug discriminations, a group of 10 naive rats was trained to discriminate saline from 10 mg/kg of pentobarbital sodium (American Pharmaceutical Co.) as previously described. Following acquisition of the discrimination, these rats were tested for the effects of the 5-HT antagonists and PCPA on the discriminability of pentobarbital. Pentobarbital was chosen since there is no compelling evidence implicating serotonin in the mode of action of this drug.

Effects of Raphe Lesions on Mescaline Discrimination

Eighteen naive rats were trained to discriminate 15 mg/kg of mescaline from saline as previously described. Following acquisition of the discrimination, the rats were given a series of three 10-min tests to ascertain the degree of

Table 5.1
Effects of Antiserotonergic Agents on Mescaline Discrimination

Pretreatment	*N*	Test compound	Percentage Mescaline appropriate responses[a]	Average total responses
None	15	Mescaline (15 mg/kg)	84.5 ± 3.6	43 ± 4.2
None	15	Mescaline (7.5 mg/kg)	55.6 ± 8.1	48.7 ± 2.4
None	10	Saline	17.2 ± 4.2	46.5 ± 3.1
Cinanserin	10	Saline	22.6 ± 3	30.6 ± 3.8
(15 mg/kg, 30 min)	10	Mescaline (15 mg/kg)	29 ± 5.9[b]	39.6 ± 5.8
Methysergide	10	Saline	31 ± 7.8	48.6 ± 5.2
(5 mg/kg, 30 min)	10	Mescaline (15 mg/kg)	49.3 ± 7.1[b]	43 ± 2.8
Cyproheptadine	10	Saline	27 ± 7.9	41.2 ± 5.7
(5 mg/kg, 30 min)	10	Mescaline (15 mg/kg)	38.6 ± 9.1[b]	33.5 ± 3.4
Xylamidine tosylate	10	Saline	26.7 ± 6	47.4 ± 3.7
(1 mg/kg, 90 min)	10	Mescaline (15 mg/kg)	81.7 ± 3.6	41.4 ± 5.9
PCPA methylester	10	Saline	38.7 ± 3.4[c]	29 ± 3.7[c]
(100 mg/kg for 3 days, tested 48 hr after last injection)	10	Mescaline (7.5 mg/kg)	83.9 ± 5.5[d]	28.2 ± 2.9[e]

[a] Each value represents the mean ± S.E.M. of the number of animals (*N*) indicated. Values are the percentage of the total responses made during a 10-min extinction test on the lever previously paired with mescaline. Test compounds were administered i.p. 15 min prior to the 10-min extinction test. Time between administration of pretreatment and test compounds is given in the "pretreatment" column.

[b] $p < .002$ different from mescaline, 15 mg/kg (2-tail Mann–Whitney U).

[c] $p < .02$ different from saline (2-tail Mann–Whitney U).

[d] $p < .025$ different from mescaline, 7.5 mg/kg (2-tail Mann–Whitney U).

[e] $p < .002$ different from mescaline, 7.5 mg/kg (2-tail Mann–Whitney U).

generalization following 15 mg/kg of mescaline, 7.5 mg/kg of mescaline and saline. Half the rats (odd-numbered subjects) received lesions of both the dorsal and medial raphe 24 hr after the last test. Control rats (even-numbered subjects) were left unoperated.

Raphe lesions were performed using a stereotaxic apparatus (David Kopf Instruments, Tujanga, California). The rats were anesthetized with pentobarbital (40 mg/kg). Raphe lesions were produced by passing 1 mA for 10 sec between a cathodal stainless-steel electrode insulated except for .5 mm at the tip and a rectal anode. With the incisor bar level with the ear bars, the coordinates used for the dorsal raphe lesion were 1 mm anterior to lambda, 1.2 mm lateral to the midline, and 6.7 mm ventral to the surface of the skull at an angle of 10 degrees to avoid damage to midsagittal sinus (Konig & Klippel, 1963). Ventral raphe lesions were made by inserting the electrode at a 10-degree angle on the contralateral side 1 mm anterior to lambda, 1.5 mm lateral to the midline, and 8.4 mm ventral to the surface of the skull.

All the rats were tested for stimulus generalization following 7.5 mg/kg of mescaline and saline 13 and 14 days after the placement of raphe lesions. Tests were performed in a counterbalanced order, with half the subjects receiving mescaline on Day 13.

The rats were sacrificed by decapitation 24 hours after the last generalization test. Brains were removed within 2 min after sacrifice, dissected on ice into various areas, weighed and frozen at −20° C until assayed 72 hr later. Brain areas consisted of pons-medulla; midbrain; thalamus-hypothalamus-hippocampus; and striatum-septum.

Serotonin Determinations

Serotonin determinations were performed using the spectrofluorometric method of Curzon and Green (1970). Various brain areas were homogenized in 5 ml of acidified *n*-butanol (0.85 ml conc. HCl per liter of butanol) using a glass homogenizer and Teflon pestle. The homogenate was centrifuged for 10 min at 3000*g* in a Sorval refrigerated centrifuge; 4 ml of the supernatant was transferred to a 35-ml glass centrifuge tube containing 20 ml washed heptane and 1.5 ml of .1% cysteine in 0.1 N HCl. The tubes were shaken on a mechanical shaker (Eberach Corp.) for 5 min and the organic layer removed by aspiration. From the remaining aqueous layer .5 ml was transferred to a 10-ml glass-stoppered centrifuge tube containing 3 ml of OPT reagent (2.0 mg *o*-phthaldehyde plus 50 mg cysteine in 10 N HCl). The tubes were placed in a boiling water bath for 15 min, and the reaction was terminated by placing the tubes in an ice bath. The contents of each tube were read at room temperature in an Amico-Bowman spectrofluorometer at activation 360 nm and fluorescence at 470 nm. Blanks consisted of boiled OPT reagent only. Serotonin levels were calculated on the basis of a 1 μg 5-HT internal standard and were not corrected for recovery. The standard was obtained by homogenizing the midbrain and striatum-septum from 2 untreated rats in 10 ml of acidified butanol and adding 1 μg of 5-HT to 5 ml of the homogenate.

Statistical Analysis

Due to the lack of homogeneity of variance between treatment groups, nonparametric statistics were chosen. Statistical analysis was performed using a two-tailed Mann–Whitney U. Spearman rank correlation coefficients with corresponding significances were calculated for 5-HT content versus percentage cue appropriate responding.

RESULTS

Acquisition of Discriminative Response Control

Mescaline can serve as a strong discriminative stimulus at a dose of 15 mg/kg (Figure 5.1). Importantly, the rats do not exhibit a preference for either

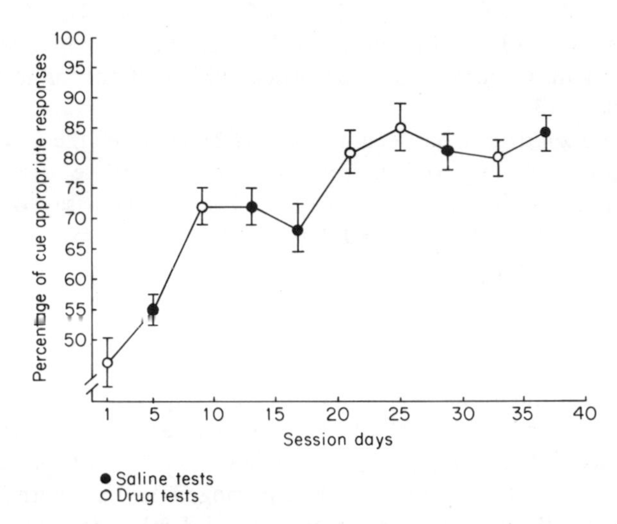

Figure 5.1. Acquisition of discrimination between mescaline (15 mg/kg) and saline administered 15 min before the session. Ordinate: Percentage of total extinction responses made during the first 5 min of the session on the mescaline-correct or saline-correct lever. Abcissa: Session days on which tests were administered. Each point is the mean ± S.E.M. of 25 rats.

lever on the first day of training. Indeed, as indicated in Figure 5.1, the rats made essentially half of their test responses on each lever during their first exposure to the drug. After about 20 days of training, the rats stabilized at a criterion of greater than 80% cue appropriate responses when tested with either mescaline or saline.

Dose-Response and Time-Course

Figure 5.2 depicts the generalization gradient produced by various doses of mescaline. A marked decrement in discrimination accuracy occurred when the subjects were tested with doses of mescaline below the 15 mg/kg training dose. The results shown in Figure 5.2 indicate that the cue produced by mescaline is a continuum rather than an all-or-none phenomenon, as evidenced by the gradual rather than abrupt decline in discriminability with incremental reductions in dosage. Doses of mescaline slightly higher than the 15 mg/kg training dose did not greatly increase the mescaline-appropriate responding, indicating a ceiling effect was reached during acquisition such that higher doses of the drug did not augment postacquisition discriminability. Doses of mescaline higher than 25 mg/kg were not tested since decreased response rates occurred with doses of mescaline greater than 25 mg/kg.

Figure 5.3 represents the generalization gradient exhibited by the rats at increasing times after injection. The peak in mescaline (15 mg/kg) appropriate responding was observed at 30 min after injection but was not significantly different from the 15-min delay used during acquisition. At 90-min postinjection, the subject's mescaline-appropriate responses were significantly lower than those observed after only 15 min, and by 360 min the rats exhibited responses characteristic of saline.

Serotonin Antagonists

Table 5.1 lists the effects of 5-HT receptor blockers on the discriminability of mescaline and saline. In addition, the average total responses made on both levers during the 10-min extinction tests are included for the different treatments. As is seen in the response column of Table 5.1, the total number of responses exhibited by the rats given mescaline is not significantly different from the number of responses following saline administration. Furthermore, the doses of 5-HT antagonists used had no influence on the overall number of responses. It is clear from Table 5.1, however, that pretreatment with the

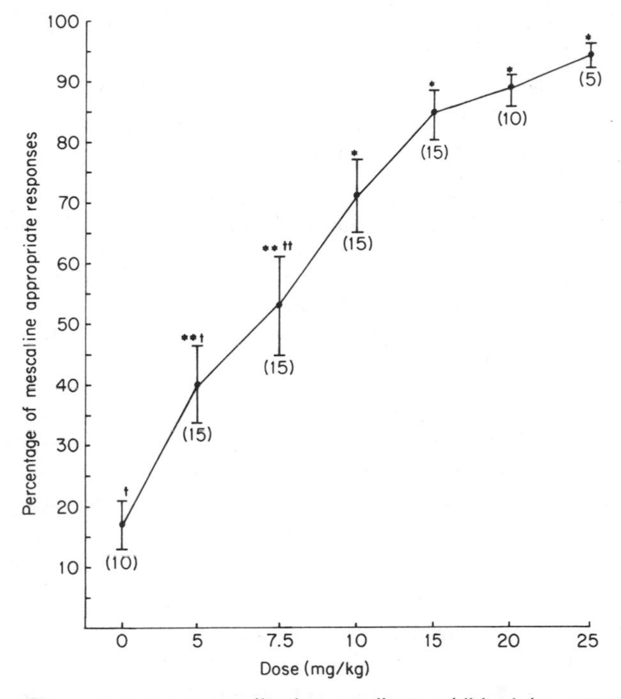

Figure 5.2. Dose–response generalization gradient exhibited by rats trained to discriminate mescaline (15 mg/kg) from saline. One 10-min test per week was interposed among four training sessions per week subsequent to acquisition. All doses of mescaline were administered 15 min before the test. Values are the mean ± S.E.M. The number of experimental animals is shown in parentheses, and the values are calculated as the percentage of total test responses made on the lever previously paired with mescaline (15 mg/kg). $^{*}p < .02$ different from saline; $^{**}p < .02$ different from saline; $^{\dagger}p < .002$ different from mescaline (15 mg/kg); $^{\dagger\dagger}p < .02$ different from mescaline (15 mg/kg).

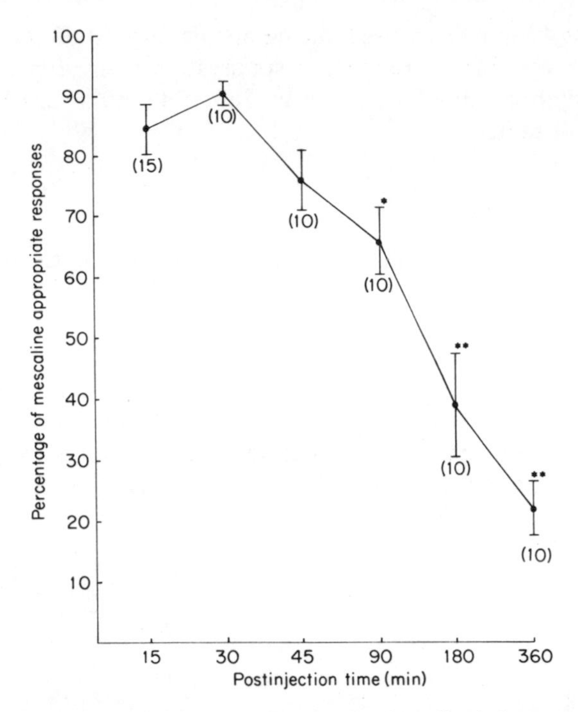

Figure 5.3. Time-course generalization in rats trained to discriminate mescaline (15 mg/kg) from saline administered 15 min before the sessions. Values are the mean ± S.E.M. of the number of animals in parentheses. Ordinate: percentage of total responses during a 10-min test on the lever previously paired with mescaline (15 mg/kg) administered 15 min before the sessions. Abcissa: time after the injection of 15 mg/kg of mescaline when the test was initiated. *$p < .02$ different from mescaline 15 min after injection; **$p < .002$ different from mescaline 15 min after injection.

central 5-HT antagonists cinanserin, methysergide, and cyproheptadine greatly reduced the discriminability of mescaline, as reflected by an increased number of test responses on the lever previously paired with saline (i.e., less than 50% mescaline appropriate responses). Indeed, following pretreatment with cinanserin, the effects of mescaline were altered to a point not significantly different from that of saline. It is of interest that while the 5-HT antagonists tended to reduce slightly the discriminability of saline (e.g., from 17% to 31% for methysergide), the disruption in saline discrimination was not statistically significant. As seen in Table 5.1, the peripheral 5-HT antagonist, xylamidine tosylate (1 mg/kg), failed to affect the stimulus properties of mescaline.

Serotonin Depletion

The results presented in Table 5.1 indicate that depletion of brain 5-HT with PCPA significantly potentiates the effects of mescaline. At a dose of 7.5

mg/kg of mescaline, the rats exhibit about 55% mescaline-appropriate responses. There was, however, a significant reduction in the discriminability of saline produced by PCPA pretreatment (Table 5.1). Furthermore, the PCPA caused a large decrease in the number of responses following both saline and mescaline administration (Table 5.1).

Pentobarbital Discrimination

Table 5.2 shows the effect of the 5-HT antagonists and PCPA on rats trained to discriminate 10 mg/kg pentobarbital from saline. Cinanserin did not alter the discriminability of either saline or pentobarbital in these animals. Similarly, methysergide, cyproheptadine, and xylamide tosylate had no detectable effect on the ability of rats to discriminate pentobarbital. PCPA consistently reduced the discriminability of saline in the pentobarbital trained rats, but because of the small number of subjects, this result was not statistically significant. More importantly, however, PCPA pretreatment had no observable effect on low-dose pentobarbital discrimination (Table 5.2).

Effects of Raphe Lesions on Mescaline Discrimination

As shown in Table 5.3, mescaline at a dose of 15 mg/kg can serve as a strong discriminative stimulus before lesioning. When tested during extinction,

Table 5.2
Effects of Antiserotonergic Agents on Pentobarbital Discrimination

Pretreatment	*N*	Test compound	Percentage pentobarbital appropriate responses[a]
None	10	Pentobarbital (10 mg/kg)	89.5 ± 2.6
None	10	Pentobarbital (5 mg/kg)	51.5 ± 3
None	10	Saline	16.9 ± 3.6
Cinanserin	5	Saline	24.2 ± 8
(15 mg/kg, 30 min)	5	Pentobarbital (10 mg/kg)	85 ± 4.5
Methysergide (5 mg/kg, 30 min)	5	Pentobarbital (10 mg/kg)	84.2 ± 7.7
Cyproheptadine (5 mg/kg, 30 min)	5	Pentobarbital (10 mg/kg)	87.4 ± 5.5
Xylamidine tosylate (1 mg/kg, 90 min)	5	Pentobarbital (10 mg/kg)	90.6 ± 3.5
PCPA methylester (100 mg/	5	Saline	33 ± 7.9
kg for 3 days, tested 48 hr after last injection)	5	Pentobarbital (5 mg/kg)	58.6 ± 3.2

[a] Each value is the mean ± S.E.M. of the number of animals (*N*) indicated. Values are the percentage of the total responses made during a 10-min extinction test on the lever previously paired with pentobarbital (10 mg/kg).

Table 5.3
Effect of Midbrain Raphe Lesions on Mescaline Discrimination[a]

Treatment	N	Percentage mescaline correct responses
	Prelesion	
Saline	18	14.3 ± 2.1
Mescaline 15 mg/kg	18	87 ± 3.1
Mescaline 7.5 mg/kg	18	55.6 ± 7.3
	Postlesion	
Unlesioned control rats		
Mescaline 7.5 mg/kg	8	58.1 ± 15
saline	8	15.4 ± 4.3
Lesioned rats		
Mescaline 7.5 mg/kg	7	89.7 ± 5[a]
Saline	8	30.4 ± 9.1

Note: Generalization exhibited by rats trained to discriminate 15 mg/kg of mescaline from saline. One 10-min extinction test per week was interposed among four training sessions per week subsequent to acquisition and before lesioning. Mescaline and saline were administered i.p. 15 min before the tests. Values are the mean ± S.E.M. of the number of animals indicated (N) and represent the percentage of total test responses made on the lever previously paired with mescaline 15 mg/kg during acquisition.

[a] $p < .02$ different from mescaline 7.5 mg/kg (2-tail Mann-Whitney U).

the rats responded greater than 85% on the lever previously paired with mescaline. Similarly, when tested following saline administration, the rats made greater than 85% saline appropriate responses (14% mescaline correct responses). At a lower dose of mescaline (7.5 mg/kg), the rats responded significantly different from both training conditions. The results summarized in Table 5.3, which were obtained from a different group of rats, are in excellent agreement with the data in Figures 5.1 and 5.2.

It can be seen from the results presented in Table 5.3 that midbrain raphe lesions greatly potentiated the discriminative properties of mescaline. When tested with 7.5 mg/kg of mescaline, the lesioned animals made nearly 90% of their responses on the mescaline (15 mg/kg) correct lever, as compared to the 58% mescalinelike responses made by the unlesioned control rats. The slight disruption in saline discrimination following raphe lesions (from 14% to 30% mescalinelike responses) was not significant.

Effects of Raphe Lesions on Rat Brain Serotonin

The effects of midbrain raphe lesions on the 5-HT content of various brain regions is depicted in Figure 5.4. The 5-HT levels in the forebrain structures were reduced approximately 50% by the lesions. A small but significant reduction in 5-HT was observed in the midbrain in which the lesions were located. The lesions had no significant effect on 5-HT levels in the pons-medulla.

Table 5.4 lists the Spearman correlation coefficients for 5-HT levels in the various brain regions versus percentage cue appropriate responding. For the control rats, a consistent negative correlation was observed between 5-HT levels and cue appropriate responding, that is, with higher 5-HT levels the rats tended to exhibit a large number of inappropriate responses. In the lesioned rats, the percentage cue appropriate responses showed a similar negative correlation with 5-HT when these rats were tested with saline. When the lesioned rats were tested with mescaline, however, a positive correlation was observed. As seen in Table 5.4, there was a statistically significant positive correlation between hippocampus-hypothalamus-thalamus 5-HT levels and percentage mescaline appropriate responding in the lesioned rats ($r_s = .85$, $p < .025$).

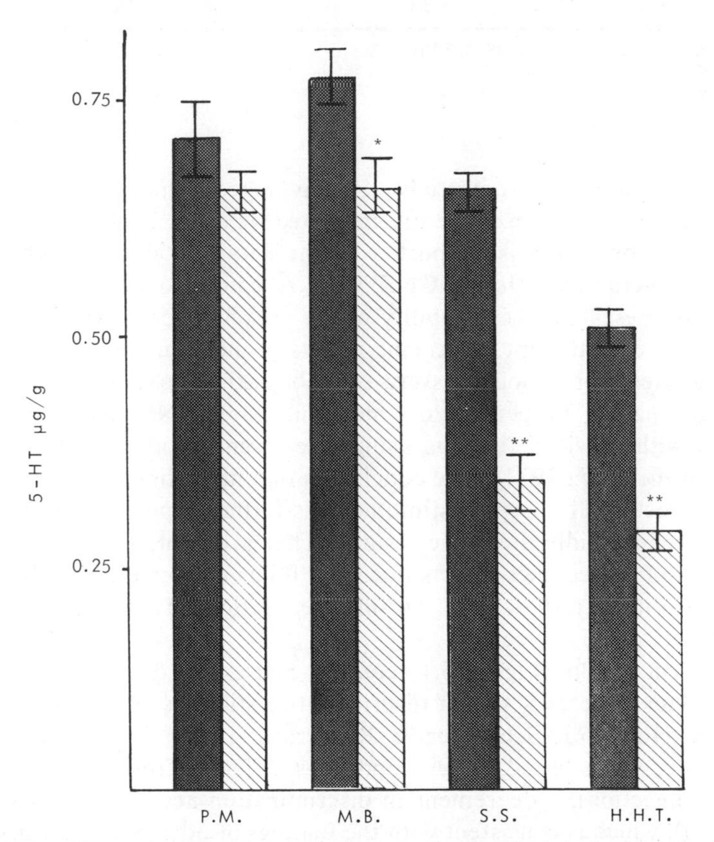

Figure 5.4. Effect of midbrain raphe lesions on the 5-HT content of four brain regions (P.M. = pons medulla; M.B. = midbrain; S.S. = striatum-septum; H.H.T. = hippocampus-hypothalamus-thalamus). Values are the mean ± S.E.M. of eight rats in both control (shaded) and lesioned (cross-hatched) groups. *$p < .05$; **$p < .001$.

Table 5.4
Correlation Coefficients: 5-HT Levels versus Cue Appropriate Responding[a]

	Control rats	
Brain area	Saline tests	Mescaline tests
P.M.	−.48	−.15
M.B.	−.55	−.38
S.S.	−.81 $p < .05$	−.43
H.H.T.	−.26	.095
	Lesioned rats	
P.M.	−.24	−.04
M.B.	−.29	.11
S.S.	−.69 $p < .05$	.46
H.H.T.	−.33	.85 $p < .025$

[a] Brain areas as in Figure 5.4.

DISCUSSION

The results of the present study strongly suggest that the discriminative-stimulus properties of mescaline are mediated by central 5-HT receptor stimulation. This conclusion is supported by the data in Table 5.1, which shows that three structurally different CNS 5-HT receptor blockers disrupt the cueing effect of mescaline. The inability of the peripheral 5-HT antagonist, xylamidine tosylate, to disrupt the cueing effects of mescaline, while blocking the peripheral effects on blood pressure (unpublished observation), suggests that the discriminative properties of mescaline are CNS-mediated. This is consistent with previous findings using direct central administration of mescaline (Browne *et al.*, 1974). The conclusion that mescaline produces its cueing effect by direct 5-HT receptor stimulation is further supported by the findings with PCPA and midbrain raphe lesions. That a potentiation rather than an antagonism of mescaline's effects occurred following depletion of brain 5-HT suggests a direct rather than an indirect action of mescaline on 5-HT receptors.

The present results (Figure 5.1) confirm and extend previous findings that mescaline can serve as a strong discriminative stimulus (Browne *et al.*, 1974; Browne & Ho, 1975; Schechter & Rosecrans, 1972a; Winter, 1973, 1974). Furthermore, when tested under conditions different from training (dose or time after injection), a decrement in discrimination accuracy occurs (Figures 5.2 and 5.3), which is consistent with the findings of other investigators (Anden *et al.*, 1974; Hirschhorn & Rosecrans, 1974a; Huang & Ho, 1974; Overton, 1972; Schechter & Rosecrans, 1971; Waters *et al.*, 1972).

The results of the 5-HT receptor blockers (Table 5.1) are consistent with biochemical evidence (Anden *et al.*, 1968, 1974) as well as with experiments on umbilical vasculature (Dyer, 1974), which indicate direct receptor stimulation by mescaline. While these antiserotonergic agents do have other pharmacological properties (Glegg & Turner, 1971; Stone, Wenger, Ludden, Stavorski, & Ross, 1961; Trottier & Malone, 1969), they are all widely accepted as 5-HT receptor blockers (Furgiuele, High, & Horovitz, 1965; Glegg & Turner, 1971; Persip & Hamilton, 1973; Winter, 1969; Stone *et al.* 1961). Hirschhorn and Rosecrans (1974b) have recently reported that neither methysergide nor cyproheptadine produced significant alterations in the discriminative stimulus properties of LSD. Since the authors had previously shown that mescaline and LSD produce similar cues (Hirschhorn & Winter, 1971; Schechter & Rosecrans, 1972a), the present results do not support their findings (Hirschhorn & Rosecrans, 1974b). Their study, however, employed a different hallucinogen (LSD), a slightly different schedule of reinforcement, and different doses of cyproheptadine and methysergide. Nevertheless, Winter (1975) has been able to replicate the antagonism of the mescaline cue with cinanserin. Perhaps LSD and mescaline produce similar cues (Hirschhorn & Winter, 1971; Schechter & Rosecrans, 1972a) but by different mechanisms.

The inability of xylamidine tosylate to disrupt the mescaline cue is consistent with observations that show that xylamidine tosylate at the dose used in the current investigation could not antagonize the rate-disrupting effects on operant behavior of a known hallucinogen, *N,N*-diethyltryptamine, but could antagonize systemically administered doses of 5-HT that produced pauses in responding (Winter, 1969).

That a potentiation of mescaline's effects occurred following PCPA pretreatment (Table 5.1) and midbrain raphe lesions (Table 5.3) suggests that the action of mescaline on 5-HT receptors is direct rather than indirect. Indeed, the PCPA pretreatment regime used has been shown to cause greater than 90% depletion of brain 5-HT (Koe & Weissman, 1966), and the midbrain raphe lesions resulted in a 50% reduction in forebrain 5-HT (Figure 5.4). In both cases, the discriminability of 7.5 mg/kg of mescaline was augmented from 55% to almost 90% mescaline (15 mg/kg) appropriate responses (Tables 5.1 and 5.3). Unlike the experiment with PCPA in which a significant reduction in the discriminability of saline and a large diminution in the number of responses was observed (Table 5.1), the effects of raphé lesions differed in that no significant effect on saline discrimination or response rate was observed. The present results with PCPA are in agreement with work by Cameron and Appel (1973), who demonstrated a potentiation of LSD discriminability after PCPA pretreatment.

The data pertaining to pentobarbital discrimination (Table 5.2) demonstrate that the effects of pharmacological manipulations on mescaline discrimination were not the result of simply rendering the rats incapable of making drug discriminations. Indeed, with the exception of PCPA, these agents did not affect pentobarbital discrimination. The disrupting effects of PCPA on saline discrimination may be caused by a behaviorally toxic manifestation of the drug since not only were the number of responses significantly reduced by the pretreatment, but also the discriminability of saline in both mescaline- and pentobarbital-trained rats was also altered. Nevertheless, PCPA importantly had no potentiating effect on low-dose pentobarbital discrimination. These results suggest that the use of the drug-discrimination paradigm provides a useful and highly specific method for evaluating the mechanisms of action of drugs.

A negative correlation was expected between appropriate responding and 5-HT levels since reduction of brain 5-HT with either PCPA or raphe lesions potentiated the effects of mescaline. This finding was consistently observed in the saline tests of both untreated and lesioned animals (Table 5.4). Statistically significant negative correlations were observed between striatum-septum 5-HT and saline-appropriate responses. When tested with mescaline, the unlesioned control rats showed negative correlations with 5-HT levels, although these findings were not significant. A differential effect was seen in the lesioned rats tested with mescaline. Here, a significant positive correlation was found between 5-HT levels and mescaline-appropriate responses (Table 5.4).

It is not clear how a reduction in brain 5-HT potentiates the discriminative properties of mescaline and other hallucinogens, but several explanations seem plausible. Since mescaline and LSD exert inhibitory effects on the firing rate of raphe neurons (Haigler & Aghajanian, 1973; Aghajanian, Haigler, & Bloom, 1972), the effects of PCPA and raphe lesions may be to augment this inhibition. This does not seem likely, however, since PCPA does not affect those raphe neurons that are depressed by LSD (Aghajanian, Graham, & Sheard, 1970). One might predict from the present studies that the rats pretreated with PCPA or with raphe lesions would have exhibited mescaline like discrimination when given saline. In addition, monoamine oxidase inhibitors depress the firing rate of raphe neurons (Aghajanian *et al.* 1970) but do not generalize to the stimulus properties of mescaline (Browne & Ho, 1975), although augmentation of mescaline discrimination has been observed after monoamine oxidase inhibition (Browne & Ho, 1975). Another possibility is that 5-HT and the hallucinogenic drugs compete for a common receptor site, and by reducing the level of 5-HT there is less competition. Support for this hypothesis is found in the recent study by Bennet and Aghajanian (1974) showing the displacement by 5-HT of *d*-LSD bound to subcellular brain fractions. Since PCPA does not totally deplete 5-HT stores, and since the raphe lesions produced only a 50% reduction in forebrain 5-HT, the possibility that mescaline might act indirectly by releasing 5-HT cannot be ruled out. Indeed, using a push–pull cannula technique to preload radioactive 5-HT, Tilson and Sparber (1972) have demonstrated that mescaline apparently releases and/or blocks the reuptake of preloaded radioactive 5-HT, while LSD inhibits the release of 5-HT. Furthermore, since a positive correlation was observed between 5-HT levels and mescaline responses after raphe lesions (Table 5.4), this might indicate an indirect action. An alternative theory of receptor sensitivity may explain the potentiation of mescaline's effects after reduction of brain 5-HT. Following chemical lesions produced by 6-hydroxydopamine to deplete brain catecholamines, a supersensitivity to norepinephrine (Segal, McAllister, & Geyer, 1974) and to a specific dopamine stimulant (apomorphine; Ungerstedt, 1971) have been observed. Perhaps when brain 5-HT is reduced, a similar supersensitivity develops to the direct or indirect actions of hallucinogenic drugs on 5-HT receptors.

Winter (1973) has recently demonstrated that a nonhallucinogenic isomer of mescaline, 2,3,4-trimethoxyphenylethylamine, has stimulus properties similar to mescaline (3,4,5-trimethoxyphenylethylamine). In preliminary findings, a generalization to the mescaline cue by 15 mg/kg of the 2,3,4-isomer was observed, and this generalization could be blocked by cinanserin. Thus, the question remains as to the exact nature of the "stimuli" produced by different drugs that exhibit stimulus generalization. Interactions between several neurochemical systems may be involved in producing these stimuli; for mescaline, at least, it appears that 5-HT is involved intimately. The present study does not rule out the possibility that other neurotransmitters may contribute to the stimulus properties of mescaline; indeed, there is a large body of data implicating the effects of mescaline on other neurochemical systems (Carlini, Sampaio, Santos, & Carlini, 1965; Clemente & Lynch, 1968; Gonzalez-Vegas, 1971; Lalley, Rossi, & Baker, 1973; Morton & Malone, 1969; Ratcliffe, 1971; Schopp, Kreutter, & Guzak, 1961; Shah, Shah, Lawrence, & Neeley, 1973; Stolk, Barchas, Goldstein, Boggan, & Freedman, 1974; Tonge & Leonard, 1971). However, preliminary studies have failed to show effects of atropine, phentolamine, propranolol, amphetamine, strychnine, pimozide, or picrotoxin on mescaline discrimination. In light of these results, the consistent effects of 5-HT antagonists observed in the present study would support further the conclusion that the cue produced by mescaline is a concomitant of CNS serotonergic receptor stimulation.

ACKNOWLEDGMENT

This work was conducted at the Texas Research Institute of Mental Sciences, Houston, Texas, and represents a portion of the author's doctoral dissertation submitted to the University of Texas Health Sciences Center at Houston.

REFERENCES

Aghajanian, G. K., Graham, A. W., & Sheard, M. H. Serotonin-containing neurons in brain: Depression of firing by monoamine oxidase inhibitors. *Science,* 1970, *169,* 1100–1102.

Aghajanian, G. K., Haigler, H. J., & Bloom, F. E. Lysergic acid diethylamine and serotonin: Direct actions on serotonin-containing neurons in rat brain. *Life Sci.,* 1972, *11,* 615–622.

Anden, N.-E., Corrodi, H., Fuxe, K., & Hokfelt, T. Evidence for a central 5-hydroxytryptamine receptor stimulation by lysergic acid diethylamide. *Brit. J. Pharmacol.*, 1968, *34*, 1–7.

Anden, N.-E., Corrodi, H., Fuxe, K., & Meek, J. L. Hallucinogenic phenylethylamines: Interactions with serotonin turnover and receptors. *European J. Pharmacol.*, 1974, *25*, 176–184.

Appel, J. B., Lovell, R. A., & Freedman, D. X. Alterations in the behavioral effects of LSD by pretreatment with *p*-chlorophenylalanine and α-methyl-*p*-tyrosine. *Psychopharmacologia*, 1970, *18*, 387–406.

Axelrod, J. The enzymatic N-methylation of serotonin and other amines. *J. Pharmacol. Exp. Therap.*, 1962, *138*, 28–33.

Barry, H., & Kubena, R. K. Discriminative stimulus characteristics of alcohol, marijuana, and atropine. In J. M. Singh, L. H. Miller, & H. Lal (Eds.), *Drug addiction: Experimental pharmacology*. Vol. I. Mt. Kisco, New York: Futura, 1972. Pp. 3–16.

Barry, H., Steenberg, M. L., Manian, A. A., & Buckley, J. P. Effects of chlorpromazine and three metabolites on behavioral responses in rats. *Psychopharmacologia*, 1974, *34*, 351–360.

Bennet, J. L., & Aghajanian, G. K. *d*-LSD binding to brain homogenates: possible relationship to serotonin receptors. *Life Sci.*, 1974, *15*, 1935–1944.

Boakes, R. J., Bradley, P. B., Briggs, I., & Dray, A. Antagonism of 5-hydroxytryptamine by LSD-25 in the central nervous system: A possible neuronal basis for the actions of LSD-25. *Brit. J. Pharmacol.*, 1970, *40*, 202–218.

Browne, R. G., Harris, R. T., & Ho, B. T. Stimulus properties of mescaline and N-methylated derivatives: Difference in peripheral and direct central administration. *Psychopharmacologia*, 1974, *39*, 43–56.

Browne, R. G., & Ho, B. T. Discriminative stimulus properties of mescaline: Mescaline or metabolite? *Pharmacol. Biochem. Behav.*, 1975, *3*, 109–114.

Cameron, O. G., & Appel, J. B. A behavioral and pharmacological analysis of some discriminable properties of *d*-LSD in rats. *Psychopharmacologia* 1973, *33*, 117–134.

Carlini, E. A., Sampaio, M. R. P., Santos, M., & Carlini, G. R. S. Potentiation of histamine and inhibition of diamine oxidase by catatonic drugs. *Biochem. Pharmacol.*, 1965, *14*, 1657–1663.

Cheng, H. C., Long, J. P., Barfknecht, C. F., & Nichols, D. E. Cardiovascular effects of 2,5-dimethoxy-4-methylamphetamine (DOM, STP). *J. Pharmacol. Exp. Therap.*, 1973, *14*, 1657–1663.

Clemente, E., & Lynch, V. dP. *In vitro* action of mescaline. Possible mode of action. *J. Pharm. Sci.*, 1968, *57*, 72–78.

Corne, S. J., & Pickering, R. W. A possible correlation between drug-induced hallucinations in man and a behavioural response in mice. *Psychopharmacologia*, 1967, *11*, 65–78.

Curzon, G., & Green, A. R. Rapid method for the determination of 5-hydroxytryptamine and 5-hydroxyindolacetic acid in small regions of rat brain. *Brit. J. Pharmacol.*, 1970, *36*, 653–655.

Dyer, D. C. Evidence for the action of *d*-lysergic acid diethylamide, mescaline and bufotenine on 5-hydroxytryptamine receptors in umbilical vasculature. *J. Pharmacol. Exp. Therap.*, 1974, *188*, 336–341.

Dyer, D. C., & Gant, D. W. Vasoconstriction produced by hallucinogens on isolated human and sheep umbilical vasculature. *J. Pharmacol. Exp. Therap.*, 1973, *184*, 366–375.

Freedman, D. X., Gottlieb, R., & Lovell, R. A. Psychotomimetic drugs and brain 5-HT metabolism. *Biochem. Pharmacol.*, 1970, *19*, 1181–1188.

Friedhoff, A. J., & Goldstein, M. New developments in metabolism of mescaline and related amines. *Ann. N.Y. Acad. Sci.*, 1962, *96*, 5–13.

Friedhoff, A. J., Schweitzer, M. W., & Miller, J. Biosynthesis of mescaline and N-acetylmescaline by mammalian liver. *Nature*, 1972, *237*, 454–455.

Furgiuele, A. R., High, J. P., & Horovitz, Z. P. Some central effects of SQ 10,643 [2′-(3,3-dimethylaminopropylthio)cinnamanilide hydrochloride], a potent serotonin antagonist. *Arch. Intern. Pharmacodyn.*, 1965, *155*, 225–235.

Gaddum, J. H. Antagonism between lysergic acid diethylamide and 5-hydroxytryptamine. *J. Physiol.* (Lond.), 1953, *121*, 15P.

Glegg, A. M., & Turner, P. Cholinergic interactions of methysergide and cinanserin on isolated human smooth muscle. *Arch. Intern. Pharmacodyn.* 1971, *191*, 301–309.

Gonzalez-Vegas, J. A. Antagonism of catecholamine inhibition of brain stem neurons by mescaline. *Brain Res.*, 1971, *35*, 264–267.

Haigler, J. H. and Aghajanian, G. K. Mescaline and LSD: Direct and indirect efforts on serotonin-containing neurons in brain. *European J. Pharmacol.*, 1973, *21*, 53–60.

Harley-Mason, J., Laird, A. H., & Smythies, J. R. Delayed clinical reactions to mescaline. *Confin. Neurol.*, 1958, *18*, 152–155.

Hirschhorn, I. D., & Rosecrans, J. A. Studies on the time course of cholinergic and adrenergic receptor blockers on the stimulus effects of nicotine. *Psychopharmacologia*, 1974, *40*, 109–120. (a)

Hirschhorn, I. D., & Rosecrans, J. A. A comparison of the stimulus effects of morphine and lysergic acid diethylamide (LSD). *Pharmacol. Biochem. Behav.*, 1974, *2*, 361–366. (b)

Hirschhorn, I. D., & Winter, J. C. Mescaline and lysergic acid diethylamide (LSD) as discriminative stimuli. *Psychopharmacologia*, 1971, *22*, 305–313.

Huang, J.-T., & Ho, B. T. Discriminative stimulus properties of *d*-amphetamine and related compounds in rats. *Pharmacol. Biochem. Behav.*, 1974, *2*, 669–673.

Jones, C. N., Hill, H. F., & Harris, R. T. Discriminative response control by *d*-amphetamine and related compounds in the rat. *Psychopharmacologia*, 1974, *36*, 374–356.

Koe, B. K., & Weissman, A. p-Chlorophenylalanine: A specific depletor of brain serotonin. *J. Pharmacol. Exp. Therap.*, 1966, *154*, 499–516.

Konig, J. R. F., & Klippel, R. A. *The rat brain: A stereotaxic atlas of the forebrain and lower parts of the brain stem.* Baltimore: Williams & Wilkins, 1963.

Kubena, R. K., & Barry, H. Generalization by rats of alcohol and atropine stimulus characteristics to other drugs. *Psychopharmacologia*, 1969, *15*, 196–206.

Kuhn, D. M., Appel, J. B., & Greenberg, I. An analysis of some discriminative properties of *d*-amphetamine. *Psychopharmacologia*, 1974, *39*, 57–66.

Lalley, P. M., Rossi, B. G., & Baker, W. W. Tremor induced by intra-caudate injections of bretylium, tetrabenzine, or mescaline: Functional deficits in caudate dopamine. *J. Pharm. Sci.*, 1973, *62*, 1302–1307.

Mandell, A. J., Buckingham, B., & Segal, D. S. Behavioral, metabolic and enzymatic studies of a brain indole (ethyl)amine N-methylating system. In B. T. Ho & W. McIssac (Eds.), *Brain chemistry and mental disease.* New York: Plenum, 1971. Pp. 37–60.

Morrison, C. F., & Stephenson, J. A. Nicotine injections as the conditioned stimulus in discrimination learning. *Psychopharmacologia*, 1969, *15*, 351–360.

Morton, J. J. P., & Malone, M. H. Investigation of adrenergic beta-receptor blockade and mescaline-induced bradycardia. *J. Pharm. Sci.*, 1969, *58*, 1169–1170.

Overton, D. A. State-dependent learning produced by depressant and atropine-like drugs. *Psychopharmacologia*, 1966, *10*, 6–31.

Overton, D. A. Differential responding in a three choice maze controlled by three drug states. *Psychopharmacologia*, 1967, *11*, 376–378.

Overton, D. A. Visual cues and shock sensitivity in the control of T-maze choice by drug conditions. *J. Comp. Physiol. Psychol.*, 1968, *66*, 216–219.

Overton, D. A. State-dependent learning produced by alcohol and its relevance to alcoholism. In B. Kissin & H. Begleiter (Eds.), *The biology of alcoholism.* Vol. II. New York: Plenum, 1972.

Persip, G. L., & Hamilton, L. W. Behavioral effects of serotonin or a blocking agent applied to the septum of the rat. *Pharmacol. Biochem. Behav.*, 1973, *1*, 139–147.

Ratcliffe, F. The effect of mescaline and bufotenine on some central actions of noradrenaline. *Arch. Intern. Pharmacodyn.*, 1971, *194*, 147–157.

Richards, D. W., Harris, R. T., & Ho, B. T. Central control of *d*-amphetamine-induced discriminative stimuli. *Proc. 3rd Ann. Mtg. Soc. Neuroscience*, 1973, *3*, 34.

Rosecrans, J. A., Goodloe, Jr., M. H., Bennet, G. J., & Hirschhorn, I. D. Morphine as a discriminative cue: Effects of amine depletors and naloxone. *European J. Pharmacol.*, 1973, *21*, 252–256.

Saavadra, J. M., & Axelrod, J. Psychotomimetic N-methylated tryptamines: formation in brain *in vivo* and *in vitro*. *Science*, 1972, *175*, 1365–1366.

Schecter, M. D. Ethanol as a discriminative cue: Reduction following depletion of brain serotonin. *European J. Pharmacol.*, 1973, *23*, 278–281. (a)

Schecter, M. D. Transfer of state-dependent control of discriminative behavior between subcutaneously and intraventricularly administered nicotine and saline. *Psychopharmacologia*, 1973, *32*, 327–335. (b)

Schecter, M. D., & Rosecrans, J. A. CNS effects of nicotine as the discriminative stimulus for the rat in a T-maze. *Life Sci.*, 1971, *10*, 821–832.

Schecter, M. D., & Rosecrans, J. A. Lysergic acid diethylamide (LSD) as a discriminative cue: Drugs with similar stimulus properties. *Psychopharmacologia*, 1972, *26*, 313–316. (a)

Schecter, M. D., & Rosecrans, J. A. Effect of mecamylamine on discrimination between nicotine and arecoline produced cues. *European J. Pharmacol.* 1972, *17*, 179–182. (b)

Schechter, M. D., & Rosecrans, J. A. Nicotine as a discriminative cue in rats: Inability of related drugs to produce a nicotine-like cueing effect. *Psychopharmacologia*, 1972, *27*, 379–387. (c)

Schechter, M. D., & Rosecrans, J. A. Atropine antagonism of arecoline-cued behavior in the rat. *Life Sci.*, 1972, *11*, 517–523. (d)

Schechter, M. D., & Rosecrans, J. A. *d*-Amphetamine as a discriminative cue: Drugs with similar stimulus properties. *European J. Pharmacol.*, 1973, *21*, 212–216.

Schopp, R. T., Kreutter, W. F., & Guzak, S. V. Neuromyal blocking action of mescaline. *Am. J. Physiol.*, 1961, *200*, 1226–1228.

Segal, D. S., McAllister, C., & Geyer, M. A. Ventricular infusion of norepinephrine and amphetamine: Direct versus indirect action. *Pharmacol. Biochem. Behav.*, 1974, *2*, 79–86.

Shah, N. S., Shah, K. R., Lawrence, R. S., & Neeley, A. E. Effects of chlorpromazine and haloperidol on the disposition of mescaline-^{14}C in mice. *J. Pharmacol. Exp. Therap.*, 1973, *186*, 297–304.

Smythies, J. R., & Sykes, E. A. The effect of mescaline upon the conditioned avoidance response in the rat. *Psychopharmacologia*, 1964, *6*, 163–172.

Stolk, J. M., Barchas, J. D., Goldstein, M., Boggan, W., & Freedman, D. X. A comparison of psychotomimetic drug effects of rat brain norepinephrine metabolism. *J. Pharmacol. Exp. Therap.*, 1974, *184*, 42–50.

Stone, C. A., Wenger, H. C., Ludden, C. T., Stavorski, J. M., & Ross, C. A. Antiserotonin-antihistaminic properties of cyproheptadine. *J. Pharmacol. Exp. Therap.*, 1961, *131*, 73–84.

Thudichum, J. W. L. *A treatise on the chemical constitution of the brain.* London: Balliere, Tindall, 1884.

Tilson, H. A., & Sparber, S. B. Studies on the concurrent behavioral and neurochemical effects of psychoactive drugs using the push–pull cannula. *J. Pharmacol. Exp. Therap.*, 1972, *181*, 387–398.

Tonge, S. R., & Leonard, B. E. Hallucinogens and non-hallucinogens: A comparison of the effects of 5-hydroxytryptamine and noradrenaline. *Life Sci.*, 1971, *10*, 161–168.

Tonge, S. R., & Leonard, B. E. Some observations on the behavioral effects of hallucinogenic drugs on rats: Potentiation by two drugs affecting monoamine metabolism. *Arch. Intern. Pharmacodyn.*, 1972, *195*, 168–176.

Trottier, R. W., Jr., & Malone, M. H. Comparative *in vitro* evaluation of cryogenine, cyproheptadine, and diphenyhydramine as antagonists of furtrethonium, histamine and serotonin. *J. Pharmacol. Sci.*, 1969, *58*, 1250–1253.

Ungerstedt, U. Postsynaptic supersensitivity after 6-hydroxydopamine induced degeneration of the nigro-striatal dopamine system. *Acta Physiol. Scand.*, 1971, *367* (Suppl.), 69–93.

Waters, W. H., Richards, D. W., & Harris, R. T. Discriminative control and generalization of the stimulus properties of *d, l*-amphetamine in the rat. In J. M. Singh, L. H. Miller, & H. Lal

(Eds.), *Drug addiction: Experimental pharmacology*. Vol. I. Mount Kisco, New York: Futura, 1972. p. 87.

Winter, J. C. Behavioral effects of N,N-diethyltryptamine: Absence of antagonism by xylamidine tosylate. *J. Pharmacol. Exp. Therap.*, 1969, *169*, 7–16.

Winter, J. C. A comparison of the stimulus properties of mescaline and 2,3,4-trimethoxyphenylethylamine. *J. Pharmacol. Exp. Therap.*, 1973, *185*, 101–107.

Winter, J. C. The effects of 3,4-dimethoxyphenylethylamine in rats trained with mescaline as a discriminative stimulus. *J. Pharmacol. Exp. Therap.*, 1974, *189*, 741–747.

Winter, J. C. Blockade of the stimulus properties of mescaline by a serotonin antagonist. *Arch. Intern. Pharmacodyn.*, 1975, *214*, 250–253.

6

Stimulus Properties of Ethanol and Depressant Drugs

Martin D. Schechter
Department of Pharmacology
Eastern Virginia Medical School

INTRODUCTION

Ethanol may be classified as a sedative–hypnotic agent since its principal action is depression of the central nervous system (CNS). Although of minor therapeutic importance, ethanol is one of the greatest social problems of mankind not only in producing chronic alcoholism but in its ability to potentiate the effects of other sedative–hypnotics when coadministered. The combination of barbiturate and nonbarbiturate hypnotic drugs with alcohol has often led to death in man (Forney & Harger, 1969).

Ethanol was shown as early as 1951 (Conger) to acquire state dependent control of discriminative responding. Since that time, ethanol has been employed in numerous animal and human state-dependent studies (Crow, 1966; Goodwin, Powell, Hill, Leiberman, & Viamontes, 1974; Holloway, 1972; Ryback, 1969; Storm & Smart, 1965). In addition, barbiturates such as pentobarbital, phenobarbital, secobarbital, and amobarbital, and nonbarbiturate hypnotics such as paraldehyde and chloral hydrate have been reported to produce control of state dependent discrimination in rats (Overton, 1972).

The behavioral technique of state-dependent conditioning consists of training an animal to make a differential response solely on the basis of its current drug condition. In essence, the drug state becomes the discriminative stimulus, and the measure of performance, on a behavioral task, is dependent on the perceived drug effect or drug cue. The relationship between the drug cue and response can be analyzed on many levels—molecular, anatomical, or neurochemical. Once responding to the differential drug states is established, antagonistic drugs may be employed to investigate the possible pharmacological mechanism of action that underlies the ability of the drug to control discriminative performance. This antagonism can be direct or indirect. Direct antagonism can be tested by pretreatment with a specific agent that has been reported to antagonize the central effects of the drug used to train the animal. Indirect antagonism refers to the use of specific depletors of biogenic amines that are presumed to mediate the effects of the drug in the CNS. By manipulating the endogenous neurotransmitters, the mediation of a specific neurotransmitter on the production of the drug cue can be analyzed.

The present study employed ethanol as the drug to control state dependent discrimination and tested direct and indirect antagonists to evaluate their effects on the production of the ethanol-produced cue. Interpretation of the results may supplement scientific knowledge as to the pharmacological mechanism of action of ethanol in the CNS.

REVIEW OF LITERATURE

Rats can be trained to make one response after the administration of a drug and to make a different response after the administration of saline, the nondrug state, by reinforcing correct responding and by punishing (or not reinforcing) incorrect responses. Once discriminative performance reaches criterion levels, novel drugs can be administered to observe whether they are perceived as the training drug, that is, whether the novel agents produce a similar drug cue. The technique of drug-cue generalization, or cross-drug transfer, has been employed using ethanol both as the training drug and as the novel test drug.

In Table 6.1, the results of several studies that tested cross-drug transfer between ethanol and barbiturates are summarized. When ethanol was used as the training drug, stimulus generalization to barbiturates, minor tranquilizers, and nonbarbiturate sedative-hypnotics was observed. In the study in which pentobarbital was the training drug, transfer to ethanol, as well as to other barbiturate and nonbarbiturate hypnotics, was observed (Krimmer & Barry, 1973a; Overton, 1964). These results would indicate that the central cues produced by effective doses of ethanol and other central depressant drugs are similar. In contrast, the central stimulants, *d*-amphetamine and bemegride, produced cues that were dissimilar to those produced by ethanol or pentobarbital (Kubena & Barry, 1969; Overton, 1964).

In another study, Krimmer and Barry (1973b) observed a synergistic effect of pentobarbital and ethanol in producing ethanol responses at doses of each that produced a saline (nondrug) response. This would indicate a depressant effect as the discriminative cue. Furthermore, a sufficiently high dose of the stimulant bemegride was reported to antagonize the pentobarbital-produced cue (Overton, 1964).

Table 6.1
Cross-Drug Transfer Studies

Training drug	Transfer observed	Transfer not observed	Reference
Ethanol	Phenobarbital Ethyl carbamate Meprobamate		Overton (1964)
Ethanol	Pentobarbital Chlordiazepoxide Chlorpromazine		Krimmer and Barry (1973b)
Ethanol	Chloral hydrate	*d*-Amphetamine	Kubena and Barry, 1969
Pentobarbital	Ethanol Phenobarbital Secobarbital Amobarbital Barbital Chloral hydrate Paraldehyde	*d*-Amphetamine Bemegride Gallamine	Overton (1964)

Although the site of central action and the behavioral effects of ethanol are fairly well established, the basis of ethanol's relationship with putative neurotransmitter substances is less clear. Feldstein (1972) has summarized the research concerning ethanol's action on concentration, uptake, and release in the brain of several chemicals believed to be neurotransmitters. The exact nature of these actions depends on such factors as the species investigated, dosage parameters, whether the investigation was *in vivo* or *in vitro*, and the specific neurochemical under consideration.

Of particular relevance to the present study is the relationship between levels of brain serotonin (5-hydroxytryptamine, 5-HT) and certain behavioral effects of ethanol. Several investigators have reported that some of the behavioral effects of ethanol can be altered by 5-HT depletion. Before outlining these results, a consideration of the location and biosynthesis of 5-HT is presented here to aid in the interpretation of its possible mediation of behavior. The highest concentration of 5-HT occurs in the gastrointestinal tract in man, with only 1–2% of total body 5-HT present within the CNS. This neurochemical is found endogenously in the brain, and because of its polar characteristics, it is generally held that 5-HT is synthesized within the brain itself rather than entering the brain via the bloodstream.

The synthesis of 5-HT is initiated by the uptake of the amino acid tryptophan, which is derived from dietary sources. Following this uptake, tryptophan is hydroxylated into 5-HTP by the enzyme tryptophan hydroxylase. The final step in serotonin synthesis is decarboxylation of 5-HTP into 5-HT. The enzyme responsible for this reaction is referred to as L-aromatic amino acid decarboxylase and is believed to be the common decarboxylase enzyme for all naturally occurring L-amino acids. Following its synthesis, serotonin is catabolized by the enzyme monoamine oxidase (MAO), which also appears to catabolize the catecholamines. The products of this reaction are 5-hydroxyindoleacetic acid (5-HIAA) and 5-hydroxytryptophol (5-HTOL), the production of the latter substance occurring in the liver.

The enzyme that seems to be of primary importance in the synthesis of 5-HT is tryptophan hydroxylase rather than 5-HTP decarboxylase. Relevant support for this hypothesis is that when tryptophan hydroxylase is inhibited, brain 5-HT synthesis rapidly decreases, whereas inhibition of 5-HTP decarboxylase has no effect on 5-HT brain levels. Also of interest in this context is the finding that 5-HTP, in contrast to the catecholamines, does not inhibit the activity of its biosynthetic enzyme, tryptophan hydroxylase. The lack of feedback inhibition suggests the absence of a control mechanism responsible for signaling slow-down of 5-HT synthesis. Such a feedback mechanism does appear to operate in catecholamine synthesis, for example, norepinephrine and dopamine can inhibit tyrosine hydroxylase (Cooper, Bloom, & Roth, 1974).

Since this study will address itself to the effects of 5-HT depletion and its interaction with ethanol, a brief summary of the behavioral effects of 5-HT will focus primarily on changes in behavior associated with depletion rather than increased concentrations of this substance. Before describing these, however, it would be useful to review certain caveats that have been presented in reviews of 5-HT and its role in behavior. Chase and Murphy (1973) have pointed out methodological weaknesses found in evaluations of the functional role of 5-HT. The first of these problems stems from the pharmacological manipulation of 5-HT levels. The use of drugs either to increase (e.g., L-tryptophan or 5-HTP) or decrease (e.g., *p*-chloroamphetamine) 5-HT may be accompanied by simultaneously altered levels of other amines or amino acids, thus rendering a 5-HT interpretation per se suspect. A second problem in 5-HT research stems from the use of lesion formation and electrical stimulation as means of altering 5-HT output. Implications drawn from the results of studies employing these techniques are questionable because both of these methods may affect fibers traversing the area under study as well as those believed to terminate there. Jacobs, Mosko, and Trulson (1975), rather than focusing on methodology, stress the paucity of knowledge surrounding the functional characteristics of 5-HT. As an example of this weakness they point out that, although 5-HT is localized within the neurons of the brainstem raphe nuclei, the afferents and efferents that comprise the raphe complex are unknown. In addition, our knowledge of the endogenous and exogenous factors that regulate the activity of raphe neurons is similarly deficient.

In any consideration of the behavioral effects of 5-HT manipulation, then, it would seem that any conclusions should be tempered by an understanding that such problems as presented by these authors always pose a potential distortion of data. In fact, Chase and Murphy (1973) consider methodological problems to be serious enough so that "the exact relationship of central indoleamines . . . cannot be said to be clear in even a single instance."

The behavioral characteristic most frequently associated with 5-HT depletion in brain is hyperactivity. For example, sleep studies with the cat have demonstrated that an 80–90% decrease in 5-HT synthesis produces total insomnia. These changes were observed whether decreased synthesis was produced by parachlorophenylalanine (PCPA), a tryptophan hydroxylase inhibitor, or with raphe lesions (Chase & Murphy, 1973). Behaviors that are associated with the effects of ethanol and that demonstrate alterations in relationship to brain 5-HT levels are most pertinent to this study and will be presented in some detail.

Myers and his associates have reported results from a series of studies that suggest that following administration of PCPA, rats' voluntary consumption of ethanol solutions is significantly reversed when compared to consumption prior to PCPA treatment. The following means of establishing ethanol preference were utilized, and the observed preference reversal was reported in each case: The preference was an original predisposition (Veale & Myers, 1970); it was acquired after exposure to stress (Myers & Cicero, 1969); by acclimation to peripherally administered ethanol (Veale & Myers, 1970); by intracerebral injections of minute quantities of ethanol into the cerebral ventricles (Myers, Evans, & Yakich, 1972). Another study used *p*-chloramphetamine (PCA) to deplete 5-HT (Frey, Magnussen, & Nielsen, 1970) and reported that treatment with this drug also reduces rats' preference for ethanol. A reduction in ethanol drinking in rats has also been reported by Opitz (1969), who treated rats with either PCPA or fenfluramine, another serotonin depletor.

A contradictory finding has been reported by Geller (1973), who observed an increase in rats' preference for ethanol following PCPA treatment. Since the dosage levels of PCPA used in Geller's and Myers' studies were different, the discrepancy between their results may be partly attributable to this difference. However, both studies found that rats' intake of ethanol is reduced following systemic injection of 5-hydroxytryptophan (5-HTP). These findings should not be regarded unequivocally since several investigators have reported that "loading doses of 5-HTP not only lead to substantial 5-HT accumulations in cells that do not ordinarily contain this indoleamine, but also interfere with catecholamines and their precursors for transport, storage, and metabolism within the CNS" (Chase & Murphy, 1973). Thus, the reduction in ethanol intake reported by Geller (1973) and Myers *et al.* (1972) may reflect 5-HTP-induced changes that indirectly influenced catecholaminergic systems.

To test the effect of indirect antagonism by depleting brain 5-HT concentrations with PCPA and the effect of direct antagonism with stimulant drugs upon the ethanol-induced cue, the present study was performed.

EXPERIMENTAL DESIGN AND RESULTS

Female, experimentally naive albino rats were housed individually with food and water provided and trained to escape from a .5 mA scrambled shock in a three-compartment chamber described previously (Schechter & Rosecrans, 1972) by discriminating between 1.5 gm/kg of ethanol and an equal volume of saline. Ethanol, in a 10% solution (w/v) of saline, and saline (.9% NaCl solution) were administered intraperitoneally 15 min before training.

To minimize the possible systematic effects of position preference, the rats were randomly divided into two equal groups. One group was required to enter the left compartment of the three-compartment chamber after ethanol administration and the right compartment after the injection of saline to escape the shock; the reverse responses were required for the other half of the rats. In each trial the rat was dropped from a height of about 20 cm into the center compartment with the shock already turned on and allowed to run freely until it entered the designated ethanol- or saline-correct compartment, at which time the shock was turned off. Each rat received 8 training trials spaced 15–30 sec apart, and the same choice of compartment was required in all trials during a given session. Rats were given 5 daily training sessions per week. The substance administered, as well as the required compartment choice, were alternated on successive days.

Since the side compartments were identical, the rat could make a correct choice only by employing its current drug state as a discriminative stimulus. Rats were trained until they made eight correct responses in the first trial of 10 consecutive sessions, that is, until they entered the correct compartment in relation to the administration of ethanol or saline.

The typical learning curve of 13 rats trained to discriminate between 10% ethanol and saline appears in Figure 6.1. Criteria performance was attained by all rats within 20 sessions, or after 10 sessions with each treatment.

Once criterion performance was maintained, the rats were tested with ethanol doses lower and higher than the 10% ethanol training dose in order to obtain a dose–response relationship. On Tuesdays and Thursdays, the rats were tested with either 10% ethanol or saline to ascertain whether criterion performance was maintained. Only the first response in each session was considered in judging a rat's response as correct or incorrect, but seven additional (training) trials were run to maintain training. On Mondays, Wednesdays, and Fridays, the rats were given either a 3, 5, 8, 12, 18, or 20% dose of ethanol (all w/v in .9% saline) and were tested 15 min later in a single trial in which shock was terminated when either compartment was entered. It was proposed that if a rat did not maintain the 80% criterion on Tuesdays and Thursdays, all data on that rat would be deleted from the results. This did not occur. All treatments, including test-training sessions with 10% ethanol and saline, were randomized throughout the experimental series.

The dose–response relationship after saline and doses of ethanol from 3 to 20% is presented in Figure 6.2. During testing–training days, 26 first responses

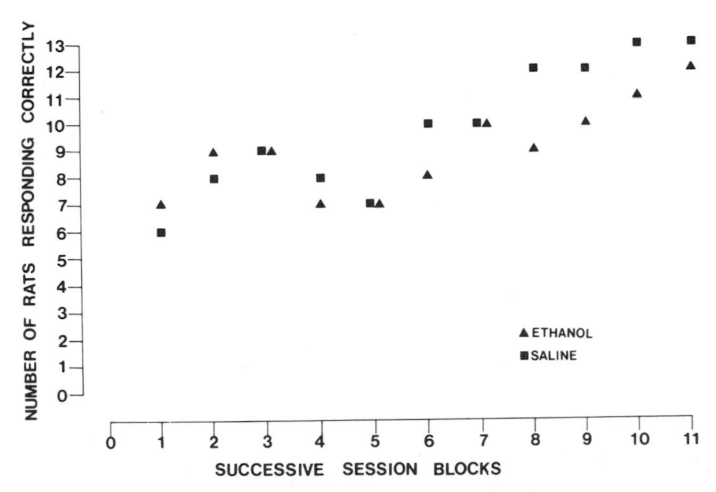

Figure 6.1. Discriminative responding following i.p. administration of ethanol, 1.5 gm/kg in 10% solution (w/v), or saline. Ordinate: number of subjects (*N* = 13) making correct response contingent upon the drug administered. Abscissa: successive session blocks; each block consists of a single administration of ethanol and saline. Reprinted by permission from: M. D. Schechter, Ethanol as a discriminative cue: Reduction following depletion of brain serotonin, *European Journal of Pharmacology*, 1973, *24*, p. 279.

were made into the ethanol-correct compartment after 208 sessions with saline (12.5%), whereas 185 of 209 first responses into the ethanol-correct side occurred after 10% ethanol administration (88.9%). After 8 trials for each of 13 rats with 3% ethanol, 27 or 104 first responses (25.9%) were into the ethanol-correct side. The percentage of correct responses increased with progressively higher doses of ethanol, the maximum discriminative performance of 87 of 91 trials (95.6%) occurring after administration of 15% ethanol. The ED_{50} seems to be between 3 and 5% ethanol. Discriminative performance after 18 and 20% ethanol was observed to decrease possibly as a result of toxicity and gross behavioral impairment at these high doses.

Following the dose–response experiments, the rats were given 10 sessions with each of the training conditions, that is, 10% ethanol and saline, to ensure that criterion performance was maintained before the onset of pretreatment with possible antagonists. Ten rats were observed to attain a level of discriminative performance of 9 out of 10 correct first-choice responses with each treatment and were subsequently exposed to three pretreatment studies.

The procedure for the pretreatment studies was the following: On alternate days, the rats were tested and trained with either 10% ethanol or saline. On every other day, at the same time of day and in the same order, the rats were injected intraperitoneally (i.p.) with the proposed antagonist drug; 15 min later, when drug effect was believed to be at its height, ethanol was administered, and 15 min later a single trial was run in which entrance into either side resulted in the termination of shock.

Each pretreatment series was treated as a block of experiments in which

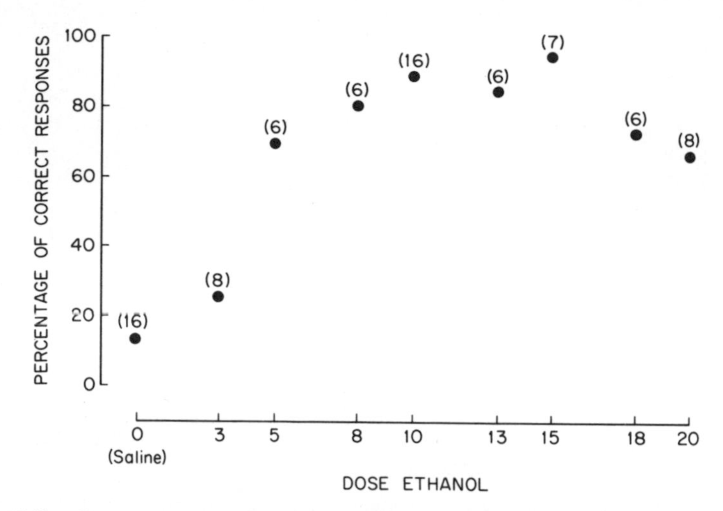

Figure 6.2. Dose–response after various doses of ethanol. Ordinate: Percentage of total responses made into the ethanol-correct compartment. Abscissa: dosage of ethanol as percentage of concentration; training dose at 10% (w/v) represents 1.5 gm of ethanol/kg. Numbers in parentheses represent the number of trials at each ethanol dose for 13 rats. Reprinted by permission from: M. D. Schechter, Effect of propranolol, d-amphetamine and caffeine on ethanol as a discriminative cue, *European Journal of Pharmacology*, 1974, *29*, p. 54.

treatments were randomized throughout. Propranolol, 1, 5, 10, and 20 mg/kg, and *d*-amphetamine sulphate, 4 mg/kg (as base), were administered before 10% ethanol or saline. In the series of experiments with caffeine, the procedure was altered slightly, and 100 mg/kg of caffeine was administered before 3% and 15% ethanol or saline. This was done to observe the possibility of an increase or decrease in discriminative performance at doses of ethanol shown to produce minimum and maximum discrimination as observed in the dose-response study (Figure 2.6). The caffeine series was controlled by testing after administering 3 and 15% ethanol doses without caffeine pretreatment.

The dosages of 100 mg/kg of caffeine and 4 mg/kg of amphetamine were selected as these dosages had been reported to interact with the effect of ethanol on the behavior of rats. Alstott and Forney (1971) observed that ethanol depresses the performance of rats in a shock-avoidance task. When 100 mg/kg of caffeine was administered with ethanol, the depression of behavior was seen to increase significantly. The authors suggested that these results are contrary to the popular belief that caffeine antagonizes ethanol's effects in man. Leonard and Wiseman (1970) reported that pretreatment with 4 mg/kg of amphetamine significantly antagonizes the depressant effects of 1.5 gm/kg of ethanol on the exploratory behavior of rats in a T-maze apparatus. Propranolol has not previously been used as an antagonist to ethanol's behavioral effects in the rat; therefore, a range of doses between 1 and 20 mg/kg of propranolol was employed in the present investigation.

The results of three series of pretreatment experiments are presented in Table 6.2. In Series 1, propranolol, at doses of 1, 5, 10, and 20 mg/kg, did

Table 6.2
Effect of Pretreatment with Propranolol, *d*-Amphetamine, and Caffeine on Discrimination of 10% Ethanol and Saline

Treatment	Number of trials[a]	Number of ethanol responses	Responses into ethanol side (%)
(1)			
Ethanol[b]	40	36	90
Saline	40	5	12.5
1 mg/kg Propranolol			
+ Ethanol	10	8	80
+ Saline	10	1	10
5 mg/kg Propranolol			
+ Ethanol	10	9	90
+ Saline	10	1	10
10 mg/kg Propranolol			
+ Ethanol	10	9	90
+ Saline	10	1	10
20 mg/kg Propranolol			
+ Ethanol	10	8	80
+ Saline	10	1	10
(2)			
Ethanol	50	45	90
Saline	50	4	8
4 mg/kg *d*-Amphetamine			
+ Ethanol	50	27	54[c]
+ Saline	50	4	8
(3)			
Ethanol	80	70	87.5
Saline	80	8	10
3% Ethanol	30	1	36.7
15% Ethanol	30	28	93.3
100 mg/kg Caffeine			
+ Saline	30	4	13.3
+ 3% Ethanol	30	10	33.3
+ 15% Ethanol	30	26	86.7

Source: M. D. Schechter, Effect of propranolol, d-amphetamine and caffeine on ethanol as a discriminative cue, *European Journal of Pharmacology,* 1974, *29,* p. 55. Reprinted by permission.

[a] $N = 10$.

[b] Ethanol dose is 10% (w/v) in saline, which is equivalent to 1.5 gm ethanol/kg.

[c] Probability of difference from ethanol score being due to chance, $p < .05$, chi-square test.

not significantly affect the rat's ability to discriminate 10% ethanol or saline. However, 4 mg/kg of *d*-amphetamine sulphate was observed to significantly ($p < .05$) antagonize the ethanol cue without altering recognition of the saline state. The correct response was made in 45 of 50 trials with 10% ethanol alone. After pretreatment with amphetamine, 27 of 50 correct first responses with ethanol were observed. Caffeine, 100 mg/kg, was shown (Series 3) to have no significant effect on the rat's ability to discriminate 3 or 15% doses of ethanol. Similarly, caffeine had no effect on saline discrimination.

Before administration of PCPA to deplete 5-HT, each of the rats was tested in five sessions with either 10% ethanol or saline, and this represents "pre-PCPA" discriminative performance. Brain 5-HT was then depleted by the oral administration of PCPA in a dose of 350 mg/kg (Schechter & Rosecrans, 1972). Drug sessions began 72 hr after PCPA administration, when depletion of brain 5-HT levels is 81–87% of normal as shown by chemical analysis (Rosecrans, Goodloe, Bennett, & Hirschhorn, 1973). Each rat received either 10% ethanol or saline at 9 A.M. and the other treatment at 3 P.M., with the order randomized over 6 consecutive weeks. The rat was tested 15 min after administration, and a response into either compartment was followed immediately by the termination of shock. Only 1 trial was run after each treatment, and the compartment entered was recorded. No training trials were conducted during this testing period, and each rat was tested with both ethanol and saline 5 times per week.

The effect of PCPA on the ethanol cue is presented in Figure 6.3. Before PCPA administration, the rats responded into the ethanol-correct compartment in 43 of 50 (86%) trials with ethanol and into the saline-correct compartment in 42 of 50 (84%) trials with saline. This is indicated as pre-PCPA per-

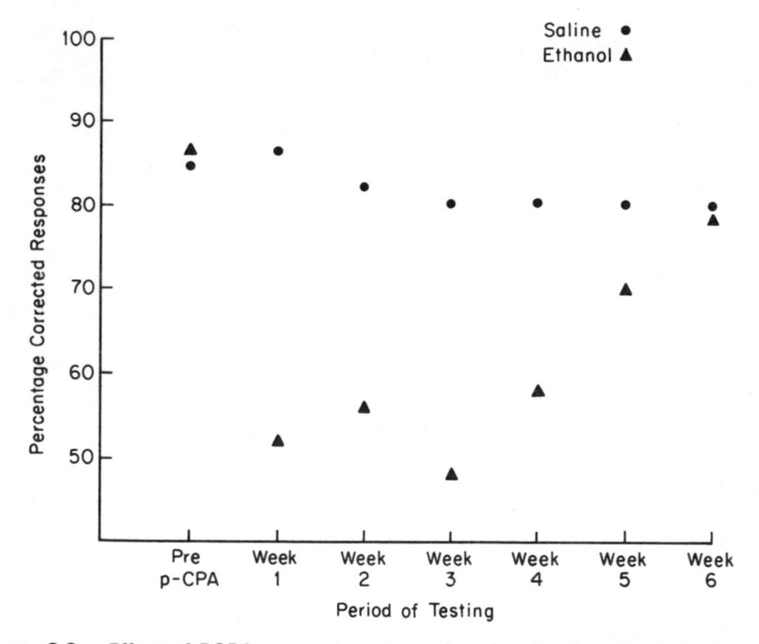

Figure 6.3. Effect of PCPA pretreatment on ethanol and saline discrimination. Abscissa: period of testing; 1 week before PCPA treatment and 6 weeks after PCPA administration. Ordinate: percentage of correct responses according to drug administered. Each point represents 10 rats given 5 administrations each of either saline or 10% ethanol. Reprinted by permission from: M. D. Schechter, Effect of propanolol, d-amphetamine and caffeine on ethanol as a discriminative cue, *European Journal of Pharmacology*, 1974, *29*, p. 55.

formance. The administration of PCPA markedly reduced the ability of the rats to discriminate the ethanol cue as seen in the first week after PCPA treatment in which 26 correct responses in 50 trials were made. The percentage of correct responses for Weeks 2–4 were, likewise, at about 50% or "chance" levels of responding. While rats were unable to detect the ethanol cue, the saline cue was unaltered. Significantly lower discrimination of ethanol than of saline ($p < .01$, t-test of means) continued until the fifth week after PCPA treatment, and at Week 6 discrimination performance after ethanol and saline was the same. At this time, rats were retrained to discriminate between ethanol and saline, and pre-PCPA performance levels were readily attained, indicating that PCPA did not produce any latent effects.

DISCUSSION

The literature pertaining to the interaction of ethanol and other drugs is voluminous (see Polacsek, Barnes, Turner, Hall, & Weise, 1972), and three drugs have recently been reported to alter the central effects of ethanol. Propranolol, a β-adrenergic blocking agent, has been observed to inhibit the respiration and narcotic effects of ethanol in mice (Smith & Hayashida, 1970) and is currently being tested as a potential blocker of the psychological and behavioral effects of ethanol in man (Mendelson, 1972; Tyrer, 1972). *d*-Amphetamine sulphate has been reported to antagonize ethanol's effects on performance and its nystagnatic effect in humans (Bernstein, Richards, Hughes, & Forney, 1965; Taylor, Wilson, Nash, & Cameron, 1964), whereas caffeine, in the form of coffee, is generally thought to be a sobering agent after excessive alcohol ingestion (Nash, 1966). The purpose of the present study was to investigate the effect of these three drugs on rats trained to discriminate the interoceptive cue produced by ethanol.

Pretreatment with *d*-amphetamine was observed to decrease significantly the discriminative ability in rats trained to discern the interoceptive cue produced by ethanol. The discriminable properties of ethanol as a state-dependent drug are discussed extensively in a study by Overton (1972) in which three rats receiving 2 gm/kg of ethanol (as 10% solution) i.p. or no drug, and trained 15 min after injection, required 15 sessions to attain a performance criterion of 8 correct first-trial choices during 10 consecutive training sessions. This time course of ethanol's discriminative effects closely matches that of the present study (Figure 6.1). The slightly longer time necessary to train the 13 rats (20 sessions versus 15 sessions) is most probably a function of the lower ethanol training dose employed. A discrepancy between Overton's investigation and the present results does, however, exist in that the former study showed that naive rats could discriminate ethanol at 2.7 and 3 gm/kg (18 and 20% solutions, respectively) faster than the 1.5 gm/kg (10%) dosage. In chronically treated rats, the two higher ethanol doses produced a decrease in discriminative performance (Figure 6.2), and the explanation offered for this result is gross toxicity produced by administration of the large ethanol doses.

Amphetamine has been reported to decrease (Miller, 1944), increase (Hughes & Forney, 1964), and have no effect on (Brown, Hughes, Forney, & Richards, 1965; Wilson, Taylor, Nash, & Cameron, 1966) ethanol's action on human experimental subjects. In rats, amphetamine (4 mg/kg) was shown to antagonize ethanol's depressant effect on exploratory behavior in a T-maze apparatus (Leonard & Wiseman, 1970). The administration of 4 mg/kg of *d*-amphetamine sulphate to rats trained to differentiate between the interoceptive cues produced by 1.5 gm/kg of ethanol and saline resulted in a significant decrease in ethanol discrimination (Table 6.2, Series 2). The simplest explanation for this effect is that the cue to which the rats respond in making a differential choice is produced by ethanol's depressant action on the CNS and that amphetamine, a central stimulant, counteracts the ethanol cue by an opposite central action. A similar explanation for ethanol-controlled state dependent behavior was previously offered by Kubena and Barry (1969). The precise mechanism and/or site of action, or the possibility of opposite actions on a brain biogenic amine, cannot be predicted from this experiment.

Like the reports on the interaction of ethanol and amphetamine, caffeine has been shown to antagonize (Graf, 1950; Nash, 1966) and to have no effect on (Forney & Hughes, 1965; Newman & Newman, 1956) ethanol's action on human performance. In rats, 100 mg/kg of caffeine and 1 gm/kg of ethanol were shown to produce significantly greater gross depressant effects than 1 gm/kg of ethanol administered alone (Alstott & Forney, 1971). In the present study, caffeine had no significant effect on discriminative performance after 3 and 15% ethanol or saline. Oral administration of 100 mg/kg caffeine 15 min prior to ethanol, 4.8 gm/kg, has been reported to depress blood alcohol levels in rats (Siegers, Strubelt, & Back, 1972). This effect was seen from 1 to 6 hr after ethanol administration and was thought to be caused by a caffeine-induced delay of stomach emptying due to relaxation of the rat's gastric musculature and, thereby, retardation of ethanol absorption. It is, therefore, possible that if the rats were trained and tested 1 or more hours after ethanol administration, the pretreatment with caffeine might have produced the reported decrease of ethanol blood and brain concentrations and, in turn, decreased discriminative performance in the state-dependent model. Nevertheless, the question: Does caffeine and/or coffee antagonize the impairment of human performance by ethanol? must await a scientifically responsible answer.

Propranolol has been used successfully in the treatment of opiate addiction (Grosz, 1972) and is currently being tested in alcohol addiction (Tyrer, 1972). In the present study, propranolol in doses of 1, 5, 10, and 20 mg/kg had no significant effect on the rat's ability to discriminate ethanol. The possibility exists that the reported efficacy of propranolol in attenuating alcohol withdrawal symptoms is produced by the drug's tranquilizing effect rather than by any interaction with ethanol.

Rats trained to discriminate between ethanol and saline states were unable to do so when treated with PCPA, a specific depletor of 5-HT. The possibility that this resulted from a general effect of PCPA toxicity can be eliminated by the observation that the cueing effect of *d*-amphetamine sulfate cannot be antagonized by the same PCPA regimen used here (Schechter & Cook, 1975).

Brain 5-HT levels have been reported to return to normal in 16 days after termination of PCPA treatment (Koe & Weissman, 1966), and the reasons for the continued reduction in the rats' ability to discriminate the ethanol cue for up to 5 weeks following PCPA treatment is not clear. A similar long-term reduction in ethanol preference has been observed by Myers and Veale (1968) for which they offer two possible explanations: (*a*) Evidence is cited that indicates that ethanol decreases the turnover rate of brain 5-HT, and the hypothesis is offered that the PCPA-treated rats reject the once-preferred ethanol in an attempt to conserve their remaining stores of 5-HT. This explanation does not apply to the observations made in the present study since the rats were not making a free choice in accepting or rejecting ethanol; (*b*) PCPA decreases tryptophan or 5-HT levels, and this alters the long-term action of ethanol on certain brain structures. It has been reported that the urinary level of 5-hydroxyindoleacetic acid, a metabolic product of 5-HT, is lower in alcoholics than in control subjects, and it was suggested that the biosynthesis of 5-HT is reduced in chronic alcoholics (Olson, Gursey, & Vester, 1960). It is tempting to speculate that chronic use of alcohol produces a decrease in brain 5-HT that, in turn, reduces central discrimination of the behavioral cues produced by continued and elevated alcohol ingestion. The proof of this hypothesis, as well as the efficacy of 5-HT manipulation in the rehabilitation of alcoholics, must wait for more extensive research into the cause-effect relationship between brain 5-HT and ethanol effect.

REFERENCES

Alstott, R. L., & Forney, R. B. Performance studies in rabbits, rats and mice after administration of caffeine or l-methylxanthine, singly and with ethanol. *Federation Proc.*, 1971, *30*, 568.

Bernstein, M. E., Richards, A. B., Hughes, F. W., & Forney, R. B. Optokinetic nystagmus under the influence of d-amphetamine and alcohol. In R. N. Harger (Ed.), *Alcohol and traffic safety.* Bloomington, Indiana: Indiana Univ. Press, 1965. P. 208.

Brown, D. J., Hughes, F. W., Forney, R. B., & Richards, A. B. Effect of d-amphetamine and alcohol on attentive motor performance in human subjects. In R. N. Harger (Ed.), *Alcohol and traffic safety.* Bloomington, Indiana Univ. Press., 1965. P. 184.

Chase, T. N., & Murphy, D. L. *Serotonin and central nervous system function.* (Report No. 6560). Bethesda, Maryland: National Institute of Mental Health, Laboratory of Clinical Science, 1973.

Conger, J. J. The effects of alcohol on conflict behaviour in the albino rat, *Quart. J. Studies Alc.*, 1951, *21*, 29.

Cooper, J. R., Bloom, F. G., & Roth, R. H. *The biochemical basis of neuropharmacology.* (2nd ed.) New York: Oxford Univ. Press, 1974. P. 175.

Crow, L. T. Effects of alcohol on conditioned avoidance responding. *Physiol. Behav.*, 1966, *1*, 89.

Feldstein, A. Effect of ethanol on neurohumoral amine metabolism. In B. Kissin & H. Begleiter (Eds.), *The biology of alcoholism.* Vol. 1. New York: Plenum Press, 1972. P. 127.

Forney, R. B., & Harger, R. N. Toxicology of ethanol. *Ann. Rev. Pharmacol.*, 1969, *9*, 379.
Forney, R. B., & Hughes, F. W. Effect of caffeine and alcohol in performance under stress of audiofeedback, *Quart. J. Studies Alc.*, 1965, *26*, 206.
Frey, H. H., Magnussen, M. P., & Nielsen, C. K. The effect of p-chloroamphetamine on the consumption of ethanol by rats. *Arch. Intern. Pharmacodyn.*, 1970, *183*, 165.
Geller, I. Effects of P-CPA and 5-HTP on alcohol intake in the rat. *Pharmacol. Biochem. Behav.*, 1973, *1*, 361.
Goodwin, D. W., Powell, B., Hill, S. Y., Leiberman, W., & Viamontes, J. Effect of alcohol on "dissociated" learning in alcoholics. *J. Nervous Mental Disease*, 1974, *158*, 198.
Graf, O. Increase of efficiency by means of pharmaceutics (stimulants). In *German aviation medicine*. Washington, D.C.: Dept. Air. Force, 1950.
Grosz, H. J. Successful treatment of a heroin addict with propranolol: Implications for opiate addiction treatment and research, *J. Indiana State Med. Assoc.*, 1972, *65*, 505.
Holloway, F. A. State-dependent effects of ethanol on active and passive avoidance learning. *Psychopharmacologia*, 1972, *25*, 238.
Hughes, F. W., & Forney, R. B. Dextro-amphetamine, ethanol and dextroamphetamine-ethanol combinations on performance of human subjects stressed with delayed auditory feedback. *Psychopharmacologia*, 1964, *6*, 234.
Jacobs, B. L., Mosko, S. S., & Trulson, M. E. The investigation of the role of serotonin in mammalian behavior. In R. R. Drucker-Colin and J. L. McCaugh (Eds.), *Neurobiology of sleep and memory*. New York: Academic Press, 1975.
Koe, B. K., & Weissman, A. p-Chlorophenylalanine: A specific depletor of brain serotonin. *J. Pharmacol. Exp. Therap.*, 1966, *154*, 499.
Krimmer, E. C., & Barry, H., III. Differential stimulus characteristics of alcohol and pentobarbital in rats. *Proc. 79th Ann. Conv. Amer. Psychol. Assoc.*, 1973, *8*, 1005. (a)
Krimmer, E. C., & Barry, H., III. Discriminability of pentobarbital and alcohol tested by two lever choice in shock escape. *Pharmacologist*, 1973, *15*, 236. (b)
Kubena, R. K., & Barry, H. Generalization by rats of alcohol and atropine stimulus characteristics to other drugs. *Psychopharmacologia*, 1969, *15*, 196.
Leonard, B. E., & Wiseman, B. O. The effect of ethanol and amphetamine mixtures on the activity of rats in a T-maze. *J. Pharm. Pharmacol.*, 1970, *22*, 967.
Mendelson, J. Propranolol against alcoholism. *Med. World News*, 1972, *13*, 13.
Miller, M. M. Amphetamine sulphate in aborting the acute alcoholic cycle, *Am. J. Psychiat.*, 1944, *100*, 800.
Myers, R. D., & Cicero, T. J. Effects of serotonin depletion on the volitional alcohol intake of rats during a condition of psychological stress. *Psychopharmacologia*, 1969, *15*, 373.
Myers, R. D., Evans, J. E., & Yakich, T. L. Ethanol preference in the rat: Interactions between brain serotonin and ethanol, acetaldehyde, paraldehyde, 5-HTP and 5-HTOL, *Neuropharmacol.*, 1972, *11*, 539.
Myers, R. D., & Veale, W. L. Alcohol preference in the rat: Reduction following depletion of brain serotonin, *Science*, 1968, *160*, 1469.
Nash, H. Psychological effects and alcohol-antagonizing properties of caffeine. *Quart. J. Studies Alc.*, 1966, *27*, 727.
Newman, H. W., & Newman, E. J. Failure of dexedrine and caffeine as practical antagonists of the depressant effect of ethyl alcohol in man, *Quart. J. Studies Alc.*, 1956, *17*, 406.
Olson, R. E., Gursey, D., & Vester, J. W. Evidence for defect in tryptophan metabolism in chronic alcoholism, *New Engl. J. Med.*, 1960, *263*, 1169.
Opitz, K. Observations in alcohol drinking rats—effect of fenfluramine. *Pharmakopschiatr. Neuropsychopharmackol.*, 1969, *2*, 202.
Overton, D. A. State-dependent or "dissociated" learning produced with pentobarbital. *J. Comp. Physiol. Psychol.*, 1964, *57*, 3.
Overton, D. A. State-dependent learning produced by alcohol and its relevance to alcoholism. In

B. Kissen & H. Begleiter (Eds.), *The biology of alcoholism.* Vol. 2. *Physiology and behavior.* New York: Plenum Press, 1972. P. 193.

Polacsek, E., Barnes, T., Turner, N., Hall, R., & Weise, C. Interaction of Alcohol and Other Drugs. (2nd ed.) Toronto: Addiction Research Foundation, 1972.

Rosecrans, J. A., Goodloe, M. H., Bennett, G. J., & Hirschhorn, I. D. Morphine as a discriminative cue: Effect of amine depletors and naloxone. *European J. Pharmacol.*, 1973, *21*, 252.

Ryback, R. S. State-dependent or "dissociated" learning with alcohol in the goldfish. *Quart. J. Studies Alc.*, 1969, *30*, 598.

Schechter, M. D., & Cook, P. G. Dopaminergic mediation of the interoceptive cue produced by d-amphetamine in the rat. *Psychopharmacologia*, 1975, *42*, 185.

Schechter, M. D., & Rosecrans, J. A. Nicotine as discriminative stimulus in rats depleted of norepinephrine and 5-hydroxytryptamine, *Psychopharmacologia*, 1972, *24*, 417.

Siegers, C. P., Strubelt, O., & Back, G. Inhibition by caffeine of ethanol absorption in rats. *European J. Pharmacol.*, 1972, *20*, 181.

Smith, A. A., & Hayashida, K. Blockade by propranolol of the respiratory and narcotic effects of ethanol. *National Research Council Comm. on Problems of Drug Dependence*, 32nd meeting, 1970, P. 687.

Storm, T., & Smart, R. G. Dissociation: A possible explanation of some features of chronic alcoholism and implications for treatment, *Quart. J. Studies Alc.*, 1965, *26*, 111.

Taylor, J. S., Wilson, L., Nash, C. W., & Cameron, D. F. The effects of ethyl alcohol and amphetamine on performance. *Can. Fed. Biol. Sci.*, 1964, *7*, 36.

Tyrer, P. Propranolol in alcohol addiction. *Lancet*, 1972, *2*, 707.

Veale, W. L., & Myers, R. D. Decrease in ethanol intake in rats following administration of p-chlorophenylalanine. *Neuropharmacol.*, 1970, *9*, 317.

Wilson, L., Taylor, J. D., Nash, C. W., & Cameron, D. F. The combined effects of ethanol and amphetamine sulphate on performance of human subjects. *Can. Med. Assoc., J.*, 1966, *94*, 478.

7

The Discriminative Stimulus Properties of *N*- and *M*-Cholinergic Receptor Stimulants[1]

John A. Rosecrans and William T. Chance
Department of Pharmacology
Medical College of Virginia
Virginia Commonwealth University

INTRODUCTION

One of the major objectives of research conducted in this laboratory has been to determine the mechanisms by which nicotine is producing its behavioral effects in experimental animals. The primary goal of this research is identification of the potent reinforcing properties of nicotine that contribute to its use in tobacco by humans. Nicotine has long been known to have a specific stimulating effect on peripheral autonomic nervous system (ANS) preganglionic synapses (nicotinic receptor sites). This effect has been shown to lead to the release of acetylcholine at the postganglionic parasympathetic (muscarinic sites) synapses and norepinephrine at the postganglionic sympathetic (adrenergic sites) synapses. Thus, in terms of a specific central nervous system (CNS) mechanism, one must first consider whether nicotine exerts an effect in the brain similar to that observed within the peripheral ANS. Specifically, it may be hypothesized that the behavioral effects resulting from the administration of nicotine may be mediated by the stimulation of specific *N*-cholinergic (nicotinic) receptors, by the release of acetylcholine at the *M*-cholinergic (muscarinic) synapses, or by the release of norepinephrine at

[1] The basic support for this research was provided by the American Medical Association Education and Research Foundation. W. T. Chance is now also being supported by a postdoctoral fellowship (USPHS Grant T22 DE00116).

the adrenergic sites. In addition to the specific hypotheses of CNS receptor stimulation, one must show that the behavioral and reinforcing properties of nicotine are mediated by central and not peripheral neural mechanisms.

To permit generalization of these results to the use of nicotine in man, the observed effects of the drug must be shown in a behavioral task that is specific to nicotine and analogous to nicotine's effects in man. Preliminary studies using a variety of tasks indicated that nicotine's behavioral effects were varied and depended to a great extent on the baseline rates of behavior in the specific tasks (Rosecrans, 1971a,b). From these studies, it became apparent that nicotine will reduce extremely high rates of baseline behavior, while increasing behavioral rates when the baseline is low. The results of these studies indicated that such tasks were inappropriate for the investigation of the mechanism of action of the drug. Thus, the need for a paradigm that allowed an animal to identify the drug effect rather than one that studied the disruptive or excitatory effects of nicotine became apparent. The discriminative stimulus procedures developed by Overton (1969) and modified by Morrison and Stevenson (1969) provided the type of paradigm required by these investigations. According to this model, a rat was required to discriminate between saline and the nicotine drug states in order to obtain food reinforcement or to escape a painful shock. Thus, the drug was the discriminative stimulus for correct-choice responses. This model of investigation of nicotine's behavioral effects has proved to be very useful and appropriate for our purposes. The behavioral response depends on the ability of the animal to detect a specific state, which is analogous to requiring human subjects to report their perceptions after receiving a specific drug.

APPROACHES AND METHODS

The discriminative–stimulus procedures involve two discrete tasks. In the first paradigm, rats were required to escape a shock by entering one area of a three-chamber box, the choice of chamber being contingent upon whether nicotine or saline was administered 10 min prior to testing (Schechter & Rosecrans, 1972a). The second task required a food-deprived rat to press one of two levers, or to enter one arm of a T-maze, for liquid food reinforcement, contingent upon which drug was administered (Hirschhorn & Rosecrans, 1974; Schechter & Rosecrans, 1971). The two-lever operant procedure has been run under several different schedules of food reinforcement, but a variable interval (VI)-15 sec has yielded the most satisfactory results. Mechanism of action studies were implemented after the rats had learned to discriminate the stimulus effects of the drug. The procedures for learning and maintaining performance involved a daily nonreinforced extinction period, which consisted of the first trial in the T-maze and escape procedures or the first 2½ of a 15-min training session using the operant procedures. These extinction periods were used to obtain learning and performance data in various studies. The results were expressed as percentage of drug choice or percentage of correct nicotine-bar responding. The criterion for learning in the T-maze was considered reached when the rats made 8 of 10 correct arm choices under each drug state. In the 2-lever operant procedure, rats were required to achieve a stable rate of responding in which at least 75% of their responses under the nicotine state were on the nicotine-correct bar. Following the injection of saline, however, stable response rates on the nicotine-correct bar were not expected to exceed 20–30%. To prevent biasing the discrimination training, drug mechanism of action studies were conducted during these 2½–5-min extinction sessions in which the rats were not reinforced.

RESULTS AND DISCUSSION

Training and Dose–Response Data

Rats generally learned to discriminate between nicotine and saline states within 20 training sessions under each drug state (Figure 7.1). The injection of nicotine has been found generally to disrupt behavior during the first few sessions. Rats soon become tolerant to the drug, however, and rates of responding under nicotine and saline do not differ after the first few sessions (Hirschhorn & Rosecrans, 1974). Dose–response studies indicate dose-related effects (Table 7.1) across the discriminable doses (100–400 μg/kg). Attempts to train rats to discriminate lower doses indicate a loss of the discrimination at 50 μg/kg. The data concerning the duration of action of nicotine have shown that the stimulus effect lasts for at least 60 min and begins to decline thereafter, terminating in about 160 min after drug administration (Figure 7.2). A study conducted in this laboratory (Murfin, 1974) also examined the influence of schedules of reinforcement on several of these response

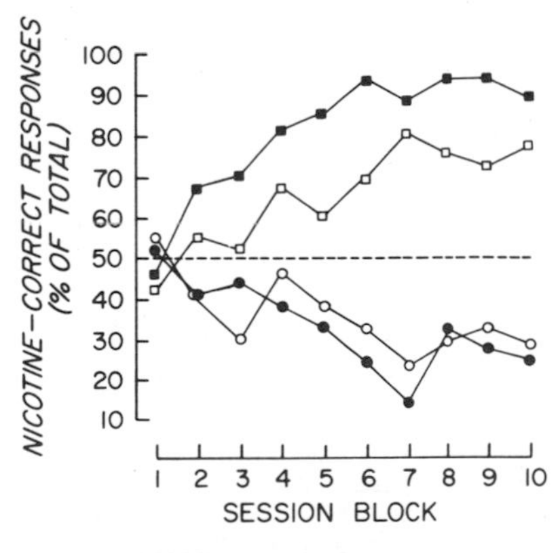

Figure 7.1. Discriminated responding following the administration of nicotine and saline. Ordinate: number of responses on the nicotine correct lever under each drug state. Two groups of rats were trained to discriminate nicotine at either 200 or 400 μg/kg. Abscissa: successive blocks of four sessions, two under each drug state (from Hirschhorn & Rosecrans, 1974, with permission).

Table 7.1
Nicotine Dose Response Generalizations in a T-Maze Procedure

Nicotine dose[a] (μg/kg)	Number of trials (replications)	Percentage of first responses into "nicotine-correct" arm
400	70	87[b]
255	28	89[b]
182	28	61[b,c]
66	28	50[b,c]
37	28	29[c]
0	70	14[c]

Source: From Schechter and Rosecrans, 1971; with permission.
[a] $N = 14$.
[b] Probability of difference from saline, $p < .001$, chi-square test.
[c] Probability of difference from 400 μg/kg, $p < .001$, chi-square test.

parameters. Three schedules of reinforcement (VI-15 sec, DRL [differential reinforcement of low response rate]-10 sec and FR [fixed ratio]-10) were studied in relation to dose–response relationships and time–duration parameters. As expected, response rates were much greater in animals trained under the FR-10 schedule, but no differences were found among the various schedules in percentage of drug responses. Furthermore, dose–response and time–duration relationships were not significantly different across the various schedules, indicating that there was no interaction between percentage of nicotine discrimination and schedule of reinforcement. Finally, no differences in response rates were found between the various doses (100–400 μg/kg) of

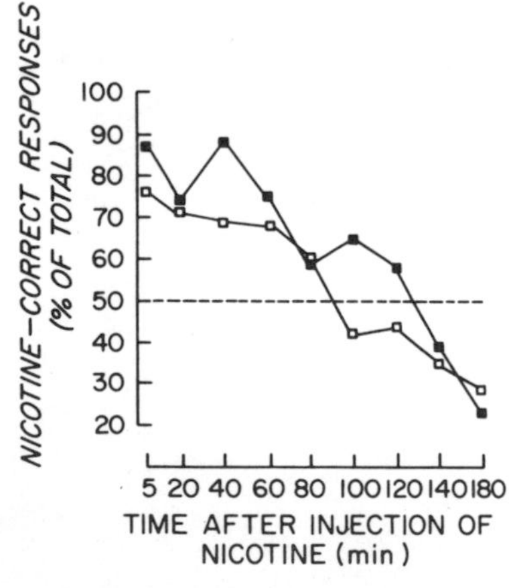

Figure 7.2. Time course of nicotine and saline discrimination (adapted from Hirschhorn & Rosecrans, 1974, with permission).

nicotine or between saline and the training dose of the drug. These studies are especially important for this research since they indicate that the stimulus can be well defined and is not affected by competing behavioral responses in the animals studied.

Specificity of the Stimulus

The demonstration of the mechanism of action of a drug requires that its stimulus properties be specific to the drug response, with little generalization to other drug states. Barry (1974) has reviewed current research showing that the degree of generalization exhibited by animals between two drug states depends primarily on the similarity of the pharmacological properties of the two drugs. Similar findings have been obtained in this laboratory, with generalization of nicotine being studied extensively. To this date, nicotine has not been shown to generalize to any other drug (Schechter & Rosecrans, 1972a). Amphetamine produces effects closest to nicotine, but our research indicates that nicotine-trained rats perceive amphetamine as different from both nicotine and saline (Chance, Murfin, Krynock, & Rosecrans, 1977). From these data, it is inferred that rats perceive amphetamine as similar to but not the same as nicotine. Peripheral injections of nicotine have been observed, however, to generalize to intraventricular injections of its major metabolite, cotinine. In this study, rats trained to discriminate nicotine at 100 μg/kg subcutaneously (s.c.) generalized the response to cotinine, while those trained at 200 μg/kg (s.c.) did not. This observation suggests that drugs with similar structures have a high probability of generalizing to each other. Thus, our research indicates that the stimulus effect of nicotine is extremely specific and presents an appropriate paradigm for interpreting how nicotine produces its CNS effects.

Receptor Specificity

Research conducted in Domino's laboratory (1967) involving electroencephalographic (EEG) activation has suggested that there are two types of specific cholinergic receptors that are not in series with each other as in the peripheral ANS. In his studies, both nicotine and the potent peripheral *M*-cholinergic agonist, aerocholine, were shown to induce EEG activation. The nicotine-induced arousal, however, could be blocked by the *N*-cholinergic antagonist, mecamylamine, but was not affected by atropine. Conversely, arousal induced by aerocholine was inhibited by atropine but was not antagonized by mecamylamine. Furthermore, the peripherally acting *N*-cholinergic antagonist, hexamethonium, and the peripherally acting *M*-cholinergic antagonist, methylatropine, were both shown to be ineffective in blocking EEG activation induced by nicotine or aerocholine. Thus, Domino has not only demonstrated the existence of two types of specific cholinergic receptors but has also presented evidence suggesting their CNS localization.

To determine whether nicotine was eliciting its stimulus properties in a manner analogous to that found by Domino, we designed a series of similar studies (Hirschhorn & Rosecrans, 1974; Schechter & Rosecrans, 1971). The results of these studies clearly indicate, in both operant and T-maze procedures, that nicotine's effects result primarily from the stimulation of a different receptor than that antagonized by atropine (Table 7.2). In these studies, mecamylamine was found to antagonize the nicotine stimulus, while hexamethonium was without effect. In a similar study, Schechter and Rosecrans (1972b) observed that the stimulus effects elicited by aerocholine were antagonized by atropine but not by methylatropine or mecamylamine. Therefore, both a receptor specificity and a CNS localization have been demonstrated for the stimulus effects of *N*- and *M*-cholinergic drugs. The specificity of these receptors was further shown in a study in which rats were trained to discriminte nicotine from aerocholine (Schechter & Rosecrans, 1972c). In this study, pretreatment with mecamylamine was also observed to block only the nicotine stimulus.

Correlations with Drug Levels

On the basis of the previously cited studies, it is apparent that nicotine stimulates a CNS cholinergic receptor different from the *M*-cholinergic receptor. To investigate these relationships further, two additional studies were designed to correlate nicotine's behavioral effects with the drug levels (Rosecrans & Schechter, unpublished observation; Hirschhorn & Rosecrans,

Table 7.2
Antagonism of the Nicotine Stimulus

Drug 1	Drug 2	Number of trials (replications)	Percentage of first responses into "nicotine-correct" arm
Saline	Saline	42	9.5[a]
Saline	Nicotine (400 μg/kg)	42	81[b]
Hexamethonium (750 μg/kg)	Saline	14	7.1
Hexamethonium (750 μg/kg)	Nicotine (400 μg/kg)	14	100
Mecamylamine (500 μg/kg)	Saline	14	4.3
Mecamylamine (500 μg/kg)	Nicotine (400 μg/kg)	14	28.6[a]

[a] Probability of difference from nicotine; $p < .001$, chi-square test, $N = 14$.
[b] Probability of difference from saline; $p < .001$, chi-square test, $N = 14$.

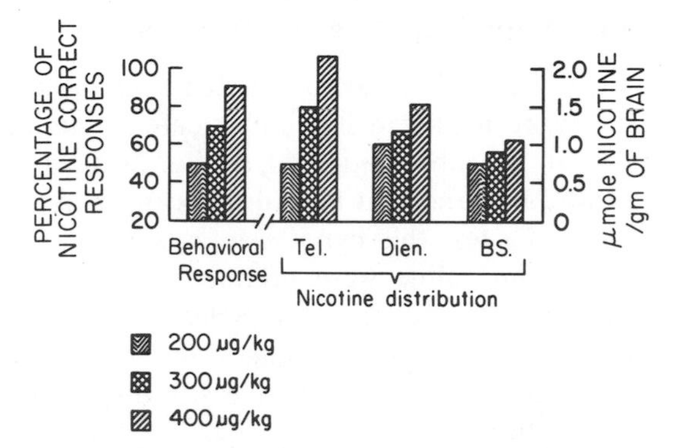

Figure 7.3. Nicotine discrimination at various dosage levels and brain area nicotine levels 5 min after drug administration. Behavioral response was measured in a T-maze, while brain area nicotine levels were determined in a different group of animals. These latter rats were maintained on a similar weight deprivation schedule (80% of weight) and received 40 s.c. injections of both various doses of nicotine and saline to simulate the same behavior conditions. *N* = 10–12 in all experimental groups.

1974). In the first study (Figure 7.3), a dose–response relationship was established that indicated high positive correlations between dosage, percentage of nicotine-correct responses, and brain nicotine levels. In the second study (Figure 7.4), both the training dosage and the duration of action were compared to brain nicotine levels. The correlations were again in the positive direction, suggesting that nicotine's stimulus effects depended on the level of nicotine in the brain at the time the behavior was measured. Thus, it seems that nicotine's behavioral effects depend on the accumulation of the drug in the brain as well as on direct stimulation of a specific receptor that can be interrupted by competitive antagonism at the receptor site.

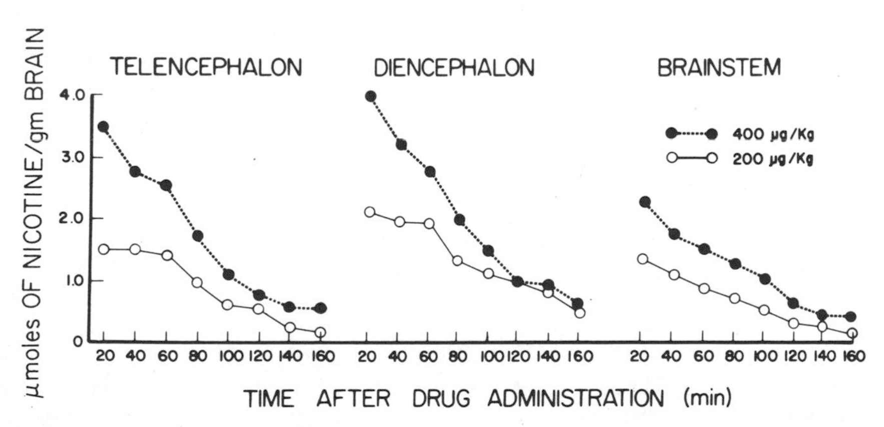

Figure 7.4. Brain levels of ^{14}C-nicotine at various times after a single injection (from Hirschhorn & Rosecrans, 1974, with permission).

Importance of Brain Norepinephrine and Dopamine

Orsingher and Fulginitti (1971) have suggested that the effects of nicotine and amphetamine may be mediated by similar catecholamine systems. They base this hypothesis on the observation of similar behavioral effects of the drugs in operant procedures and that both drugs can be antagonized by catecholamine antagonists. To test this hypothesis, rats that had been trained to discriminate nicotine from saline were pretreated with alpha-methyl paratyrosine (AMPT) prior to the test session (Schechter & Rosecrans, 1972d). Pretreatment with AMPT was found to antagonize the nicotine stimulus effect (Figure 7.5), which suggests that the nicotine stimulus depended on a catecholamine system. That this effect was specific to a catecholamine system was shown by the lack of inhibition by pretreatment with parachloraphenylalanine, an antagonist of brain serotonin, on the stimulus effect of nicotine. Further investigations with various adrenergic blocking agents and AMPT have given us the opposite results (Hirschhorn & Rosecrans, 1974). In this study, such drugs as propranolol and dibenamine produced no disruption of the stimulus effect of nicotine, indicating that neither alpha nor beta receptors were involved. We have had the same difficulty with other amine-synthesis inhibitors, and recent research suggests that the differences in these results may be caused by the type of task used to assess the stimulus effects of the drug. Thus,

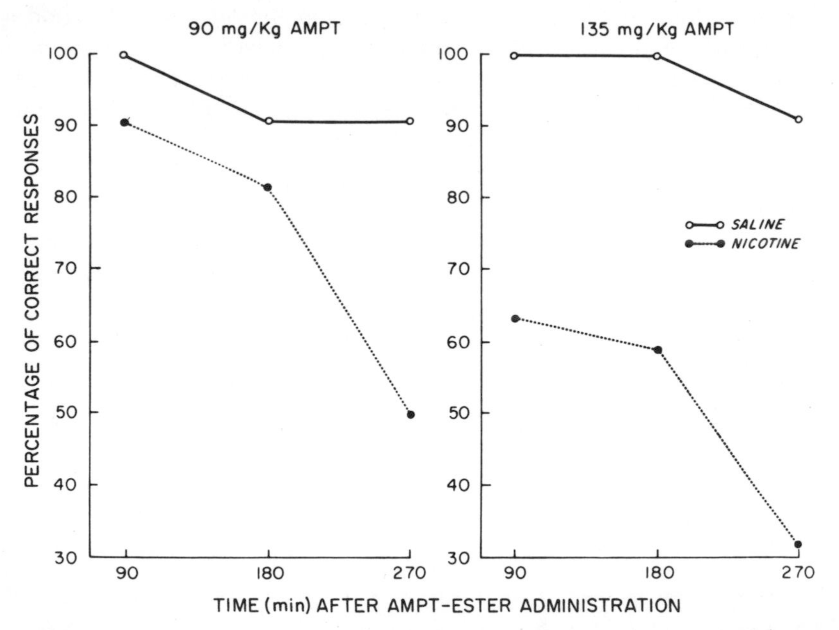

Figure 7.5. Effect of pretreatment with 90 and 135 mg/kg AMPT-ester on the discrimination of nicotine (400 μg/kg) and saline. Each point represents 12 trials with each of six rats receiving both saline and nicotine twice at each dose and at each time after AMPT-ester (from Schechter & Rosecrans, 1972c, with permission).

Table 7.3
Nicotine and *d*-Amphetamine Stimulus ED_{50} Doses in Catecholamine Depletions

Experimental group	DA/NE[a]	*d*-Amphetamine (μg/kg)	Nicotine[b] (μg/kg)
Control	3.31	221	85
DA ↓	1.99	308	97
NE ↓	4.84	426[c]	142

[a] μg/g DA/μg/g NE; DA ↓ and NE ↓ significant at $p < .05$.
[b] ED_{50} doses calculation from the regression lines, probit versus log doses.
[c] Analysis of variance of dose–response data indicated $p < .01$.

the blockade of the nicotine stimulus with amine depletors seems to be more likely when the stimulus effects are measured in tasks involving escape from or avoidance of shock (three-chamber box), while the blockade is less likely to occur when the task is involved with food reinforcement (operant procedures).

To elaborate the possible influence of catecholamine systems on the stimulus effects of nicotine, a study was designed in which rats were depleted of either dopamine (DA) or norepinephrine (NE) by the administration of 6-OH-dopamine to rats at 14 days of age (Elchisak & Rosecrans, 1973). In these studies, the mean effective doses (ED_{50}) of both nicotine and amphetamine were increased in catecholamine-depleted rats (Rosecrans, Elchisak, & Schechter, 1976). NE depletion reduced the ED_{50} dose for both amphetamine and nicotine, but this effect was significant only in *d*-amphetamine-trained rats (Table 7.3). In reviewing these findings, we hypothesize that NE may be specifically involved with only the *d*-amphetamine stimulus effects. Furthermore, it is postulated that the reduction of the stimulus effect of nicotine by NE depletion is not specific and that it may be related to a reduced arousal level produced by the NE depletion, yielding a decrease in sensitivity to the discrimination stimulus effects.

CNS Sites of Action

The results of experiments reported by Domino (1967) suggest that the reticular activating system may be a specific site of action for the effects of nicotine on EEG arousal. Other investigators (Nelson, Pelley, & Goldstein, 1975), however, have suggested that the hippocampus may be the critical site of nicotine-induced EEG arousal. These authors suggest that chronic treatment with nicotine causes a shift of subcortical mechanisms of EEG arousal from the reticular activating system to the hippocampus. To investigate the effects of nicotine applied directly to brain tissue and to test the site-of-arousal hypothesis, we designed a series of studies in which cannulae were implanted into the dorsal hippocampus of DA-depleted and NE-depleted rats. These rats

Table 7.4
Discriminability of Nicotine from Saline in Catecholamine Depleted Rats: Peripheral and Hippocampal Injection

Experimental Group	Peripheral[a] injection (400 μg/kg)	Hippocampal[a] injection (1 μg/area)
Control ($N = 5$)	79% (5)[b]	33% (10)[b]
NE ↓ ($N = 4$)	65% (5)	23% (6)
DA ↓ ($N = 4$)	58% (5)	— (7)

[a] Discriminability = percentage of nicotine-correct responding after nicotine subtracted by percentage of nicotine-correct responding after saline.

[b] Values in parentheses represent replications in these rats.

had previously been trained to discriminate 400 μg/kg of nicotine versus saline, and this discrimination was maintained by peripheral administration of the drugs after surgery. The results (Table 7.4) indicate that these rats were able to discriminate 1 μg of nicotine from saline when each of these substances were injected bilaterally into the hippocampus. The discrimination was reduced in NE-depleted rats and was not observed in DA-depleted rats. These studies involving intracranial sites of drug discrimination are very preliminary, but they indicate at least one brain site that may be involved in responding to the stimulus properties of nicotine. The reduced effect of the nicotine stimulus in catecholamine-depleted rats provides additional support for the involvement of NE and/or DA in this discrimination. Since only a small number of subjects were used in this study, the results are viewed as suggestive. These data, however, indicate a new direction for research in the further elaboration of the nicotine stimulus effect.

SUMMARY AND CONCLUSIONS

The studies reviewed here clearly indicate how the stimulus properties of a potent psychoactive drug may be used to learn how such a drug may be producing its CNS effects. Nicotine provides an excellent model. The drug elicits a very specific effect on *N*-cholinergic receptors, does not generalize to other CNS stimulants, and its stimulus effects are well correlated with brain levels. The degree of receptor specificity is further emphasized by the facts that the stimulus effects of the drug can be discriminated from *M*-cholinergic stimulus effects, and only pretreatment with mecamylamine has been shown to disrupt this stimulus effect. Ancillary studies have indicated that other amine systems may not be directly involved with the stimulus effect of nicotine but can alter the sensitivity of the CNS to this stimulus. More recent studies have provided information indicating that the hippocampus may be an important CNS location for the stimulus effect of the drug.

All the experimental evidence indicates that nicotine possesses very potent and specific stimulus properties, which could constitute the main reinforcing quality of tobacco. Although experimental animals cannot easily be made to self-administer nicotine, the drug does provide an extremely potent stimulus to behavior. Thus, the dissociative properties of the drug may reinforce its use in society. These interpretations, however, must be delivered with caution, for it is well known that the use of tobacco, like any social behavior, may be maintained by a vast array of reinforcers.

ACKNOWLEDGMENTS

The senior author (J. A. R.) is grateful to Drs. M. D. Schechter and I. D. Hirschhorn for their research contributions; both authors were also supported by A.M.A. Eduation Research Foundation postdoctoral fellowships.

REFERENCES

Barry, H., III. Classification of drugs according to their discriminable effects in rats. *Federation Proc.*, 1974, *33*, 1814–1824.

Chance, W. T., Murfin, D., Krynock, G. & Rosecrans, J. A. A description of the nicotine stimulus and tests of its generalization to amphetamine. *Psychopharmacology* (in press).

Domino, E. F. Electroencephalographic and behavioral arousal effects of small doses of nicotine: A neuropsychological study. *Ann. N.Y. Acad. Sci.*, 1967, *142*, 216–244.

Elchisak, M. A., & Rosecrans, J. A. Effect of central catecholamine depletions by 6-hydroxydopamine on morphine antinociception in rats. *Res. Comm. Chem. Path. Pharmacol.*, 1973, *6*, 349–352.

Hirschhorn, I. D., & Rosecrans, J. A. Studies on the time course and the effect of cholinergic and adrenergic receptor blockers on the stimulus effect of nicotine. *Psychopharmacologia*, 1974, *40*, 109–120.

Morrison, C. F., & Stephenson, J. A. Nicotine injections as the conditioned stimulus in discrimination learning. *Psychopharmacologia*, 1969, *15*, 351–360.

Murfin, D. A parametric study of the effects of nicotine's discriminative control on tests of dose transfer and time duration for three schedules of reinforcement. Unpublished master's thesis, Virginia Commonwealth Univ., Richmond, Virginia, 1974.

Orsingher, O. A., & Fulginitti, S. Effects of *alpha*-methyl tyrosine and adrenergic blocking agents on the facilitating action of amphetamine and nicotine on learning in rats. *Psychopharmacologia*, 1971, *19*, 231–240.

Overton, D. A. Control of T-maze choice by nicotinic, antinicotinic and antimuscarinic drugs. *Proc. Annual Am. Psychol. Assoc. Conv.*, 1969, *77*, 869.

Nelson, J. M., Pelley, K., & Goldstein, L. Protection by nicotine from behavioral disruption caused by reticular formation stimulation in the rat. *Pharmacol. Biochem. Behav.*, 1975, *3*, 749–754.

Rosecrans, J. A. Effects of nicotine on brain area 5-hydroxytryptamine function in male and female rats separated for differences of activity. *European J. Pharmacol.* 1971, *16*, 123–127. (a)

Rosecrans, J. A. Effects of behavioral arousal and brain 5-hydroxytryptamine function in female rats selected for differences in activity. *European J. Pharmacol.*, 1971, *14*, 29–37. (b)

Rosecrans, J. A., Elchisak, M. A., & Schechter, M. D. Dopamine and psychoactive drugs, a further evaluation. *Federation Proc.*, 1976, *60*, 506.

Schechter, M. D., & Rosecrans, J. A. CNS effect of nicotine as the discriminative stimulus for the rat in the T-maze. *Life Sci.*, 1971, *10*, 821–832.

Schechter, M. D., & Rosecrans, J. A. Nicotine as a discriminative cue in rats: Inability of related drugs to produce a nicotine-like cueing effect. *Psychopharmacologia*, 1972, *27*, 379–387. (a)

Schechter, M. D., & Rosecrans, J. A. Atropine antagonism of aerocholine-cued behavior in the rat. *Life Sci.*, 1972, *11*, 517–523. (b)

Schechter, M. D., & Rosecrans, J. A. Effect of mecamylamine on discrimination between nicotine- and aerocholine-produced cues. *European J. Pharmacol.* 1972, *17*, 179–182. (c)

Schechter, M. D., & Rosecrans, J. A. Nicotine as a discriminative stimulus in rats depleted of norepinephrine and 5-hydroxytryptamine. *Psychopharmacologia*, 1972, *24*, 417–429. (d)

8

The Discriminative Stimulus Properties of Cannabinoids: A Review[1]

Robert L. Balster and Robert D. Ford[2]
Pharmacology Department
Medical College of Virginia
Virginia Commonwealth University

There has been an increase in research on the biological effects of marihuana and *Cannabis* plant constituents. Because at self-administered doses most of the effects of marihuana are on the central nervous system (CNS), there is particular concern with its behavioral effects. Our understanding of the complex chemistry of marihuana has coincided with the emergence of sophisticated research methodologies in behavioral pharmacology. As a consequence, the behavioral effects of cannabinoids, especially in animals, are widely under investigation. This chapter will review one aspect of this work.

Not until the mid-1960s was the chemistry of marihuana substantially elucidated. In 1964, Gaoni and Mechoulam isolated and identified Δ^9-tetrahydrocannabinol (Δ^9-THC), and it was shown to mimic the behavioral effects of marihuana in man and laboratory animals. In 1966, Hively, Mosher, and Hoffman isolated and identified the isomer Δ^8-THC, which also was found to have marihuana-like activity. Two other cannabinoids, cannabinol and cannabidiol, are also present in relatively large amounts in the plant, and much smaller amounts of about 30 other cannabinoids have been isolated and

[1] Preparation of this manuscript and research from our laboratory was supported by USPHS Grant DA-00490.

[2] Robert D. Ford was a postdoctoral fellow supported by USPHS Training Grant T-22 DA-00128. Present address: Pharmacology Department, Michigan State University.

identified. Although cannabinol and cannabidiol have little behavioral activity, not much is known about the activity of these less abundant cannabinoids (Hollister, 1974). It is now generally believed that Δ^9-THC is responsible for most of the psychological effects of marihuana, while Δ^8-THC is present in lesser amounts in the plant and is less potent than its isomer (Hollister, 1974; Mechoulam, 1970).

THE BEHAVIORAL PHARMACOLOGY OF CANNABINOIDS

The effects of Δ^9-THC on behavior have been extensively studied in many animal species. The acute effects of Δ^9-THC are in general depressant, although mixed depressant and stimulant effects can occur, particularly in higher species. Δ^9-THC is reported to decrease spontaneous motor activity in fish (Gonzalez, Matsudo, & Carlini, 1971), baby chicks (Abel, McMillan, & Harris, 1972), gerbils (Grunfeld & Edery, 1969), mice (Holtzman, Lovell, Jaffe, & Freedman, 1969), rats (Kubena & Barry, 1970), dogs (Grunfeld & Edery, 1969), and monkeys (Scheckel, Boff, Dahlen, & Smart, 1968). However, in mice and rats, and more markedly in dogs and monkeys, high doses of Δ^9-THC can produce an initial period of increased motor activity followed by a more prolonged depression of behavior. Along with these general effects on activity, Δ^9-THC characteristically produces ataxia in the dog (Dewey, Jenkins, O'Rourke, & Harris, 1972) and an overt behavioral syndrome in the mouse (Christensen *et al.,* 1971) that is marked by hypersensitivity to sound ("popcorn effect") and in rhesus monkeys (Mechoulam, 1970) that is marked by tameness and a characteristic crouched posture ("thinker position").

Δ^9-THC has been shown to disrupt learned behaviors in numerous species. (See review by Carlini, 1973.) It seems, however, that Δ^9-THC may be readily distinguished from other psychoactive drugs only by its effects on responding to avoid electric shock in rats and monkeys and in particular by its effects on short-term memory in nonhuman primates. The effects of Δ^9-THC on the behavior of rats and monkeys maintained by nondiscriminated schedules of electric shock presentation are unusual (Carlini, 1973; Scheckel *et al.,* 1968). In contrast to most drugs, Δ^9-THC decreases avoidance responding at low doses and increases this responses at high doses. Delayed matching to sample tasks have often been used to assess the short-term memory of humans and nonhuman primates. Δ^9-THC is reported to specifically decrease the accuracy of matching to sample performance when delay times are fairly long (30 sec or more). Unlike a wide variety of other drugs tested, Δ^9-THC does not at the same doses decrease the rate of responding or the accuracy at short delay times (Ferraro, 1972; Zimmerberg, Glick, & Jarvik, 1971).

Most of these effects of Δ^9-THC on behavior have been used as screening tests for Δ^9-THC-like activity. Most frequently used is the dog ataxic test and, more recently, measures of overt behavior of the mouse and rhesus monkey.

Because of the size of the animal or the large number of animals used, these procedures require large amounts of drug, a major problem in testing plant constituents and metabolites that are available in small amounts. Also, the specificity of these tests for predicting Δ^9-THC-like effects in man has not been adequately validated. In addition, behavioral tolerance to Δ^9-THC has been demonstrated to occur in the mouse (McMillan, Dewey, & Harris, 1971), the rhesus monkey (Harris, Waters, & McLendon, 1972), and to a marked and rapid degree in the dog (Dewey *et al.*, 1972). On the other hand, Δ^9-THC has a specific effect on short-term memory of monkeys, and tolerance to this effect develops slowly (Ferraro & Grilly, 1974).

In this review, we shall examine the use of marihuana and its plant constituents as discriminative stimuli. We shall describe the experimental procedures that have demonstrated stimulus control by cannabinoids and look at evidence that bears on tolerance development to this property of cannabinoids. We shall point out the specificity of stimulus control by cannabinoids as evidence by transfer studies. Lastly, we shall discuss research with humans on the stimulus properties of marihuana.

DISCRIMINATIVE CONTROL OF BEHAVIOR BY CANNABINOIDS

A great many tasks and species have been used to study the discriminative-stimulus properties of cannabinoids. Although it makes comparisons more difficult, the diversity shows that stimulus control by these drugs is not task- nor species-specific. Discriminative control of maze performance or operant behavior has been demonstrated in rats, gerbils, pigeons, and rhesus monkeys.

Jarbe and Henriksson and their associates have studied stimulus control by cannabinoids of maze performance in rats (Henriksson & Jarbe, 1972; Jarbe and Henriksson, 1973, 1974; Jarbe, Johansson, & Henriksson, 1976), and gerbils (Jarbe, Johansson, & Henriksson, 1975). For rats, they initially used a T-shaped water maze in which swimming to the left side led to an escape ladder under one drug condition and swimming to the right resulted in escape under the other drug condition. After an incorrect choice, the animals encountered a barrier but were allowed to swim back to the correct arm (correction procedure).

Figure 8.1 shows acquisition data using this procedure, with hashish, synthetic Δ^9-THC, or Δ^8-THC as the training stimuli. For Δ^9-THC (5 mg/kg) and Δ^8-THC (10 mg/kg), the nondrug stimuli were intraperitoneal (I.P.) vehicle injections. Hashish (2.5 gm) was administered by inhalation. The nondrug stimulus for hashish was chervil smoke. All three groups evidenced greater than 75% correct first-trial choices by the tenth session.

Jarbe *et al.* (1975) have also used T-maze escape to study the discriminative-stimulus properties of Δ^9-THC in the gerbil. Instead of water, they used electric shock to motivate escape behavior. Again, a left–right position discrimination was based on the presence or absence of the training drug. Five

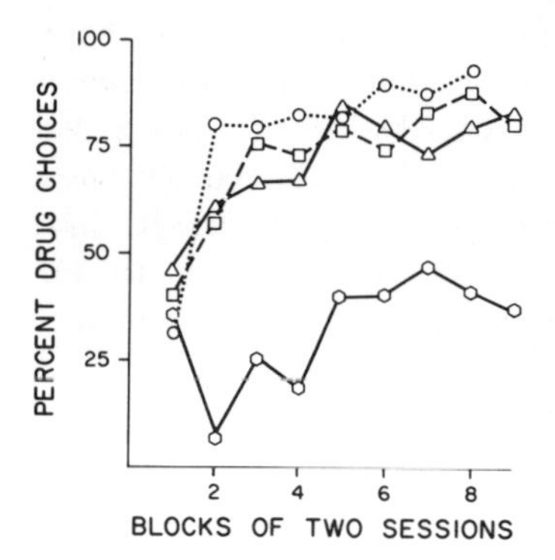

Figure 8.1. Acquisition of discriminative control of T-maze performance by cannabinoids. Training stimuli: △——△, 5 mg/kg of Δ^9-THC versus vehicle; □---□, 10 mg/kg of Δ^8-THC vehicle; ○---○, hashish versus chervil smoke; ○——○, control (chervil smoke versus vehicle). Adapted from Jarbe and Henriksson (1974).

trials per session and a correction procedure were used. Jarbe *et al.* studied the effect of training dose of Δ^9-THC on the rate of acquisition. Animals trained with 8, 12, or 16 mg/kg did not show different acquisition rates, but all three groups learned faster than did animals trained with either .5 or 2 mg/kg. In each case, vehicle was the nondrug stimulus. For the three larger-dose groups, acquisition was very rapid. If a criterion of 8 correct first-trial choices out of 10 consecutive training sessions is used, these animals inititated the criterion performance within 4–5 sessions.

With rats in a more recent study, Jarbe *et al.* (1976) used a similar T-maze task with shock to motivate escape. They again studied the rate of acquisition of drug-stimulus control as a function of dose of both Δ^9-THC and Δ^8-THC. As in the gerbil study, they found acquisition to be dose-related. The results in gerbil and rat are similar to previous studies that have also found rate of acquisition of drug discriminations to be a function of training dose of a number of other drugs (e.g., Hill, Jones, & Bell, 1971; Overton, 1974).

Other studies of discriminative control of behavior by cannabinoids have used operant behavior by rats, rhesus monkeys, or pigeons. In operant terminology, these procedures involve two-component multiple schedules in which each component is associated with a drug stimulus. The differential responding associated with each stimulus can be either two different reinforcement schedules (e.g., fixed-ratio and extinction) or the same schedule on two different response manipulanda. Drug-controlled multiple schedules differ from typical schedules under exteroceptive stimulus control only in that the components change from day to day (to allow the drug stimulus to dissipate) rather than within daily sessions.

Barry and Kubena have used a one-lever approach-avoidance conflict task to study stimulus control by a number of psychoactive drugs, including cannabinoids (Barry & Kubena, 1972; Kubena & Barry, 1972). Details of the procedure are given in Kubena and Barry (1969). The approach component (associated with one drug condition) consisted of a fixed-ratio (FR) 5 schedule of food reinforcement. During the avoidance component (associated with the other drug condition), food was withheld, and every fifth lever press was followed by grid shock. Five-minute sessions were divided equally between food and shock sessions in an unsystematic sequence. For one-half of the rats, the approach component was associated with drug injections; for the other half, the avoidance component was associated with drug. Vehicle injections were the nondrug stimulus for each group. For purposes of data analysis, an animal was considered to have made an approach response if it completed 5 or more lever presses within the 5-min session and an avoidance response if it made less than 5. Barry and Kubena (1972) used 4 mg/kg of Δ^9-THC and vehicle injections as the training stimuli for this task. Within 15 sessions (8 food, 7 shock), the animals made more than 90% drug-appropriate approach responses and more than 70% drug-appropriate avoidance responses. Further training increased the percentage of correct avoidance responding to more than 90%. Acquisition of discriminative control by this dose of Δ^9-THC was roughly comparable to the rate of acquisition achieved by either 1200 mg/kg of ethanol versus vehicle or by 10 mg/kg of atropine versus vehicle.

Other operant studies of stimulus control by marihuana constituents utilized a two-manipulanda choice paradigm in which reinforced responding on one manipulandum was associated with one drug condition and reinforced responding on the other was associated with the other drug condition. The studies differ principally on the schedule by which responding was reinforced. Bueno and Carlini (1972) and Hirschhorn and Rosecrans (1974) studied the effects of tolerance on the discriminative stimulus properties of Δ^9-THC. The results of these studies will be discussed later. The training procedure for the two studies was similar, differing principally in that the former used an FR-10 schedule of reinforcement on the correct lever, the latter a variable-interval (VI) 15-sec. schedule. Responding on the incorrect lever had no programmed consequence in either study.

We also use a two-lever choice procedure with rats. As we have reported (Ford & Balster, 1975), we studied the effects of various metabolites of Δ^9-THC and other cannabinoid congeners of Δ^9-THC in rats trained to discriminate 3 mg/kg of Δ^9-THC from vehicle. Responding on the correct lever was reinforced on an FR-10 schedule of milk presentation. After 30 sessions of discrimination training, every third session began with a 2.5-min period of extinction on both levers. This allowed us to check the stimulus control exerted by the training dose of Δ^9-THC or vehicle. During these check periods, the animals consistently made more than 85% of their responses on the drug-appropriate lever. Test sessions were run periodically in which the animals received an injection of a test drug. In these sessions, the animals were removed from the chamber after the 2.5-min extinction period. This allowed evaluation of the extent of transfer of discriminative control from Δ^9-THC to other cannabinoids. These results will be discussed in another section.

Ferraro, Gluck, and Morrow (1974) used a complex variation of the two-lever choice procedure to study the discriminative stimulus properties of Δ^9-THC in rhesus monkeys. The clearest difference in their procedure from the two-lever choice procedures described above lies in the consequences of incorrect lever responding. In the other studies, the incorrect lever is programmed for extinction, but in Ferraro *et al.*, incorrect responding led to the illumination of a green light associated with a schedule of reinforcement in which the frequency of reinforcement was low (a random interval [RI] of 2 min). Correct-lever responding led to illumination of a red light associated with a richer reinforcement schedule (RI—30 sec). Using Δ^9-THC (3 to 4 mg/kg, oral [*p.o.*]) or vehicle as the stimuli, they trained monkeys to press the lever associated with the red light more than 80% of the time. Training required 55–89 sessions. After this extensive training, the discrimination remained relatively stable from day to day.

Recently, Henriksson, Johansson, and Jarbe (1975) studied the discriminative stimulus properties of Δ^9-THC in the pigeon with a two-key operant task. Responding on each key was reinforced on an FR 15 schedule of grain presentation. Discrimination was acquired within 30 15-min training sessions with a training dose of .25 mg/kg.

In summary, most studies of the discriminative-stimulus properties of cannabinoids have used a two-manipulanda operant task with rats or pigeons, although T-maze performance has also been employed with rats and gerbils. One study has used rhesus monkeys on a complex variant of the two-lever choice task. Except for one study with Δ^8-THC and combusted hashish smoke and one with marihuana extract, Δ^9-THC was the training drug for all of the research. For operant studies, the dosages of Δ^9-THC used for training were similar (3 or 4 mg/kg, i.p. or *p.o.*). As is usually the case (but cf. Overton, 1973), escape procedures routinely involved somewhat higher training doses (5 mg/kg), but Jarbe *et al.* (1975) found that with extended training in the shock-motivated T-maze escape task doses as low as .5 or 2 mg/kg can develop discriminative control in the gerbil.

TOLERANCE

It now seems established that tolerance develops more slowly to the discriminative-stimulus effects of drugs than to most of their disruptive effects on behavior. The most prevalent evidence for this comes from the numerous studies based successfully on the prolonged use of the discriminative-stimulus effects of drugs produced by fixed doses. This is most striking in the case of Δ^9-THC and marihuana since a marked and rapid tolerance develops to most of their disruptive effects on behavior (McMillan *et al.*, 1971).

Bueno and Carlini (1972) showed that the administration of a marihuana extract can still function as a discriminative stimulus for bar-passing behavior in rats tolerant to the disruptive effects of marihuana on rope-climbing behavior. On the other hand, Jarbe and Henriksson (1973) reported some evidence for mild tolerance to the discriminative properties of Δ^9-THC. Rats pretreated with 5 mg/kg of Δ^9-THC for 18 days developed differential responding to Δ^9-THC in a water maze more slowly than did rats preteated with vehicle. However, similar to the study of Bueno and Carlini (1972), Δ^9-THC still acquired discriminative-stimulus control after complete tolerance had developed to its effects on open-field behavior.

Hirschhorn and Rosecrans (1974), using a two-lever discrimination task, trained one group of rats to discriminate morphine from vehicle and another to distinguish Δ^9THC from vehicle. After a daily experimental session, rats were given supplementary doses of the respective drugs that were 2–8 times the training dose. After 2 months of this treatment, rats still consistently discriminated morphine or Δ^9-THC from vehicle, although a mild tolerance is suggested by a slight reduction of stimulus control. In conclusion, tolerance to the discriminative-stimulus properties of cannabinoids, if it occurs at all, develops at a slower rate than to other behavioral effects of these drugs.

TRANSFER STUDIES

One of the powerful research applications using drugs as discriminative stimuli utilizes the phenomenon of stimulus generalization. Behavior brought under the discriminative control of a drug stimulus can also occur in the presence of other drug stimuli. The more the test stimulus resembles the training stimuli, the greater the transfer of stimulus control. A function that relates the degree of stimulus control to different values of test stimuli is termed a generalization gradient. Generalization can occur along quantitative and qualitative dimensions. For drug stimuli, changes along the quantitative dimension correspond to dosage changes of the training drug. The generalization gradient, then, represents the dose-response curve in which the percentage of drug choices is expressed as a function of dose.

In the case of the qualitative dimension, the underlying continuum is less clear. Perhaps the easiest way to define this dimension is to consider it as changes in the physical structure of the drug. For example, a generalization gradient could be obtained for successive hydroxylations of the A ring of the THC molecule. In other words, generalization gradients along the qualitative dimension correspond to structure–activity relationships. This type of research is difficult since the relevant changes in the physical structure of cannabinoids that might show a relationship to behavior are poorly understood.

The logical extension of changes along the qualitative dimension leads to molecules representative of other classes of drugs. One of the important research questions concerning cannabinoids is the degree to which they resemble drugs from other classes of psychoactive compounds or whether they represent a distinct class of their own. Generalization tests are well suited to answer such questions since the overwhelming preponderance of research shows that transfer of stimulus control is very specific, only occurring between drugs with very similar pharmacological activities, which produce very similar subjective effects in man. For reviews of the literature supporting this conclusion, see Overton (1973, 1974), Barry (1974), and Schuster and Balster (1977).

Generalization to Other Doses of the Training Drug

The substitution of various doses of Δ^9-THC during test sessions in animals trained to discriminate a single dose of Δ^9-THC from vehicle results in a generalization gradient. This is illustrated in Figure 8.2 by work in our laboratory. Rats were trained in a two-lever discrimination task with 3 mg/kg of Δ^9-THC and vehicle as the training stimuli. Generalization tests with doses lower than 3 mg/kg resulted in a dose-related decrease in the percentage of responses on the lever associated with Δ^9-THC during training. The dose higher than the training dose also resulted in a high percentage of Δ^9-THC choices. This sigmoid generalization gradient in which only doses less than one-half the training dose show significantly poorer stimulus control is consistent with generalization gradients obtained with other classes of drugs. (See Schuster & Balster, 1977, for review.) Similar dose-generalization gradients have been obtained with Δ^9-THC in the rat in a variety of other tasks, including the water maze (Jarbe & Henriksson, 1974), conflict procedure (Barry & Kubena, 1972; Kubena & Barry, 1972), and the shock escape maze (Jarbe *et al.*, 1976). The studies described earlier with gerbils (Jarbe *et al.*, 1975), rhesus monkeys (Ferraro *et al.*, 1974), and pigeons (Henriksson *et al.*, 1975) also obtained dose-response curves for stimulus control that are similar to those in the rat.

In addition to varying dosage as a means of manipulating the quantitative dimension, one can also vary the time between drug administration and test session to produce a time–effect curve. Since Δ^9-THC is metabolized and excreted over the course of many hours, varying the time before the session should result in a generalization gradient similar to the dose–effect curve. Fer-

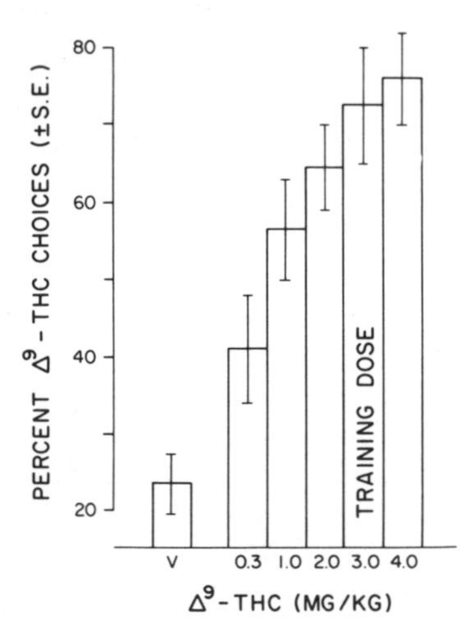

Figure 8.2. Generalization gradient for doses of Δ^9-THC in rats trained to discriminate Δ^9-THC from vehicle in a two-lever operant task. $N = 7$.

raro *et al.* (1974) varied oral administration time from .5 to 16.5 hr in generalization test sessions with rhesus monkeys trained to discriminate Δ⁹-THC from vehicle given 2.5 hr before the session. Although it differed quantitatively somewhat from animal to animal, an orderly gradient was obtained with respect to its general shape. At .5 and 1 hr, presumably before substantial drug absorption occurred, the animals made predominantly vehicle-lever choices. Between 1 and 10.5 hr, when drug levels would be relatively high, they made drug choices; after 10.5 hr, by which time substantial metabolism of Δ⁹-THC would be expected, vehicle choices again occurred. Jarbe *et al.* (1976) found a similar relationship between administration time and stimulus control using both Δ⁹-THC and Δ⁸-THC as training stimuli in different groups of rats.

Generalization from Δ⁹-THC to Other Cannabinoids

As stated, generalization tests can also be made along the qualitative dimension. In this section, we will consider generalization to other cannabinoids, both metabolites of marihuana ingredients and various congeners of Δ⁹-THC. Then we will look at transfer tests with other psychoactive drugs.

There are two important research questions to which transfer of stimulus-control studies of cannabinoids can be addressed. The first concerns the search for active metabolites, the second the development of a screening procedure to assess THC-like activity of other cannabinoid congeners. Before reviewing research relevant to these issues, we must ask, How do we know that the stimulus properties of naturally occurring cannabis products are interchangeable with synthetic Δ⁹-THC? Three studies bear on this question. Kubena and Barry (1972), using their approach–avoidance conflict procedure, trained rats to discriminate 4 mg/kg of Δ⁹-THC from vehicle. At appropriate doses, a marihuana-extract distillate and an alcohol–marihuana extract both elicited a high percentage of Δ⁹-THC responses when given during generalization tests. Jarbe and Henriksson (1974) trained three different groups of rats to discriminate between Δ⁹-THC, Δ⁸-THC or hashish smoke, and control. Generalization tests of these three training stimuli showed them to be interchangeable at equieffective doses. Barry and Krimmer (1975) also verified the equivalence of crude manihuana extract to Δ⁹-THC in a two-lever discrimination task in rats.

One of the unresolved questions about the pharmacology of THC is whether Δ⁹-THC is converted to active metabolites in the body. 11-Hydroxylation of the A ring is generally considered to be the first major metabolic transformation of Δ⁹-THC (Nilsson, Agurell, Nilsson, & Ohlssen, 1970), and this compound (11-OH-Δ⁹-THC) has been shown to be behaviorally active in animals (Kosersky, McMillan, & Harris, 1974) and to produce Δ⁹-THC-like subjective effects in man (Perez-Reyes, Timmons, Davis, & Wall, 1972; Lemberger, Martz, Rodda, Forney, & Rowe, 1973; Hollister, 1974). There is also evidence that Δ⁹-THC is converted to 8-hydroxy metabolites, of which there are two forms, an 8α-OH-Δ^9-THC and an 8β-OH-Δ^9-THC (Ben-Zvi, Mechoulam, Edery, & Porath, 1971); the corresponding 8,11-dihydroxy metabolites (8α,11-diOH-Δ^9-THC and 8β,11-diOH-Δ^9-THC) can also be found (Wall, Brine, Pitt, & Perez-Reyes, 1972). A corresponding series of hydroxylations occur with Δ^8-THC. Tests for stimulus generalization to some of these metabolites have been conducted in our laboratory and are summarized in Table 8.1. The 11-OH metabolites clearly produce Δ^9-THC-like responding. Similar results with 11-OH-Δ^8-THC have been reported by Barry and Krimmer (1975). The other metabolites at the dosages tested result in vehicle responding, which indicates that they lack Δ^9-THC-like CNS effects.

One problem with generalization testing between different drugs concerns the choice of dosage for comparison. Those familar with the psychophysical literature on stimulus generalization along a qualitative dimension will remember that in order to compare qualitatively different stimuli, it is necessary to equate them on intensity since relative intensity changes also result in generalization decrement. In vision, for example, equieffective stimuli are obtained from spectral sensitivity curves. In the case of drug stimuli, intensity corresponds to dosage (or perhaps more specifically to drug concentration at critical receptor sites), and equieffective drug stimuli would be equipotent. One of the important aspects of comparing different drugs for stimulus control is the necessity of choosing effective doses since the absence of generalization could simply reflect a difference in drug potency. In a two-lever discrimination task, as used in our laboratory, one way to determine the effective dosage of the test drug is to give successively higher doses until no responding occurs on either lever. If no generalization occurs at any dosage, one can reasonably conclude that the difference is qualitative rather than quantitative. Unfortunately, such small quantities of the minor metabolites (8-OH and diOH compounds) were available that doses whose effectiveness could be shown by decreasing overall response rates were not tested. It is still possible

Table 8.1
Stimulus Generalization from Δ^9-THC to Various Metabolites in Rats[a]

Drug[b]	Dose range tested (mg/kg)	Transfer
11-OH-Δ^9-THC	.1–1	yes
11-OH-Δ^8-THC	.1–1	yes
8α-OH-Δ^9-THC	15–60	no
8β-OH-Δ^9-THC	3–15	no
8α,11-diOH-Δ^9-THC	12	no
8β,11-diOH-Δ^9-THC	12	no

[a] Training stimuli: 4 mg/kg Δ^9-THC and vehicle.
[b] Dibenzopyran numbering system.

that at very high doses generalization to Δ^9-THC may have occurred. The problem of using behaviorally inactive doses because of limited supplies of Δ^9-THC metabolites and congeners has hampered most research studies of these compounds.

We have also tested a number of Δ^9-THC congeners for stimulus generalization to Δ^9-THC to determine whether other plant constituents have Δ^9-THC-like activity. This research strategy is also applicable to the search for therapeutic agents among the cannabinols. Δ^9-THC or related cannabinoids are potentially useful in the treatment of glucoma (Hepler, Frank, & Ungerleider, 1972), anorexia (Hollister, 1971), and rejection reactions in transplant patients (Levy, Munson, Harris, & Dewey, 1975). In addition, there is some interest in the antitumor properties of cannabinoids (Harris, Carchman, & Munson, 1975), and some can produce antinociception (Bloom, Dewey, Harris, & Brosius, 1975). For these uses, it would be advantageous to develop cannabinoids devoid of Δ^9-THC-like subjective effects at therapeutic doses. We believe that generalization tests for stimulus control presently are the most sensitive and specific tests for this side effect of cannabinoids. If, in a series of cannabinoids tested in this manner, a behaviorally potent compound that failed to transfer to Δ^9-THC at any dose were found, it might represent a new class of psychoactive drug with possible psychotherapeutic benefits.

Transfer tests with various cannabinoids have been conducted in three laboratories (Barry & Kubena, 1972; Jarbe & Henriksson, 1974; Ford & Balster, 1975). Table 8.2 summarizes the results of these studies. For compounds that showed transfer to Δ^9-THC, the results are easy to interpret. Except for potency differences, these compounds produce qualitatively similar stimulus effects, and they might be predicted to produce Δ^9-THC-like subjective effects in man. In the case of no transfer, the data are more difficult to interpret. As stated earlier, lack of stimulus generalization may simply reflect the use of an ineffective dose; the best one can say is that a substantial potency difference exists between Δ^9-THC, which still shows stimulus generalization when tested at dosages one-third the training dosage, and these congeners. In the absence of data demonstrating that behaviorally effective dosages were used (e.g., that overall response rates were affected), conclusions concerning qualitative differences may not be warranted. It is important to point out, however, that drug-stimulus generalization occurs often at dosages lower than those needed to produce overall behavioral disruption. Only one of the compounds we tested was available in sufficient quantities to give it in behaviorally active dosages. At 100 mg/kg cannabidiol significantly lowered overall response rate during test sessions, but at no dosages did the distribution of these responses favor the Δ^9-THC-lever choice. Thus, at least with cannabidiol, there is strong evidence for a qualitatively different behavioral activity at very high dosages.

These observations might help explain the ambiguous results obtained by Jarbe and Henriksson (1974) with high doses of cannabidiol. At 150 mg/kg there was a slight tendency for drug responding to occur, although there were

Table 8.2
Stimulus Generalization from Δ^9-THC to Various THC Congeners in Rats

Drug[a]	Dose range tested (mg/kg)	Transfer
Barry and Kubena (1972). Approach–avoidance conflict. Training stimuli: 4 mg/kg Δ^9-THC and vehicle.		
Δ^{6a}-THC	4	yes
Cannabinol	10	no
Cannabidiol	10–20	no
Jarbe and Henriksson (1974). T-shaped water maze. Training stimuli: 5 mg/kg Δ^9-THC and vehicle.		
Δ^8-THC	1.25–10	yes
Cannabinol	10–60	no
Cannabidiol	15–150	?
Ford & Balster (1975). Two-lever discrimination. Training stimuli: 3 mg/kg Δ^9-THC and vehicle.		
9-*nor*-Δ^8-THC	10–60	yes
9-*nor*-9α-OH hexahydrocannabinol	1–10	?
9-*nor*-9β-OH hexayhdrocannabinol	0.1–1	yes
Cannabidiol	30–100	no
Abnormal cannabidiol	30–100	no

[a] Dibenzopyran numbering system.

never more than 50% drug choices. If a behaviorally disruptive dose had been administered, the animals might have responded indiscriminately, resulting in near-chance (50%) performance. Data on swimming time or other measures of behavioral disruption were unfortunately not included in their report.

In summary, stimulus-generalization studies show great promise as a screening procedure to measure Δ^9-THC-like behavioral activity of various congeners of THC. Dose for dose they are more sensitive than other behavioral measures currently available, and when limited drug supply is a problem (as often is the case with these difficult to synthesize cannabinoids), this sensitivity can be a significant advantage. An even greater advantage is the specificity of the procedure. Inactive cannabinoids do not transfer, and as will be discussed in the next section, other classes of psychoactive drugs also do not transfer stimulus control from Δ^9-THC.

Lack of Generalization from Δ^9-THC to Other Psychoactive Drugs

The logical extension along the qualitative dimension of stimulus generalization is to other classes of drugs. Stimulus-generalization tests with other psychoactive drugs in animals trained to discriminate Δ^9-THC from vehicle have had remarkably consistent results. There are no drugs, other than the cannabinoids discussed previously, whose stimulus properties transfer to Δ^9-THC. These studies are summarized in Table 8.3. The results of each of these studies are presented by the authors with varying degrees of completeness. In one case (Barry & Kubena, 1972), data on the extent of generalization are given, but in others the results are only summarized. In spite of the cursory nature of some of these data, it seems clear, since many drugs were tested in more than one laboratory with essentially similar results, that none of these compounds show any evidence for transfer to Δ^9-THC.

The discriminative stimulus properties of Δ^9-THC are specific: They do not transfer to a wide variety of psychoactive drugs. The question we raised earlier concerning evidence for the use of active dosages also applies to these data, however. With the exception of Barry and Kubena (1972), none of these transfer studies presented measures (e.g., overall response rate or latency measures), which would show unequivocally that effective doses were used. The usefulness of including these other performance measures should by now be clear. Barry and Kubena (1972) present data that bear on the effectiveness of the dosages they studied. At some higher dosages of various drugs, animals often failed to make approach responses regardless of which drug condition was paired with approach during training. Failure to complete the five lever presses, which constituted an approach response when give a test drug, even when approach was paired with vehicle during training, is evidence that these doses disrupted responding, hence were behaviorally active. Since in one-half

Table 8.3
Drug and Dosages (mg/kg) That Fail to Show Transfer of Stimulus Control from Δ^9-THC

Barry and Kubena (1972)	
Pentobarbital (10)	Cocaine (10–20)
Alcohol (500–1000)	Atropine (5–10)
Chlordiazepoxide (15)	Scopolamine (.25)
Chlorpromazine (2–4)	LSD (.1–.2)
Morphine (4–8)	Mescaline (15–30)
Methamphetamine (1)	Dimethyltryptamine (2–4)
Jarbe and Henriksson (1974)	
Pentobarbital (12.5–18.75)	Psilocybin (3)
Alcohol (1000–2000)	Amphetamine (2.5–5)
Phenobarbital (60)	Cocaine (5–20)
Diazepam (3.75–7.5)	Yohimbine (7.5)
Chlorpromazine (5)	Tacrine (2–5)
Nortryptiline (30)	Morphine (1.25–10)
Atropine (25–150)	Levallorphan (20–60)
Scopolamine (10–100)	Phencyclidine (1–2)
Ditran (3.125–12.5)	
Jarbe *et al.* (1975)	
Pentobarbital (10–20)	Atropine (10)
Barry and Krimmer (1975)	
Chlordiazepoxide (5–10)	Pentobarbital (10)
Alcohol (1000)	

their animals approach was paired with Δ⁹-THC and in the other half paired with vehicle, a behaviorally disruptive dose of a test drug would result in overall 50% drug responding. Barry and Kubena found this to be the case with the higher dosages of alcohol, chlorpromazine, morphine, atropine, scopolamine, LSD, mescaline, and DMT. The drugs tested in other studies were generally within the dosage range in which one would expect behavioral effects.

2 × 2 DESIGN STUDIES

As an alternative to training subjects to discriminate between Δ^9-THC and vehicle injections, it is also possible to train two different groups of animals—one under each of the conditions—and then split each group and test for transfer of training from drug to nondrug conditions and vice versa. Failure to transfer training from one drug condition to the other could be considered a case of stimulus generalization decrement. The theoretical implications of considering these studies within a stimulus-control framework have been discussed elsewhere (Schuster & Balster, 1977). Overton (1974) has discussed the problems of data interpretation with this research design. Nevertheless, despite its limitations, the 2 × 2 design is the only procedure that has ever been used with human subjects to test for drug-stimulus control. First we will consider two animal studies of Δ^9-THC as one of the drug conditions.

Henriksson and Jarbe (1971) used the 2 × 2 design to determine the effects of both Δ^9-THC (7.5 mg/kg) and Δ^8-THC (15 mg/kg) on acquisition and retention of a shuttle-box avoidance task with rats. They found retention substantially decreased in animals whose drug conditions changes from training to test sessions. The dissociation was symmetrical; that is, a switch from drug to nondrug (D–ND) resulted in roughly the same generalization decrement as did the switch from nondrug to drug (ND–D) conditions. There was no retention deficit in either control group (D–D or ND–ND). Herring (1972), on the other hand, failed to find evidence for state-dependent learning (SDL) in rats on a discriminative lever-press avoidance task. Regardless of training conditions, Δ^9-THC (4 mg/kg) interfered with retention on the day after training. It is difficult to explain these results since Δ^9-THC actually facilitated acquisition.

Three studies have been conducted using the 2 × 2 design with human subjects in which one of the drug conditions was produced by smoking marihuana. Rickles, Cohen, Whitaker, and McIntyre (1973) studied acquisition and retention of paired associates in four groups of "social" users of marihuana. Marihuana interfered with acquisition of the paired associates. In fact, some subjects were dropped from the study for failing to reach criterion during marihuana intoxication. In spite of this difference in acquisition, the subjects trained and tested with marihuana (D–D) had the greatest recall, the ND–ND,

D–ND, and ND–D groups having respectively less recall. These results may be interpreted as evidence for stimulus control by marihuana in human subjects.

The other two marihuana studies using the 2 × 2 design with human subjects involved a number of different tests: in both, evidence was found for state-dependent effects on some tests but not on others. Stillman, Weingartner, Wyatt, Gillan, and Eich (1974) found generalization decrement on a four-way picture-choice test in which subjects had to select the correct target picture on each of 16 slides containing four pictures and on a picture-arrangement task. On the other hand, a free-association task in which subjects were asked to recall the verbal free associations they had made to a series of five pictures on the training day and a word-recall sorting task showed no evidence of SDL. Hill, Schwin, Powell, and Goodwin (1973) also found the state-dependent effects of marihuana to be task-specific. As did Stillman *et al.* (1974), they found word-recall and word-association tasks to be insensitive to SDL with marihuana, but they did find an effect on tests requiring the recall of the order of various items.

In summary, cannabinoids have shown state-dependent effects using the 2 × 2 design in rodents and human beings. For many purposes, this design is less satisfactory for studying drug stimulus control than is a discrimination task (Overton, 1974; Schuster & Balster, 1977), and it is hoped that in spite of the relative technical difficulty of a discrimination task in man, such studies will be carried out in human subjects in the future.

REFERENCES

Abel, E. L., McMillan, D. E., & Harris, L. S. Tolerance to the behavioral and hypothermic effects of l-Δ^9-tetrahydrocannabinol in neonatal chicks. *Experientia*, 1974, *28*, 1188–1189.

Barry, H., III. Classification of drugs according to their discriminable effects in rats. *Federation Proc.*, 1974, *33*, 1814–1824.

Barry, H., III, & Krimmer, E. C. Discriminative Δ^9-tetrahydrocannabinol stimulus tested with several doses, routes, intervals and related compounds. *Federation Proc.*, 1975, *34*, 743.

Barry, H., III, & Kubena, R. K. Discriminative stimulus characteristics of alcohol, marihuana and atropine. In J. M. Singh, L. H. Miller, & H. Lal (Eds.), *Drug addiction: Experimental pharmacology*. Mt. Kisco, New York: Futura, 1972. Pp. 3–16.

Ben-Zvi, L., Mechoulam, R. M., Edery, H., & Porath, G. 6β-Hydroxy-Δ^1-tetrahydrocannabinol: Synthesis and biological activity. *Science*, 1971, *174*, 951–952.

Bloom, A. S., Dewey, W. L., Harris, L. S., & Brosius, K. Brain catecholamines and the antinociceptive action of ±9-*nor*-9β-OH-hexahydrocannabinol. *Soc. Neuroscience*, 1975, *5*, 242.

Bueno, O. F. A., & Carlini, E. A. Dissociation of learning in marihuana tolerant rats. *Psychopharmacologia*, 1972, *25*, 49–56.

Carlini, E. A. Acute and chronic behavioral effects of *Cannabis sativa*. *Proc. 5th Intern. Congr. Pharmacol.*, 1973, *1*, 31–43.

Christensen, H. D., Freudenthal, R. J., Gidley, J. T., Rosenfeld, R., Boegli, G., Testino, L., Brine, D. R., Pitt, C. G., & Wall, M. E. Activity of Δ^8-and Δ^9-tetrahydrocannabinol and related compounds in the mouse. *Science*, 1971, *175*, 778–779.

Dewey, W. L., Jenkins, J., O'Rourke, T., & Harris, L. S. The effects of chronic administration of trans-Δ⁹-tetrahydrocannabinol on behavior and the cardiovascular system of dogs. *Arch. Intern. Pharmacodyn.*, 1972, *198*, 118–131.

Ferraro, D. P. Effects of Δ^9-tetrahydrocannabinol on simple and complex learned behaviors in animals. In M. F. Lewis (Ed.), *Current research in marihuana.* New York: Academic Press, 1972. Pp. 49–95.

Ferraro, D. P., Gluck, J. P., & Morrow, C. W. Temporally-related stimulus properties of Δ^9-tetrahydrocannabinol in monkeys. *Psychopharmacologia*, 1974, *35*, 305–316.

Ferraro, D. P., & Grilly, D. M. Effects of chronic exposure to Δ^9-tetrahydrocannabinol on delayed matching-to-sample in chimpanzees. *Psychopharmacologia*, 1974, *37*, 127–138.

Ford, R. D., & Balster, R. L. The discriminative stimulus properties of Δ^9-tetrahydrocannabinol: Generalization to some metabolites and derivatives. *Federation Proc.*, 1975, *34*, 743.

Gaoni, Y., & Mechoulam, R. Isolation, structure and partial synthesis of an active component of hashish. *J. Am. Chem. Soc.*, 1964, *80*, 1646–1647.

Gonzalez, S. C., Matsudo, V. K. R., & Carlini, E. A. Effects of marihuana compounds on the fighting behavior of siamese fighting fish (Betta splendens). *Pharmacol.*, 1971, *6*, 186–190.

Grunfeld, Y., & Edery, H. Psychopharmacological activity of the active constituents of hashish and some related cannabinoids. *Psychopharmacologia*, 1969, *14*, 200–210.

Harris, L. S., Carchman, R. A., & Munson, A. E. Structure-antitumor activity of cannabinoids. *Pharmacologist*, 1975, *17*, 265.

Harris, R. T., Waters, W., & McLendon, D. Behavioral effects in rhesus monkeys of repeated intravenous doses of Δ^9-tetrahydrocannabinol. *Psychopharmacologia*, 1972, *26*, 297–306.

Henriksson, B. G., & Jarbe, T. U. C. The effect of two tetrahydrocannabinols (Δ^9-THC and Δ^8-THC) on conditioned avoidance learning in rats and its transfer to normal state conditions. *Psychopharmacologia*, 1971, *22*, 23–30.

Henriksson, B. G., & Jarbe, T. U. C. Δ^9-Tetrahydrocannabinol used as a discriminative stimulus for rats in position learning in a T-shaped water maze. *Psycho. Sci.*, 1972, *27*, 25–26.

Henriksson, B. G., Johansson, J. O., & Jarbe, T. U. C. Δ^9-Tetrahydrocannabinol produced discrimination in pigeons. *Pharmacol. Biochem. Behav.*, 1975, *3*, 771–774.

Herring, B. The effect of *l*-Δ^1-trans-tetrahydrocannabinol on learning and retention of avoidance performance in rats. *Psychopharmacologia*, 1972, *26*, 401–406.

Hepler, R. S., Frank, J. M., & Ungerleider, J. T. Pupillary constriction after marijuana smoking. *Am. J. Ophthamol.*, 1972, *74*, 1185–1190.

Hill, H. E., Jones, B. E., & Bell, E. C. State-dependent control of discrimination by morphine and pentobarbital. *Psychopharmacologia*, 1971, *22*, 305–313. 1971.

Hill, S. Y., Schwin, R., Powell, B., & Goodwin, D. W. State-dependent effects of marihuana on human memory. *Nature*, 1973, *243*, 241–242.

Hirshhorn, I. D., & Rosecrans, J. A. Morphine and Δ^9-tetrahydrocannabinol: Tolerance to the stimulus effects. *Psychopharmacologia*, 1974, *36*, 242–253.

Hiveley, R. L., Mosher, W. A., & Hoffman, F. W. Isolation of trans-Δ^9-tetrahydrocannabinol from marihuana. *J. Am. Chem. Soc.*, 1966, *88*, 1832–1833.

Hollister, L. E. Hunger and appetite after single doses of marihuana, alcohol, and dextroamphetamine. *Clin. Pharmacol. Therap.*, 1971, *12*, 44–49.

Hollister, L. E. Structure-activity relationships in man of cannabis constituents, and homologs and metabolites of Δ^9-tetrahydrocannabinol. *Pharmacol.*, 1974, *11*, 3–11.

Holtzman, D., Lovell, R. A., Jaffe, J. H., & Freedman, D. X. *l*-Δ^9-Tetrahydrocannabinol: Neurochemical and behavioral effects in the mouse. *Science*, 1969, *163*, 1464–1467.

Jarbe, T. U. C., & Henriksson, B. G. Open field behavior and acquisition of discriminative response control in Δ^9-THC tolerant rats. *Experientia*, 1973, *29*, 1251–1253.

Jarbe, T. U. C., & Henriksson, B. G. Discriminative response control produced with hashish, tetrahydrocannabinols (Δ^8-THC and Δ^9-THC), and other drugs. *Psychopharmacologia*, 1974, *40*, 1–16.

Jarbe, T. U. C., Johansson, J. O., & Henriksson, B. G. Δ^9-Tetrahydrocannabinol and pentobarbital as discriminative cues in the Mongolian gerbil (Meriones unguiculatus). *Pharmacol. Biochem. Behav.*, 1975, *3*, 403–410.

Jarbe, T. U. C., Johansson, J. O., & Henriksson, B. G. Characteristics of tetrahydrocannabinol (THC) produced discrimination in rats. *Psychopharmacologia*, 1976, *48*, 181–187.

Kosersky, D. S., McMillan, D. E., & Harris, L. S. Δ^9-Tetrahydrocannabinol and 11-hydroxy-Δ^9-tetrahydrocannabinol: Behavioral effects and tolerance development. *J. Pharmacol. Exp. Therap.*, 1974, *189*, 61–65.

Kubena, R. K., & Barry, H., III. Two procedures for training differential responses in alcohol and nondrug conditions. *J. Pharm. Sci.*, 1969, *58*, 99–101.

Kubena, R. K., & Barry, H., III. Interactions of Δ^1-tetrahydrocannabinol with barbiturates and methamphetamine. *J. Pharmacol. Exp. Therap.*, 1970, *185*, 101–107.

Kubena, R. K., & Barry, H., III. Stimulus characteristics of marihuana components. *Nature*, 1972, *235*, 397–398.

Lemberger, L., Martz, R., Rodda, B., Forney, R., & Rowe, H. Comparative pharmacology of Δ^9-tetrahydrocannabinol and its metabolite, 11-OH-Δ^9-tetrahydrocannabinol. *J. Clin. Invest.*, 1973, *52*, 2411–2417.

Levy, J. A., Munson, A. E., Harris, L. S., & Dewey, W. L. Effects of Δ^9-THC on the immune response of mice. *Federation Proc.*, 1975, *34*, 782.

Mechoulam, R. Marihuana chemistry. *Science*, 1970, *168*, 1159–1169.

McMillan, D. E., Dewey, W. L., & Harris, L. S. Characteristics of tetrahydrocannabinol tolerance. *Ann. N.Y. Acad. Sci.*, 1971, *191*, 83–99.

Nilsson, I. M., Agurell, S., Nilsson, J. L. G., & Ohlssen, A. Δ^1-Tetrahydrocannabinol: Structure of a major metabolite. *Science*, 1970, *168*, 1228–1229.

Overton, D. A. State-dependent learning produced by addicting drugs. *In* S. Fisher & A. M. Freedman (Eds.), *Opiate addiction: Origins and treatment*. Washington, D.C.: Winston, 1973. Pp. 61–74.

Overton, D. A. Experimental methods for the study of state-dependent learning. *Federation Proc.*, 1974, *33*, 1800–1813.

Perez-Reyes, M., Timmons, M. C., Davis, K. H., & Wall, E. M. Intravenous injection in man of Δ^9-tetrahydrocannabinol and 11-OH-Δ^9-tetrahydrocannabinol. *Science*, 1972, *177*, 633–634.

Rickles, W. H., Jr., Cohen, M. J., Whitaker, C. A., & McIntyre, K. E., Marijuana induced state-dependent verbal learning. *Psychopharmacologia*, 1973, *30*, 349–354.

Scheckel, C. L., Boff, E., Dahlen, P., & Smat, T. Behavioral effects in monkeys of racemates of two biologically active marijuana constituents. *Science*, 1968, *160*, 1467–1469.

Schuster, C. R., & Balster, R. L. The discriminative stimulus properties of drugs. In T. Thompson & P. B. Dews (Eds.), *Advances in behavioral pharmacology*. Vol. 1. New York: Academic Press, 1977. Pp. 85–138.

Stillman, R. C., Weingartner, H., Wyatt, R. J., Gillin, J. C., & Eich, J. State-dependent (dissociative) effects of marihuana on human memory. *Arch. Gen. Psychiat.*, 1974, *31*, 81–85.

Wall, M. E., Brine, D. R., Pitt, C. G., & Perez-Reyes, M. Synthesis of 11-hydroxy-Δ^9-tetrahydrocannabinol and other physiologically active metabolites of Δ^8- and Δ^9-tetrahydrocannabinol. *J. Am. Chem. Soc.*, 1972, *94*, 8578–8581.

Zimmerberg, B., Glick, S. D., & Jarvik, M. E. Impairment of recent memory by marihuana and THC in rhesus monkeys. *Nature* 1971, *233*, 343–345.

9

Dual Receptor Mediation of the Discriminative Stimulus Properties of Pentazocine[1]

James B. Appel, Donald M. Kuhn, and Francis J. White
Behavioral Pharmacology Laboratory
Department of Psychology
University of South Carolina

INTRODUCTION

Attempts to delineate the mechanisms by which drugs alter ongoing behavior are often restricted by the nonspecificity of the behavioral "assay" used to study a particular drug effect. That is, many drugs not closely related either pharmacologically or chemically (e.g., *d*-amphetamine and LSD) decrease food intake, increase locomotor activity, and decrease rate of bar-pressing behavior maintained by fixed-ratio schedules of positive reinforcement. One method that appears to be somewhat more specific than any of these behavioral tests involves the measurement of the ability of drugs to act as discriminative stimuli. In this situation, animals (rats, gerbils, monkeys) are trained to make one response (e.g., press lever A) in the presence of a particular stimulus (Drug A) and to make a spatially different response in the presence of a different stimulus (usually the Drug A vehicle or a different drug, B). Thus, the drug or, more accurately, the state induced by the drug, is the direct object of study. When this is the case, it is generally found that the stimulus properties of numerous drugs are unique, and even low doses of such compounds as *d*-amphetamine are readily distinguishable from those of other drugs, for example LSD (Cameron & Appel, 1973).

[1] Preparation of this chapter was supported by USPHS Research Grants MH-24,333 and MH-24,593 from the National Institute of Mental Health.

In addition to promising a greater degree of sensitivity and specificity than most behavioral assays, probably because detection by the animal of drug-induced states or interoceptive cues is differentially reinforced, the ability of certain drugs to act as discriminative stimuli can be used for other purposes: for example, (*1*) to classify psychoactive compounds (Barry, 1974) and, with varying degrees of success; (*2*) to predict the extent to which any given drug will produce state-dependent learning (SDL) (Overton, 1968) or be addicting (Overton, 1972, 1973), reinforcing (Harris, 1975), or hallucinogenic (Hirschhorn & Rosecrans, 1974a; Kuhn, Greenberg, & Appel, 1976; Winter, 1974).

Another way in which the ability of drugs to act as discriminative stimuli has been used with great success in our own and other laboratories (Browne & Ho, 1975; Hirschhorn & Rosecrans, 1974a,b; Schechter & Cook, 1975) is in the delineation of specific neurochemical mechanisms that might determine drug action. That is, the involvement of a putative neurotransmitter or a neurotransmitter receptor in the mediation of the stimulus properties of a drug can often be specified accurately by using various pharmacological agents to alter the integrity of a particular transmitter system. For example, the *d*-amphetamine cue seems to be based on the ability of amphetamine to release dopamine (DA) from the terminals of DA-containing neurons (Cooper, Bloom, & Roth, 1974). Drugs that specifically block DA receptors (haloperidol, pimozide) block the *d*-amphetamine cue (Huang & Ho, 1973; Schechter & Cook, 1975), as does depletion of brain DA with the tyrosine hydroxylase inhibitor alpha-methyl-*p*-tyrosine (Kuhn, Appel, & Greenberg, 1974; Schechter & Cook, 1975). A role for norepinephrine (NE) was ruled out when it was demonstrated that neither alpha- nor beta-adrenergic receptor blockers, nor depletion of NE with dopamine-beta-oxidase inhibitors such as disulfiram, affected the *d*-amphetamine cue (Huang & Ho, 1973; Schechter & Cook, 1975).

Similarly, we have been studying the discriminative stimulus properties of narcotics, narcotic agonists, and narcotic antagonists and the relationship of these properties to underlying neurochemical events (Kuhn *et al.*, 1976). When we began these experiments, the only narcotic that had received any attention was morphine (Hill, Jones, & Bell, 1971; Hirschhorn & Rosecrans, 1974a,b; Rosecrans, Goodloe, Bennet, & Hirschhorn, 1973). This seemed surprising in view of the hypothesis that the discriminative stimulus properties and abuse potential of many drugs may be related (Overton, 1972, 1973).

Since clinical evidence had shown the weak benzomorphan antagonist (or, more accurately, mixed agonist–antagonist) and analgesic, pentazocine (Talwin), to be subjectively similar to morphine (Jasinski, Martin, & Hoeldtke, 1970; Keats & Telford, 1964) and said to produce physical dependence (Jasinski *et al.*, 1970; Kemp, 1968; Sandoval & Wong, 1973), we decided to investigate this compound. Our aims were (*1*) to test the relationship between discriminability and physical dependency by determining whether or not pentazocine is discriminable from its vehicle; (*2*) to compare directly the discriminable properties of pentazocine to those of morphine; (*3*) to analyze the effects of such "pure" narcotic antagonists as naloxone and naltrexone on the discriminable properties of both pentazocine and morphine; and (*4*) to determine whether pentazocine interacts in an agonistic or antagonistic manner with the morphine stimulus cue.

Recently, we extended our research to include (*5*) a more extensive characterization of the pentazocine cue by comparing it to other mixed narcotic agonist–antagonists (cyclazocine and levallorphan); and (*6*) a study of the neurochemical nature of the pentazocine cue. Since small doses of naloxone (e.g., .4 mg/kg) completely antagonize the discriminative stimulus properties of morphine (Rosecrans *et al.*, 1973) but only partially antagonize those of pentazocine (Kuhn *et al.*, 1976), we reasoned that the behavioral effects of mixed agonists–antagonists are mediated by two different sets of receptors in the central nervous system (CNS), a hypothesis first proposed by Martin (1967). That is, certain effects such as "analgesia" (Blumberg, Dayton, & Wolf, 1966; O'Callaghan & Holtzman, 1975), which are completely blocked by naloxone, are mediated by narcotic receptors (Pert & Snyder, 1973, 1974), while other behavioral (facilitation of locomotor activity) and neurochemical (depletion of brain catecholamines) effects are not blocked by naloxone (Holtzman & Jewett, 1972), and may therefore be mediated by some other receptor or set of receptors.

The results of several recent neurochemical experiments suggest that this second receptor system involves the catecholamines, especially DA. That is, brain DA is markedly reduced by pentazocine (Berkowitz, 1974; Holtzman, 1974; Holtzman & Jewett, 1972), while striatal homovanillic acid, the primary DA metabolite, is significantly increased by the same compound (Ahtee & Kaariainen, 1973). Pentazocine also accelerates the depletion of DA after inhibition of tyrosine hydroxylase (Paalzow, Paalzow, & Stalby, 1974). These data suggest that pentazocine may cause the release of striatal DA—an action that might be at least a partial basis for the discriminable cue of this drug.

While this chapter will briefly review our earlier research on the discriminative stimulus properties of pentazocine (Kuhn *et al.*, 1976), it will concentrate on our most recent attempts to specify the precise neurochemical nature of the pentazocine cue. Thus, we will present further evidence to support the hypothesis that mixed narcotic agonists–antagonists such as pentazocine exert their effects by interacting directly with narcotic receptors as well as interacting indirectly with dopamine receptors.

METHODS

Subjects

Male albino rats of Sprague–Dawley strain were used. They were experimentally naive and about 110 days old at the start of the experiment. All animals were maintained at 80% of their expected free-feeding weight by restricting water intake to 5 min after each daily experimental session. Lab chow was always freely available in individual home cages that were maintained in a room of relatively constant temperature (21–23°C) and humidity (40–50%).

Apparatus

Four commercially available, 2-lever rat chambers (BRS/LVE Model 143-24) housed within sound- and light-attenuating outer shells were used. Each chamber contained 2 levers that were mounted at opposite ends of one wall. A single dipper that delivered .01 ml of tap water was positioned between the 2 levers. Each chamber was illuminated by 28-V white house light. All programming and recording was done by solid-state and electromechanical equipment located in an adjoining room.

Training Procedure

Rats were initially shaped to respond on a continuous or fixed-ratio 1 (FR-1) schedule of reinforcement on one lever, with the other lever removed from the chamber. The schedule of reinforcement was gradually increased from FR-1 to FR-10 on each lever independently. Bar-press training on FR-10 continued on a double-alternation schedule, that is, 2 days on the left lever followed by 2 days on the right level until rates of responding stabilized. After an equal number of days of FR-10 responding on each lever, a variable-interval component averaging 1 min in length (VI-1) was added to the FR component such that each animal had to emit 10 responses after the interval elapsed to obtain reinforcement. Training on this Tandem VI-1 FR-10 schedule of reinforcement continued until response rates stabilized and until each animal had an equal number of days of the Tandem VI-1 FR-10 schedule on each lever. At this point, drug-discrimination training was begun.

Drug-Discrimination Procedure

Rats were injected intraperitoneally (i.p.) with either *dl*-pentazocine (10 mg/kg in Experiment 15; mg/kg in Experiment 2) or its vehicle 30 min before each session and were placed in the chambers with both levers present. For half of the rats, responding on the right lever after drug administration was arbitrarily deemed correct; responding on the left lever after drug injection was correct for the other half. Responding on the opposite lever after vehicle injection was correct for each group, respectively. To prevent superstitious chaining from the incorrect to the correct lever, a punishment contingency was added to the FR component after discrimination training began. That is, an incorrect response during the FR sequence reset the stepper so that 10 successive *correct* responses were required for reinforcement. Incorrect responding during the VI component was recorded but had no programmed consequences.

Training sessions were always 30 min long and were conducted at the same time each day, Monday through Friday.

To assess the degree of stimulus control exerted over choice behavior by the drug, rats were injected with either pentazocine or its vehicle and were tested for lever choice during a 5-min extinction period. It is necessary that testing be carried out in extinction both to avoid further learning and to avoid presenting animals the additional cue of the reinforcing stimulus. With some schedules of reinforcement (e.g., DRL) rats can learn to distinguish test days from normal training days and subsequently do not respond further when, on completion of the schedule requirement, reinforcement is not delivered (Cameron & Appel, 1973). We have essentially avoided this problem by using a variable, low-density schedule of reinforcement (Tandem VI-1 FR-10).

Before beginning Experiment 2, we made several procedural modifications in an effort to enhance drug-discrimination learning. That is, discrimination training as well as bar-press shaping was carried out on a double-alternation schedule (not random) and pentazocine and vehicle injections were started *during* initial shaping. Thus, the rats were trained to bar-press (FR-1) on the specified lever, with the other lever removed as before. Two days of a particular drug-lever pairing were followed by 2 days of the vehicle-"other" lever pairing until response rates on FR-10 stabilized on both levers, after which both levers were placed in the chamber. Drug injections were given earlier in training (actually from Day 1 of behavioral shaping) to minimize the behaviorally disruptive effects of the drug once discrimination training began. In this manner, responding on each lever was always paired with the appropriate drug stimulus instead of responding on both levers being associated with the nondrug state (shaping) until drug injections began and both levers were presented.

Experimental Design

Discriminability, that is, degree of stimulus control exerted by pentazocine, is expressed as the percentage of total responding that occurs on the treatment-related lever during extinction. Thus, to state that the rats were 90% correct after pentazocine or vehicle means that 90% of all responding in extinction occurred on the pentazocine-related or vehicle-related lever, respectively. When attempts to modify the discriminability of the drug were made, the results are expressed in terms of the percentage responding on the drug-related lever.

Specific attempts to block or mimic pentazocine under study are usually not carried out in our laboratory until choice behavior is at least 85% correct. Testing for blockade of pentazocine, or any drug for that matter, simply involves injecting the potential antagonist at some specified time before or after the training drug is given and then testing the rats for lever choice during extinction.

Treatments designed to mimic the training drug are given in place of the training drug, and the rats are tested for lever choice. The treatment time and dose parameters are specified when appropriate.

RESULTS

Experiment 1

The results of the first experiment have been reported elsewhere, in part (Kuhn *et al.*, 1976). They can be described briefly as follows: Figure 9.1 shows pentazocine (10 mg/kg) is discriminable from its vehicle. The acquisition of this discrimination is fairly slow (12 to 13 days) but comparable to that of numerous other drugs (Kuhn *et al.*, 1976); asymptotic level of performance is about 85%.

Figure 9.2 shows the dose-response relationship for pentazocine and the dose–response transfer of pentazocine to morphine. Since the slopes of these log-probit plots do not depart significantly from parallel, it was concluded that the stimulus properties of appropriate doses of these two compounds are identical.

When the rats were pretreated with a small dose of naloxone (.4 mg/kg) and subsequently given equipotent doses of pentazocine or morphine, the morphine cue was blocked to a significantly greater extent than was the pentazocine cue (Table 9.1). Combination injections of morphine and pentazocine cue (dis-

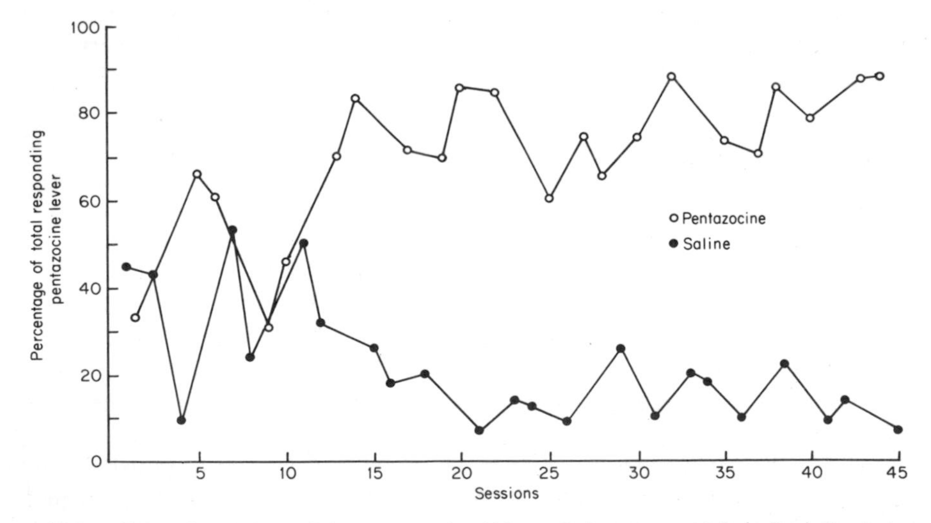

Figure 9.1. Acquisition of the pentazocine (10 mg/kg) versus vehicle (saline) discrimination (Experiment 1). All injections were given 30 min before testing for lever choice.

Table 9.1
Effects of Pentazocine, Morphine, and Naloxone on Lever Choice

Treatment	Doses (mg/kg)	Replications	Percentage of total responding on pentazocine lever	Comparison	
Pentazocine	10		81.2[a]		
Saline			12.4[a]		
Morphine	7.5	4	87.7	(1) versus (3)	NS
Naloxone	.4	2	8.1	(4) versus (2)	NS
Morphine and pentazocine	1.7 + 4.2[b]	4	93.9	50%	$<.001$
Pentazocine and naloxone	10 + .4	1	41.8	(6) versus (1)	$<.01$
Morphine and naloxone	7.5 + .4	1	16.1	(7) versus (3)	$<.01$
				(6) versus (7)	$<.05$

Source: Reprinted from the *Journal of Pharmacology and Experimental Therapeutics,* 1976, Vol. *196,* p. 125. Courtesy of Williams & Wilkins Company, Baltimore, Maryland.

[a] Average of last five training sessions.

[b] ED_{50} values estimated according to Litchfield and Wilcoxon (1949).

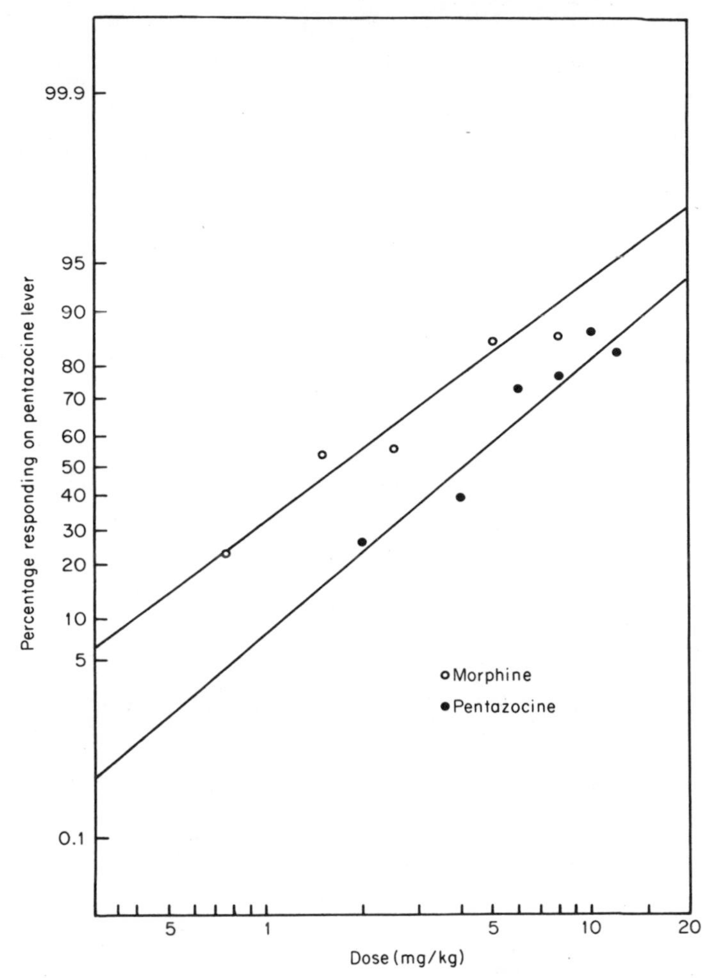

Figure 9.2. Dose–response curves for pentazocine and morphine in rats trained with pentazocine (10 mg/kg) versus no drug (saline). Doses of each drug were administered in random order 30 min before testing. Each point for pentazocine represents 3 determinations in each of 8 rats; each point for morphine represents 4 determinations in the same 8 rats. All data are based on responding during extinction. The data are plotted according to the method of Litchfield and Wilcoxon (1949). Reprinted from the *Journal of Pharmacology and* Experimental Therapeutics, 1976, Vol. *196*, p. 124, with the permission of Williams & Wilkins Company, Baltimore, Maryland.

criminable ED_{50} doses) produced a synergistic (additive) effect. Thus, the data suggest that pentazocine and morphine share some common sites of action. Since the stimulus properties of both pentazocine and morphine were blocked to some degree by naloxone, we assume that the stimulus properties of these two drugs are mediated in part by stimulation of narcotic receptors. This reasoning accounts for the similarity of some of the properties of pentazocine and morphine. However, the ability of naloxone to block the morphine cue to a much greater extent than the pentazocine cue indicated that pentazocine may exert some of its effects through a set of receptors that are different from those that mediate the effects of morphine.

Experiment 2

In view of the data from the first experiment, a second experiment was designed to ascertain information concerning the nature of other possible receptors that might mediate the pentazocine cue. Since it had been shown that pentazocine releases striatal dopamine, we decided to try to block the pentazocine cue with the DA-receptor blocker haloperidol (Anden, Rubenson, Fuxe, & Hokfelt, 1967) and to examine the effects on the pentazocine-vehicle discrimination of a wide range of doses of the narcotic-receptor blocker naltrexone.

A new group of rats were trained to discriminate 15 mg/kg of pentazocine from its vehicle. Using a somewhat more efficient training procedure than was used in Experiment 1, we were able to obtain 80% correct responding on the eighth day of training. As can be seen in the acquisition curve of Figure 9.3, accuracy rapidly increased to more than 95% and has remained at this level for a period of several months.

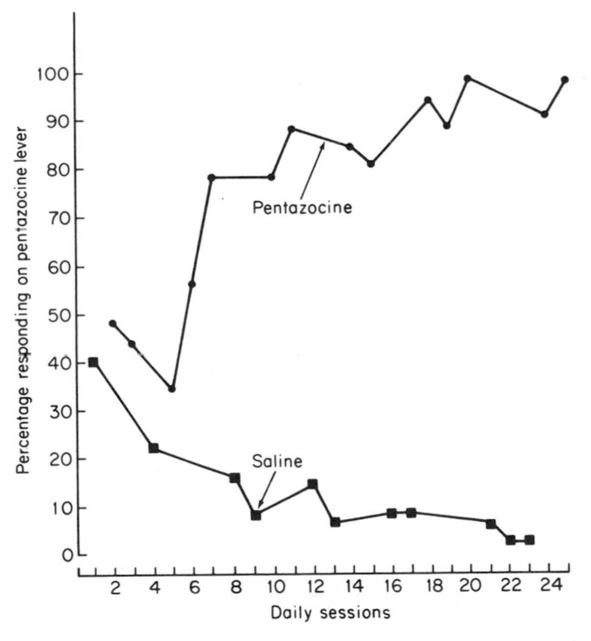

Figure 9.3. Acquisition of the pentazocine (15 mg/kg) versus vehicle (saline) discrimination (Experiment 2).

Naltrexone was administered 15 min after pentazocine (15 min prior to behavioral testing). As can be seen in Figure 9.4, naltrexone only partially blocked the pentazocine cue over a wide range of doses tested (1–20 mg/kg), that is, percentage responding on the pentazocine cue ranged from about 40 to 50%.

Figure 9.4 also demonstrates that haloperidol in doses ranging from .05 to .50 mg/kg given 1 hr before pentazocine partially blocked the pentazocine cue, to about 60%. As with naltrexone, the dose–response curve for haloperidol was relatively flat. When *both* haloperidol and naltrexone were given in combination with pentazocine, an almost complete reversal of the pentazocine cue was seen. The rats' response on the pentazocine lever was only 18% of their total response output.

Various other blocking agents were then administered in an attempt to discern the specificity of the pentazocine cue. These included both alpha- and beta-adrenergic blocking agents, as well as an anticholinergic, an antiserotonergic, and an antihistaminic agent. As can be seen in Table 9.2, none of these agents had a significant effect on the pentazocine discrimination.

In a further attempt to demonstrate the role of dopamine receptors in the pentazocine cue, apomorphine, a drug that directly stimulates dopamine recep-

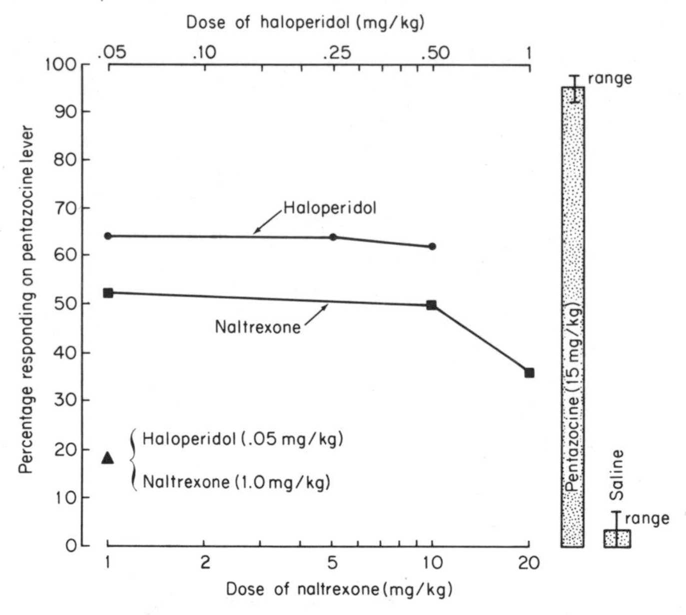

Figure 9.4. Effects of pretreatment with the narcotic antagonist, naltrexone (1, 10, and 20 mg/kg), the DA receptor antagonist, haloperidol (.05, .25, and .50 mg/kg), or a combination of haloperidol and naltrexone on the pentazocine versus saline discrimination (Experiment 2). The mean performance and range during the last 5 days of training prior to pretreatment are indicated by the bar graphs to the right of the dose–response curves.

Table 9.2
Effects of Receptor-Blocking Agents on a Pentazocine (15 mg/kg) versus Vehicle (Saline) Discrimination

Agent	Dose (mg/kg)	Time	Percent of total responding on pentazocine lever	Standard error
Phentolamine	1	15 min after pentazocine	98.7	1.2
	5	15 min after pentazocine	95.4	3.8
	5	60 min before pentazocine	100	
Propranolol	5	15 min after pentazocine	96.1	3.6
	10	15 min after pentazocine	91.3	4
	20	60 min before pentazocine	92.6	7
Methysergide	5	15 min before pentazocine	98.9	.8
	10	15 min after pentazocine	99.6	.2
Atropine	1	15 min before pentazocine	97.6	1.6
	1	15 min after pentazocine	98.8	.4
Cyproheptadine	1	60 min before pentazocine	99.2	.1

tors (Anden *et al.*, 1967), was tested for possible transfer to pentazocine. Injections of 1 mg/kg of apomorphine produced 34% pentazocine responding. This agrees with the blockade data since it was seen that haloperidol reduced pentazocine responding by about 35%.

DISCUSSION

These data indicate that the discriminative stimulus effect of pentazocine is mediated by two sets of receptors. Partial blockade by naltrexone confirms and extends our earlier results that suggest involvement of narcotic receptors. This conclusion is strengthened by studies with optical isomers of pentazocine. Since we used the racemic (*d,l*) form of the drug to establish the discrimination, and since opiate receptors are stereospecific, we tested our rats with both the *d*- and *l*-isomers of pentazocine during extinction. We found that the discriminative properties reside completely in the *l*-isomer. Choice responding after 15 mg/kg of *l*-pentazocine was 99% on the pentazocine lever while responding on the pentazocine lever after 15 mg/kg of *d*-pentazocine was 10%.

That haloperidol partially blocked the pentazocine cue suggests that pentazocine and related compounds also act (probably indirectly) through DA receptors. Several studies suggest that in addition to their ability to mediate similar drug-behavior interactions, narcotic and DA receptors may be related anatomically. That is, their sites of action are in the striatum. Opiate receptors seem to be localized in the caudate-putamen (Kuhar, Pert, & Snyder, 1973; Pert & Snyder, 1973; Pert, Kuhar, & Snyder, 1975); we have already seen that pentazocine has numerous effects on the striatal dopaminergic system. (See Introduction.) Thus, our behavioral data are consistent with both anatomical and neurochemical data.

It is, of course, conceivable that the additive blocking effects of naltrexone and haloperidol (Figure 9.3) might not reflect blockade of two different types of receptors but might instead represent *more* DA-receptor or *more* narcotic-receptor blockade than either haloperidol or naltrexone could produce alone. This seems unlikely for several reasons. First, the dose–response curves for blockade of the pentazocine cue by both naltrexone and haloperidol are relatively flat over a wide dose range. Second, we are aware of no reports in the literature in which a narcotic antagonist acts as a dopamine-receptor blocker; similarly, haloperidol, at a dose 10 times larger than ours, does not block narcotic receptors (Pert & Snyder, 1975).

RECENT ADVANCES

We proposed originally (Kuhn *et al.*, 1976) that the stimulus properties of pentazocine may be related to its addictive properties. Recent evidence suggests that this is not true. That is, both cyclazocine and levallorphan have stimulus properties similar to those of pentazocine; animals respond on the pentazocine lever following administration of these compounds and do so in a dose-related manner. Since neither cyclazocine nor levallorphan is known to produce physical dependence, it seems that the ability of a drug to serve as a discriminative stimulus is not sufficient evidence for judging its abuse potential.

Our data also suggest that the stimulus properties of pentazocine (in rats) do not reflect its psychotomimetic properties (in man). Since the psychotomimetic effects of pentazocine in man appear to be associated exclusively with the *d*-isomer (Still, 1975), and since the discriminative properties of the drug reside completely in the *l*-isomer, the stimulus properties of pentazocine cannot be related to its psychotomimetic properties.

ACKNOWLEDGMENTS

We thank Dr. H. Blumberg of Endo Laboratories for the gifts of naloxone and naltrexone, Dr. F. C. Nachod of Winthrop Laboratories for his gift of pentazocine, Mrs. Cheryl Funderburk for typing the manuscript, and the Williams & Wilkins Company, Baltimore, Maryland for permitting us to reprint material originally published in the *Journal of Pharmacology and Experimental Therapeutics*.

REFERENCES

Ahtee, L., & Kaariainen, I. The effect of narcotic analgesics on the homovanillac acid content of rat nucleus caudatus. *European J. Pharmacol.*, 1973, *22*, 206.

Anden, N. E., Rubenson, A., Fuxe, K., & Hokfelt, T. Evidence for dopamine receptor stimulation by apomorphine. *J. Pharm. Pharmacol.*, 1967, *19*, 627.

Barry, H., III. Classification of drugs according to their discriminable effects in rats. *Federation Proc.*, 1974, *33*, 1814.

Berkowitz, B. A. Effects of the optical isomers of the narcotic analgesic pentazocine on brain and heart biogenic amines. *European J. Pharmacol.*, 1974, *26*, 359.

Blumberg, H., Dayton, H. B., & Wolf, P. S. Counteraction of narcotic antagonist analgesics by the narcotic antagonist naloxone. *Proc. Soc. Exp. Biol. Med.*, 1966, *123*, 755.

Browne, R. G., & Ho, B. T. Role of serotonin in the discriminative stimulus properties of mescaline. *Pharmacol. Biochem. Behav.*, 1975, *3*, 429.

Cameron, O. G., & Appel, J. B. A behavioral and pharmacological analysis of some discriminable properties of d-LSD. *Psychopharmacologia*, 1973, *33*, 117.

Cooper, J. R., Bloom, F. E., & Roth, R. H. *The biochemical basis of neuropharmacology.* (2nd ed.) New York: Oxford Univ. Press, 1974.

Harris, R. Relationship of discriminative and reinforcing stimulus functions of drugs. Paper presented at the 22d annual convention, Southwest Psychological Association, Houston, April, 1975.

Hill, H. E., Jones, B. E., & Bell, E. C. State-dependent control of discrimination by morphine and pentobarbital. *Psychopharmacologia*, 1971, *22*, 305.

Hirschhorn, I. D., & Rosecrans, J. A. A comparison of the stimulus effects of morphine and lysergic acid diethylamide (LSD). *Pharmacol. Biochem. Behav.*, 1974, *2*, 361. (a)

Hirschhorn, I. D., & Rosecrans, J. A. Morphine and Δ^9-tetrahydrocannabinol: Tolerance to the stimulus effects. *Psychopharmacologia*, 1974, *36*, 243. (b)

Holtzman, S. G. Interactions of pentazocine and naloxone on the monoamine content of discrete regions of the rat brain. *Biochem. Pharmacol.*, 1974, *23*, 3029.

Holtzman, S. G., & Jewett, R. E. Some actions of pentazocine on behavior and brain monoamines in the rat. *J. Pharmacol. Exp. Therap.*, 1972, *181*, 346.

Huang, J.-T., & Ho, B. T. The role of monoamines in discriminative response control by d-amphetamine. Paper presented at the 3rd annual meeting of the Society for Neurosciences, San Diego, 1973.

Jasinski, D. R., Martin, W., & Hoeldtke, R. Effects of short- and long-term administration of pentazocine in man. *Clin. Pharmacol. Therap.*, 1970, *11*, 385.

Keats, A. S., & Telford, J. Narcotic antagonists as analgesics: Clinical aspects. *Advan. Chem. Ser.*, 1964, *49*, 170.

Kemp, W. Abuse liability and narcotic antagonism of pentazocine: Report of two cases. *Diseases Nervous System*, 1968, *29*, 599.

Kuhar, M. J., Pert, C. B., & Snyder, S. H. Regional distribution of opiate receptor binding in monkey and human brain. *Nature*, 1973, *245*, 447.

Kuhn, D. M., Appel, J. B., & Greenberg, I. An analysis of some discriminative properties of d-amphetamine. *Psychopharmacologia*, 1974, *39*, 57.

Kuhn, D. M., Greenberg, I., & Appel, J. B. Stimulus properties of the narcotic antagonist pentazocine: Similarity to morphine and antagonism by naloxone. *J. Pharmacol. Exp. Therap.*, 1976, *196*, 121–127.

Litchfield, J. T., & Wilcoxon, F. A simplified method of evaluating dose-effect experiments. *J. Pharmacol. Exp. Therap.*, 1949, *96*, 99.

Martin, W. R. Opioid antagonists. *Pharmacol. Rev.*, 1967, *19*, 463.

O'Callaghan, J. P., & Holtzman, S. G. Quantification of the analgesic activity of narcotic antagonists by modified hot-plate procedure. *J. Pharmacol. Exp. Therap.*, 1975, *192*, 497.

Overton, D. A. Dissociated learning in drug states (state-dependent learning). In D. Efron (Ed.), *Psychopharmacology: A Review of Progress 1957–1967.* Washington, D.C.: U.S. Govt. Printing Office, 1968.

Overton, D. A. State-dependent learning produced by alcohol and its relevance to alcoholism. In

B. Kissen & H. Begleiter (Eds.), *The biology of alcoholism: Physiology and behavior.* (Vol. II). New York: Plenum, 1972.

Overton, D. A. State-dependent learning produced by addicting drugs. In S. Fisher & A. Freedman (Eds.), *Opiate addiction: Origins and treatment.* Washington, D.C.: V. H. Winston, 1973.

Paalzow, G., Paalzow, L., & Stalby, B. Pentazocine analgesia and regional rat brain catecholamines. *European J. Pharmacol.*, 1974, *27,* 78.

Pert, C. B., Kuhar, M. J., & Snyder, S. H. Autoradiographic localization of the opiate receptor in rat brain. *Life Sci.*, 1975, *16,* 1849.

Pert, C. B., & Snyder, S. H. Opiate Receptor: Demonstration in nervous tissue. *Science,* 1973, *179,* 1011.

Pert, C. B., & Snyder, S. H. Opiate receptor binding of agonists and antagonists affected differentially by sodium. *Mol. Pharmacol.*, 1974, *10,* 868.

Pert, C. B., & Snyder, S. H. Identification of opiate receptor binding in intact animals. *Life Sci.*, 1975, *16,* 1623.

Rosecrans, J. A. Goodloe, M. H., Bennet, G. J., & Hirschhorn, I. D. Morphine as a discriminative cue: Effects of amine depletors and naloxone. *European J. Pharmacol.*, 1973, *21,* 252.

Sandoval, R. G., & Wong, R. I. H. Characteristics of pentazocine dependence in hospitalized patients after naloxone administration. *Psychopharmacologia,* 1973, *30,* 205.

Schechter, M. D., & Cook, P. G. Dopaminergic mediation of the interoceptive cue produced by d-amphetamine in rats. *Psychopharmacologia,* 1975, *42,* 165.

Still, C. N. Pain and pentazocine: Problems of control. *Southern Med. J.*, 1975, *68,* 805.

Winter, J. C. Hallucinogens as discriminative stimuli. *Federation Proc.*, 1974, *33,* 1825.

10

Stimulus Properties of Narcotic Analgesics and Antagonists

I. D. Hirschhorn

Department of Pharmacology
New York Medical College
Valhalla, New York

INTRODUCTION

Narcotic analgesics (also called opiates) are drugs with pharmacological properties like those of morphine (Jaffe & Martin, 1975). They include both natural and synthetic compounds with widely varying chemical structures. Although several are in wide clinical use, all have serious side-effects that limit their use. All of the narcotic analgesics produce tolerance and physical dependence upon repeated administration. Some are subject to compulsive self-administration, or addiction. Addiction is not inextricably related to physical dependence, but physical dependence may sometimes be a contributing factor in the development of addiction. Another important factor is the ability of a drug to produce morphinelike subjective effects, such as euphoria. Much research has been directed toward developing drugs that are effective in alleviating severe pain but have a low in liability for producing addiction.

Some narcotic-antagonist analgesics exhibit the desired separation of analgesic activity and addiction liability. Narcotic antagonist analgesics antagonize the pharmacological effects of morphine and other narcotic analgesics when given in combination with these drugs, but they also have narcotic-agonist (morphinelike) activity of their own. Nalorphine, for example, is a drug in this category that has analgesic efficacy and a low addiction liability (Lasagna & Beecher, 1954b). The low addiction liability is predictable from the nature of physical dependence and the subjective effects produced by this drug. The abstinence syndrome that follows the termination of chronic nalorphine administration lacks the intense craving for the drug that often accompanies morphine withdrawal (Martin & Gorodetzky, 1965). The subjective effects differ from those produced by morphine in that they include dysphoria and psychotomimetic effects, and human subjects perceive them generally as unpleasant (Haertzen, 1970). Unfortunately, these same unpleasant subjective effects also render nalorphine unsuitable for clinical use.

Although subjectively perceived effects of both narcotic analgesics and antagonists are prominent and important, there is obviously no direct method for assessing such effects in animals. Drug discrimination procedures, however, permit us to determine whether animals can distinguish the effects of a drug from no drug or from a second drug, and they may, therefore, be a relevant behavioral variable in animals for drugs that produce prominent subjective effects in man. This chapter will discuss some investigations of the stimulus properties of narcotic analgesics and antagonists.

GENERAL METHODS

Subjects

Male Sprague–Dawley rats (Flow Research Animals, Dublin, Virginia), without previous drug or behavioral experience, were housed in individual home cages in which they had unlimited access to water. An adjusted amount of a commercial rat chow was given following each experimental session to maintain them at 70–80% of their free-feeding weights.

Behavioral Procedures

The discrimination training and testing procedures were similar to those described by Hirschhorn and Winter (1971). Rats were trained to press both levers of a standard operant test chamber (Lehigh Valley Electronics) by successive approximations. Sweetened condensed milk diluted 1:2 with tap water was the reinforcer. During discrimination training, each daily session was preceded by one of two treatments. Following Treatment A, only responses on Lever A were reinforced; after Treatment B, Lever B was correct. For one-half of the animals of any group, Lever A was the right lever and Lever B was the left lever; for the remaining animals, these conditions were reversed. Experimental sessions were of 15-min duration. During the first 4 discrimination training sessions, Treatments A and B were alternated daily (ABA . . .), and each response on the appropriate lever resulted in delivery of reinforcement (FR 1). During subsequent sessions, Treatments A and B were alternated every other day (AABBA . . .). No responses on either lever were reinforced during the first 2.5 min of these sessions, and during the remaining 12.5 min, one of two reinforcement schedules, a variable interval of 15 sec (VI 15) or a fixed ratio of 4 (FR 4), was in effect. Only responses made during the initial unreinforced 2.5-min period were used to calculate discrimination accuracy.

Stimulus generalization and drug interaction experiments were performed in test sessions. These were 2.5-min sessions in which no responses were reinforced. After discriminated responding was established and stable, test sessions were interposed among the regular training sessions. At least one, usually 3, regular training sessions were given between any two test sessions.

Drugs

The drugs used in these experiments were morphine sulfate, pentazocine, cyclazocine, nalorphine HCl, lysergic acid diethylamide (LSD) tartrate, atropine sulfate, cyproheptadine HCl, methysergide, methadone HCl, meperidine HCl, and naloxone HCl. All drug doses refer to these salts or bases with the exception of LSD, which is referred to as the free base (.15 μmoles of LSD = 72 μg). Cyproheptadine HCl was dissolved in polyethoxylated vegetable oil and then diluted with saline (Hirschhorn & Rosecrans, 1974a). Pentazocine and cyclazocine were dissolved in a small amount of 8.5% lactic acid, resulting in a final pH of 4–5. All other drugs were dissolved in .9% sodium chloride. All drugs were injected in a constant volume of 1 ml/kg.

RESULTS AND COMMENTS

Morphine as a Discriminative Stimulus: Tolerance

When Treatments A and B were morphine sulfate and saline, discriminated responding readily occurred. Figure 10.1 shows the development of discriminated responding in 1 group of 5 rats. Early in discrimination training, they made the same percentage of their responses on the morphine-correct lever after both saline and morphine treatment (20% in the first block of 4 sessions). A consistent pattern of discriminated responding appeared by the fourth session block, and after 10 blocks they were making at least 80% of their responses in the initial 2.5-min unreinforced period on the rewarded lever.

Since, with repeated administration, tolerance develops to many of the pharmacological effects of morphine, it is of interest to determine whether tolerance also develops to the stimulus properties of morphine. During discrimination training, the rats received morphine 2–3 days per week, but no decrease in the accuracy of discrimination was observed; rather, discrimination initially increased, then became stable. To determine whether a more intense schedule of morphine administration would produce tolerance, the rate of administration and dose of morphine were increased. The same discrimination training schedule was continued in sessions subsequent to those shown in Figure 10.1. This time, however, supplementary subcutaneous injections of morphine were given in increasing doses from 20 mg/kg to 160 mg/kg, 7 days

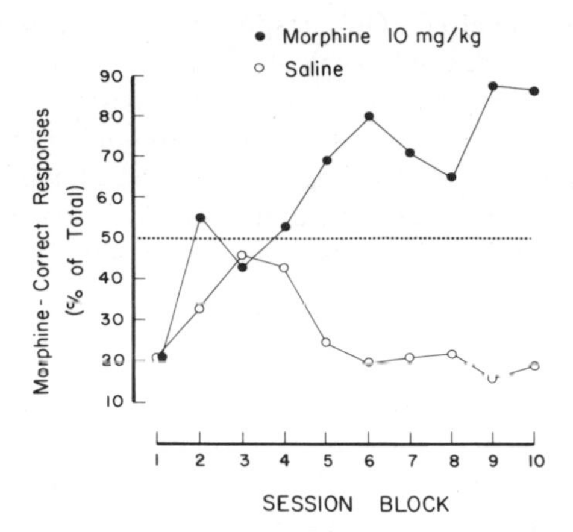

Figure 10.1. Discriminated responding with morphine and saline as discriminative stimuli. Morphine or saline was injected, i.p., 45 min before the experimental session. Each point represents the mean of 1 determination in each of 5 animals. Ordinate: Number of responses in the first 2.5 min of the session (unreinforced) on the morphine-correct lever expressed as a percentage of total responses. Abcissa: successive blocks of four sessions each. Two morphine and two saline sessions comprise each block (from Hirschhorn & Rosecrans, 1974b).

a week, over a period of about 2 months. On weekdays, the supplementary injections immediately followed the daily experimental sessions. Doses of 80 and 160 mg/kg were divided into two equal injections, the second given about 6 hr after the first. This relatively short time interval was chosen to minimize the possibility of residual morphine being present at the experimental session on the following day.

Throughout this treatment, the rats continued to discriminate morphine from saline, but the accuracy of discrimination, as measured by the difference in the pattern of responding during morphine and saline sessions, was reduced to about one-half of baseline discrimination ($p < .05$, Duncan's range test; Table 10.1). Although some tolerance was observed, the degree of tolerance did not increase in an orderly fashion as the supplementary dose of morphine was increased. A confounding factor in this experiment, which may have masked the appearance of a greater degree of tolerance, is that discrimination training continued during the period of chronic morphine treatment. Thus, it is possible that the animals learned to discriminate diminished effects of morphine.

Antagonism of Morphine Discrimination with Other Drugs

The effects of various other drugs on the discrimination of morphine are summarized in Table 10.2. Naloxone, a narcotic antagonist with few pharmacological effects of its own (Blumberg, Dayton, George, & Rappaport,

Table 10.1
Effect of Chronic Morphine Administration on the Discrimination of Morphine

Dose of morphine (mg/kg)[b]	Sessions	Responses on morphine-correct lever (% of total)[a]		
		Morphine	Saline	Morphine-saline
0 (prechronic)	16	76.8 ± 6.4	19.2 ± 2.6	57.6 ± 6.9
20	8	72.1 ± 9.1	29.8 ± 4.4	47.5 ± 10.6
40	8	67.9 ± 5.8	34.8 ± 6.7	30.1 ± 9.2[c]
80	8	57.6 ± 6.6	23.9 ± 3.7	33.7 ± 8.4[c]
160	16	67.8 ± 5.9	19.3 ± 5.	48.5 ± 7.6
0 (postchronic)	16	87.6 ± 3.	24.5 ± 3.3	63.1 ± 4.2

[a] Mean for five rats ± S.E.M.
[b] Dose is in addition to 10 mg/kg discriminative stimulus dose of morphine.
[c] Difference from baseline discrimination (mean of 16 sessions prior to and 16 sessions following chronic morphine treatment) at $p < .05$ according to Duncan's range test.
Source: Adapted from Hirschhorn and Rosecrans (1974b).

1961) specifically antagonizes the effects of morphine and other narcotic analgesics. If the stimulus property of morphine is a specific opiate effect, then naloxone should, similarly, block morphine discrimination. It was found that the percentage of morphine-correct responses decreased as the dose of naloxone was increased, and with the highest dose the animals responded as though they had been given saline. In an attempt to delineate neurochemical correlates of morphine discrimination, two serotonin (5-hydroxytryptamine [5-HT]) antagonists, cyproheptadine and methysergide, and a cholinergic

Table 10.2
Effect of Other Drugs on the Discrimination of Morphine

Drug (dose, mg/kg)[a]	Responses on morphine-correct lever (% of total)[b]			
Naloxone	(0) 96.5 ± 2.7	(.1) 43.8 ± 14.3	(.2) 39 ± 9.5	(.4) 18.3 ± 5.4
Cyproheptadine	(0) 98.4 ± 1.3	(3) 97.5 ± 2.5		
Methysergide	(0) 85.1 ± 10.6	(3) 98.5 ± 1.5	(6) 87.7 ± 8.8	(10) 100 ± 0
Atropine	(0) 76.6 ± 7.6	(.5) 83.3 ± 16.7	(1) 88.9 ± 7	(2) 91.2 ± 5.5

Adapted from Hirschhorn and Rosecrans (1974a).
[a] Morphine (10 mg/kg) was injected 45 min before the session. The other drugs were injected, respectively, 30, 60, 35, and 15 min before the session. Atropine was injected s.c.; all other drugs were given i.p.
[b] Mean of one determination in each of six rats ± S.E.M.

antagonist, atropine, were given in combination with morphine, but none of these treatments antagonized the morphine cue.

Stimulation Generalization Tests

The effect of morphine that is the basis of its stimulus properties is not known. One might question whether it is a specific opiate effect or a general action that is shared by many drugs, for example, general behavioral depression. Evidence that the morphine stimulus is a specific opiate effect is provided by the naloxone antagonism of morphine discrimination. Another method for testing the specificity of the morphine stimulus is stimulus-generalization tests. If the morphine stimulus is a specific opiate effect, then only narcotic analgesics would be expected to produce responding appropriate for morphine treatment in animals trained to discriminate morphine from saline.

Five narcotic analgesics were tested for stimulus generalization in rats previously trained to discriminate morphine from saline. These can be divided into three categories (Jaffe & Martin, 1975): (*1*) Methadone and meperidine are morphinelike analgesics. Although both are synthetic compounds with chemical structures considerably different from morphine, they produced physical and subjective effects qualitatively similar to those of morphine. (*2*)

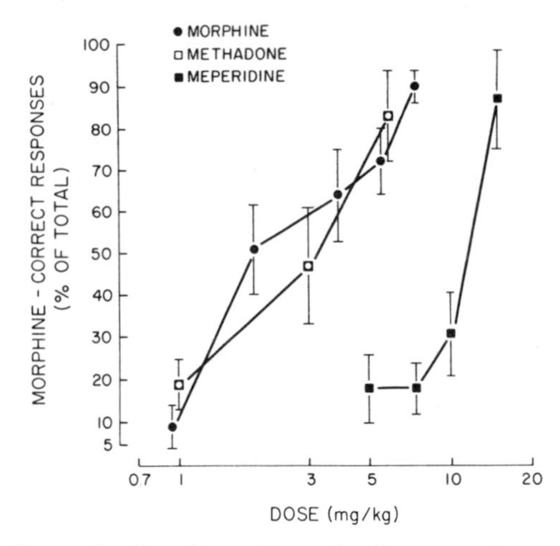

Figure 10.2. Generalization of morphine stimulus properties to narcotic analgesics. Rats were trained to respond differentially to morphine (7.5 mg/kg) and saline. Then, test sessions of 2.5 min duration, during which no responses were reinforced, were interposed among the regular training sessions. The rats received various doses of narcotic analgesics in test sessions. Morphine, methadone, and meperidine were injected, i.p., 45, 30, and 15 min, respectively, before the experimental session. Ordinate: Number of responses on the morphine-correct lever expressed as a percentage of total responses. Abcissa: Dose of test drug plotted on a log scale. Each point is the mean of one determination in each of 9 animals. Vertical lines indicate + S.E.M. (from Hirschhorn & Rosecrans, 1976).

Table 10.3
Generalization of Morphine Stimulus Properties to Narcotic Antagonists

Drug (dose, mg/kg)[a]	Responses on morphine-correct lever (% of total)[b]			
Nalorphine	(1) 41.6 ± 11.1	(10) 48.6 ± 8.6	(20) 53.6 ± 9.9	(40) 28.4 ± 8.5
Cyclazocine	(.25) 33.6 ± 8	(.50) 22.5 ± 5.8	(1) 42.8 ± 8.6	
Pentazocine	(5) 30.6 ± 5.5	(10) 60.4 ± 10.7	(20) 73.0 ± 8.7	

Source: Adapted from Hirschhorn and Rosecrans (1976).
[a] All drugs were injected, i.p., 30 min before the session.
[b] Mean of one determination in each of 9 animals ± S.E.M.

Nalorphine and cyclazocine are narcotic-antagonist analgesics. They have both morphine-antagonist and morphinelike (agonist) pharmacological properties, including analgesic efficacy. They produce subjective effects in man that, instead of euphoria, are characterized by dysphoria and include a high incidence of psychotomimetic effects. (*3*) Pentazocine is a narcotic-antagonist analgesic but with pharmacological properties different from those of nalorphine and cyclazocine. The agonist activity of pentazocine is more potent relative to its antagonist activity than is the case for the latter two compounds, and the subjective effects are more like those of morphine.

The results of stimulus generalization tests with methadone and meperidine are unambiguous (Figure 10.2). When various doses of morphine were given, a typical dose–response relationship was obtained, that is, the percentage of total responses on the morphine-correct lever varied directly with the dosage of morphine. Methadone and meperidine produced similar curves, the highest dose of each producing the same percentage of morphine-correct responses as did the training dose of morphine. Methadone was seen to be equipotent with morphine, while higher doses of meperidine were required to produce an equivalent response. The results with nalorphine and cyclazocine are also unambiguous (Table 10.3). There was no evidence of stimulus generalization to these drugs. Yet, doses of each drug within the range tested here can serve as discriminative stimuli when paired with saline (20 mg/kg of nalorphine; .5 mg/kg of cyclazocine; Hirschhorn, in press). The results obtained with pentazocine are less clear-cut. Although a dose-related trend of stimulus generalization is apparent, the highest dose tested (20 mg/kg) failed to produce as high a percentage of morphine-correct responses (73%) as the training dose of morphine (90%). Since 10 mg/kg of pentazocine is a strong discriminative stimulus when paired with saline (Hirschhorn, in press), 20 mg/kg should be adequate to produce stimulus generalization.

Since nalorphine and cyclazocine produce psychotomimetic effects in man, it is of interest to determine the degree of similarity of the stimulus properties

Table 10.4
Generalization of LSD Stimulus Properties to Narcotic Analgesics and Antagonists

Drug (dose, mg/kg)[a]	Responses on LSD-correct lever (% of total)[b]					
LSD	(0) 22.2 ± 3	(.0125) 24 ± 14.2	(.025) 40.2 ± 18.2	(.050) 46.6 ± 14.7	(.075) 73 ± 9.2	(.10) 88.1 ± 4.6
Methadone	(1) 7.8 ± 2.6	(3) 14.2 ± 9.5	(6) 5.3 ± 5.3			
Meperidine	(5) 22.2 ± 15.8	(7.5) 17 ± 13.1	(10) 6.5 ± 5.2			
Nalorphine	(1) 19.1 ± 9.1	(10) 8 ± 2.8	(20) 34.8 ± 12.3	(40) 31.3 ± 19.1		
Cyclazocine	(.25) 8 ± 3.5	(.50) 45.2 ± 17.7	(1) 63 ± 16.8			
Pentazocine	(5) 10 ± 5	(10) 13.2 ± 10	(20) 22.2 ± 17.2			

Source: Adapted from Hirschhorn & Rosecrans (1976).

[a] All drugs were injected i.p. LSD and meperidine were injected 5 and 15 min, respectively, before the session; all other drugs were injected 30 min before the session.

[b] Mean of one determination in each of five animals ± S.E.M.

of these drugs to the stimulus properties of a psychotomimetic such as LSD. The same 5 drugs that were tested for stimulus generalization in animals trained to discriminate morphine from saline were tested for stimulus generalization in animals trained to discriminate LSD from saline. Only cyclazocine regularly produced a dose-related increase in the percentage of LSD-correct responses (Table 10.4). Complete generalization was not obtained, however, since the highest dose of cyclazocine that did not abolish responding (1 mg/kg), produced only 63% LSD-correct responses, compared with 88% LSD-correct responses obtained with .1 mg/kg of LSD.

DISCUSSION

The data of these experiments indicate that the stimulus properties of morphine are a specific opiate effect. Evidence for this comes from the antagonism of morphine discrimination by the specific narcotic antagonist naloxone and from stimulus generalization tests. The antagonism to morphine discrimination by naloxone has been replicated in several laboratories and with several different discrimination tasks (Rosecrans, Goodloe, Bennett, & Hirschhorn, 1973; Shannon & Holtzman, 1975; Winter, 1975). Naloxone does not interfere with drug discrimination per se since it has no effect on the discriminability of LSD (Hirschhorn & Rosecrans, 1974a) or ethanol (Winter, 1975).

Stimulus-generalization tests permit us to determine the degree of similarity of the stimulus properties of other drugs to a training drug. If the stimulus properties of morphine are a specific opiate effect, then we would expect only narcotic analgesics to produce morphinelike responding in animals trained to discriminate morphine from saline. This is what has been observed. Although morphine is a general behavioral depressant, this action is apparently not the basis for discrimination since morphine discrimination does not generalize to ethanol, another general depressant (Winter, 1975). Other nonopiates to which morphine discrimination does not generalize are LSD (Hirschhorn & Rosecrans, 1974a), loperamide (an antidiarrheal drug), haloperidol (an antipsychotic; Gianutsos & Lal, 1975), and thebane (Shannon & Holtzman, 1975). Similarly, the narcotic-antagonist analgesics, nalorphine and cyclazocine, which differ from morphine in their pharmacological properties, did not generalize from morphine (Table 10.3).

In the present experiments, unambiguous generalization was obtained only with methadone and meperidine, narcotic analgesics with pharmacological properties much like morphine (Figure 10.2). Stimulus generalization to the narcotic antagonist, pentazocine, was less definitive (Table 10.3). The relative potencies of the drugs to which generalization was observed are in general agreement with the relative potencies for analgesia and subjective effects in man. Methadone, approximately equipotent to morphine in terms of analgesia and subjective effects (Denton & Beecher, 1949; Isbell, Wikler, Eisenman, Daingerfield, & Frank, 1948), was found to be also equipotent in stimulus properties. Meperidine, which is one-sixth to one-tenth as potent as morphine in man (Lasagna & Beecher, 1954a; Houde & Wallenstein, 1958), and pentazocine, which is one-half to one-fourth as potent (Beaver, Wallenstein, Houde, & Roger, 1966; Keats & Telford, 1964) had similar lower potencies in the present experiments.

Of the five narcotic analgesics and antagonists tested, only cyclazocine produced evidence for stimulus generalization in rats trained to discriminate LSD from saline (Table 10.4). Since only partial generalization was observed, that is, the highest dose of cyclazocine produced a lower percentage of LSD-correct responses from the training dose of LSD, this result is difficult to interpret. One possible interpretation is that the stimulus properties of these two drugs are similar but have some differences. This could be analogous to the situation in man, where both cyclazocine and LSD produce similar psychotomimetic phenomena, but the overall patterns of subjective effects are clearly differentiable (Haertzen, 1970). The different results obtained with nalorphine and cyclazocine in the LSD generalizations suggest that the two drugs have differentiable stimulus properties. This would be inconsistent with their highly similar clinical effects (Haertzen, 1970). However, a direct comparison of these two drugs is necessary before a definitive statement can be made as to whether their stimulus properties are indeed differentiable.

Most behavioral criteria of drug effects are an impairment of "normal" (nondrugged) behavior, for example, loss of righting reflex, decrease in locomotor activity, or a disruption of schedule-controlled operant behavior. Tolerance to these measures appears as a return toward a baseline of nondrugged performance. Drug stimuli, in contrast, are not an impairment of normal behavior, and tolerance would be manifested in a different way, as an impairment of discrimination. Therefore, a behavioral adaptation to a constant drug effect, sometimes referred to as behavioral tolerance, which can occur with a behavioral impairment, is inconsistent with drug stimuli.

Tolerance develops to many of the effects of morphine in both animals and man, although tolerance does not develop equally to all of the varied effects of this drug. For example, in man, profound tolerance develops rapidly to the analgesic and subjective effects, but there is observed very little tolerance to pupillary constriction or decreased intestinal motility (Jaffe & Martin, 1975). Although the procedure of training an animal to discriminate a drug from saline requires repeated drug administration, 2–3 times per week in the present experiments, we observed no tolerance during discrimination training. Supplementary daily injections of morphine up to 16 times the training dose for 2 months resulted in only a limited tolerance. Because discrimination training continued during chronic drug administration, a tolerance-induced diminution of the discriminable effects of morphine may have been partially masked by the learning of a more difficult discrimination. It should be possible to determine experimentally whether a more profound tolerance will occur if chronic drug administration is given in the absence of discrimination training.

In the present experiments, we were unable to affect morphine discrimination with serotonergic or cholinergic antagonists. This does not rule out the possibility, however, that the morphine stimulus may be mediated by one of these neurotransmitters. There are several difficulties with attempts to specifically antagonize the effects of neurotransmitters in the central nervous system. Although it is often assumed that drugs that antagonize the effects of neurotransmitters in the periphery have the same effect in the brain, this is not necessarily the case. Haigler and Aghajanian (1974) have reported that cyproheptadine and methysergide, both of which block the action of serotonin in the periphery and cross the blood–brain barrier, failed to antagonize the effects of 5-HT in the brain.

The observation that morphine is a very efficacious stimulus supports Overton's (1973) hypothesis that drugs that are readily discriminable from the no-drug condition are the drugs most subject to nonmedical self-administration. Overton further suggests that there may even be a causal relationship between the two phenomena. The finding that the rarely abused narcotic-antagonist analgesics, nalorphine and cyclazocine, are also very efficacious stimuli is inconsistent with this hypothesis. One of the most important factors in the determination of the abuse potential of a narcotic is the degree to which it produces morphinelike subjective effects. Since the results of stimulus generalization tests suggest that the degree of similarity of subjectively perceived effects is in general agreement with the degree of similarity of the stimulus properties in animals, it seems reasonable to suggest that a comparison of the stimulus properties of narcotic agents with the stimulus properties of morphine may prove useful for evaluating the potential for abuse of these drugs.

REFERENCES

Beaver, W. T., Wallenstein, S. C., Houde, R. W., & Roger, A. A. A comparison of the analgesic effects of pentazocine and morphine in patients with cancer. *Clin. Pharmacol. Therap.* 1966, *7*, 740–751.

Blumberg, H., Dayton, H. B., George, M., & Rappaport, D. N. N-allyl-noroxymorphine: A potent narcotic antagonist. *Federation Proc.*, 1961, *20*, 311.

Colpaert, F. C., Lal, H., Niemegeers, C. J. E., & Janssen, P. A. J. Investigations on drug produced and subjectively experienced discriminative stimuli-one. *Life Sci.*, 1975, *16*, 705–716. (a)

Colpaert, F. C., Niemegeers, C. J. E., Lal, H., & Janssen, P. A. J. Investigations on drug produced and subjectively experienced discriminative stimuli-two. *Life Sci.*, 1975, *16*, 717–728. (b)

Denton, J., & Beecher, H. K. New Analgesics II. A clinical appraisal of methadone and its isomers. *J. Am. Med. Assoc.*, 1949, *141*, 1146–1148.

Gianutsos, G., & Lal, H. Effect of loperamide, haloperidol and methadone in rats trained to discriminate morphine from saline. *Psychopharmacologia*, 1975, *41*, 267–270.

Haertzen, C. A. Subjective effects of narcotic antagonists cyclazocine and nalorphine on the Addiction Research Center Inventory (ARCI). *Psychopharmacologia*, 1970, *18*, 366–377.

Haigler, H. J., & Aghajanian, G. K. Peripheral serotonin antagonists: failure to antagonize serotonin in brain areas receiving a prominent serotonergic input. *J. Neural Transmission,* 1974, *35,* 257–273.

Hirschhorn, I. D. Pentazocine, cyclazocine, and nalonphine as discriminative stimuli. *Psychopharmacology,* in press.

Hirschhorn, I. D., & Rosecrans, J. A. A comparison of the stimulus effects of morphine and lysergic acid diethylamide. *Pharmacol. Biochem. Behav.,* 1974, *2,* 361–365. (a)

Hirschhorn, I. D., & Rosecrans, J. A. Morphine and Δ^9-tetrahydrocannabinol: Tolerance to the stimulus effects. *Psychopharmacologia,* 1974, *36,* 243–253. (b)

Hirschhorn, I. D., & Rosecrans, J. A. Generalization of morphine and lysergic acid diethylamide (LSD) stimulus properties to narcotic analgesics. *Psychopharmacology,* 1976, *47,* 65–69.

Hirschhorn, I. D., & Winter, J. C. Mescaline and lysergic acid diethylamide (LSD) as discriminative stimuli. *Psychopharmacologia,* 1971, *22,* 64–71.

Houde, R. W., & Wallenstein, S. C. Minutes of the 19th Meeting, Committee on Drug Addiction and Narcotics, App. D., NRC-NAS, 1958.

Isbell, H., Wikler, A., Eisenman, A. J., Daingerfield, M., & Frank, K. Liability of addiction to 6-dimethylamino-4-4-diphenyl-3-neptanone (methadone, "amidone," or "10820") in man. *Arch. Internal Med.,* 1948, *82,* 362–392.

Jaffe, J. H., & Martin, W. R. Narcotic analgesics and antagonists. In L. S. Goodman & A. Gilman (Eds.), *The pharmacological basis of therapeutics.* (5th ed.). New York: Macmillan, 1975.

Keats, A. S., & Telford, J. Studies of analgesic drugs VIII. A narcotic antagonist analgesic without psychotomimetic effects. *J. Pharmacol. Exp. Therap.,* 1964, *143,* 157–164.

Lasagna, L., & Beecher, H. K. The analgesic effectiveness of codeine and meperidine (demerol). *J. Pharmacol.,* 1954, *112,* 306–311. (a)

Lasagna, L., & Beecher, H. K. The analgesic effectiveness of nalorphine and nalorphine-morphine combinations in man. *J. Pharmacol.,* 1954, *112,* 356–363. (b)

Martin, W. R., & Gorodetzky, C. W. Demonstration of tolerance to and physical dependence on n-allylnormorphine (nalorphine). *J. Pharmacol. Exp. Therap.,* 1965, *150,* 437–442.

Overton, D. A. State dependent learning produced by addicting drugs. In S. Fisher & A. M. Freedman (Eds.). *Opiate addiction: Origins and treatment.* Washington, D.C.: V. H. Winston, 1973.

Rosecrans, J. A., Goodloe, M. H., Jr., Bennett, G. J., & Hirschhorn, I. D. Morphine as a discriminative cue: Effects of amine depletors and naloxone. *European J. Pharmacol.,* 1973, *21,* 252–256.

Shannon, H. E., & Holtzman, S. G. Evaluation of the discriminative effects of morphine in the rat. *Federation Proc.,* 1975, *34,* 3207.

Winter, J. C. The stimulus properties of morphine and ethanol. *Psychopharmacologia,* 1975, *44,* 209–214.

11

Attributes of Discriminative Pentobarbital Stimulus Immediately after Intravenous Injection[1]

Edward C. Krimmer and Herbert Barry, III
School of Pharmacy
University of Pittsburgh

Many drugs have been shown to function as discriminative signals through little-understood properties that seem to produce distinctive mental states and control the learning of differential responses (Barry, 1974). The site of this action is presumed to be the central nervous system (CNS), and the changes in the CNS that mediate this phenomenon have been investigated by a variety of methods. Pentobarbital is the prototypical compound for demonstrating the phenomenon (Overton, 1966). The largest number of studies thus far have been with pentobarbital, which has strong discriminative stimulus attributes, generalizing at least partially to other central sedatives (Barry, 1974).

This chapter reviews the research findings for discrimination learning using pentobarbital and attempts to identify equivalence or difference among other agents of the same general pharmacological classification. Questions of general pharmacological effects, applicable to the actions of pentobarbital as a discriminative signal, are effective doses, onset and duration of stimulus attributes, and how these vary or co-vary with other pharmacological actions. The other actions have been induced by tests with CNS stimulants to antagonize the sedative actions and by pretreatments to alter neurotransmitter levels.

[1] Preparation of this paper was supported by U.S. Public Health Service Post-Doctoral Research Fellowship DA-2376 (to E. C. K.), from the National Institute on Drug Abuse, by Research Scientist Development Award K2-MH-5921 (to H. B.), from the National Institute of Mental Health, and by Research Grant MH-13595 (to both authors), from the National Institute of Mental Health.

This chapter also reports some new data on training discriminative responses at a very short interval after intravenous injection of the drug or control condition. This procedure is especially interesting because of the possibility that an immediate effect, experienced by the animals with the intravenous (i.v.) route, may be qualitatively different from the effects produced by the more frequently used routes of administration. Most drug discrimination studies use onset times of various durations, usually 5 min or longer, intended to coincide with the maximal drug effect. Brain levels of the drug have generally peaked or stabilized at these intervals, but the intervals also allow time for the subject to adapt to the drug effect. The intraperitoneal (i.p.) route of administration, used in most studies, requires several minutes for complete absorption. A recent report on drug discrimination after intravenous injection (Ando, 1975) used a 5-min interval, which is several minutes after the initial drug effect is felt.

SURVEY OF PRIOR FINDINGS

A recent review (Barry, 1974) indicates that sedatives comprise a major category of drugs determined to have discriminable properties. In spite of different training techniques or reinforcers, the results are generally in agreement.

The initial study of the discriminative properties of pentobarbital was reported in an abstract by Overton (1961). Rats were trained to make left or right turns for shock escape in a T-maze, depending on whether they had been injected with pentobarbital. This drug has received a large proportion of the subsequent effort from workers in this important area of psychopharmacology, although many other drugs have been tested, especially in conjunction with the rapid increase in drug-discrimination research during the last several years.

Overton (1964) continued to study this phenomenon by training rats to make differential (left or right) turns to escape shock in a T-maze on the basis of a prior i.p. injection of pentobarbital (25 mg/kg) or saline administered 15 min before the test. Two trials were given on each day (1 each in the morning and afternoon), and learning was evident by 10 trials. The high dose (25 mg/kg) used in this study probably contributed to the rapid learning curve. This was substantiated in a second experiment that repeated the task using four new groups trained to discriminate one dose of pentobarbital (25, 20, 15, or 10 mg/kg) from saline. The fifth group received no drug treatment and served as a control to determine whether rats could learn discriminative responses on the basis of the time of day (morning or afternoon) during which they were tested. This fifth group did not acquire differential responding, but the learning time decreased for the other four groups as the dose increased.

This work by Overton (1964) was important not only because it was the first major report in a new area but because it also included an attempt to elucidate the discriminable actions of pentobarbital. The results indicated that in comparing pentobarbital (20 mg/kg) with saline, discrimination learning was slower if the differential signals were sensory stimuli (light versus dark), gallamine (a drug producing muscle flaccidity approximating that of pentobarbital), tetraethylammonium (a drug stimulating ganglia of the autonomic nervous system), or various combinations of food satiation or deprivation.

An obvious question is whether pentobarbital might alter the rat's perception, for example, diminishing shock effect or blurring vision, which would serve as discriminative signals. Overton (1968) addressed this issue by showing that when rats were trained to escape shock in a T-maze using pentobarbital (15 mg/kg) versus nondrug, no difference in learning was found between a group with higher shock level in the pentobarbital condition and another group with higher shock level in the nondrug condition. He also found that blinding rats at various stages of training had no effect on the rate at which they learned to discriminate between the pentobarbital and nondrug condition.

As the drug dose increases, discrimination learning becomes more rapid (Overton, 1964), until limited by ataxia or other impairment of behavioral capability at high doses. Krimmer (1974b) reported on rats trained to discriminate pentobarbital (10 mg/kg) from saline in a food-rewarded task. In a test with the higher dose of 15 mg/kg, a few of the animals were unable to respond, but a majority did respond and with a higher frequency of drug choices than at the training dose of 10 mg/kg. The same dosage increase had a different outcome, however, for animals given similar training in a shock-escape situation, in which all animals could be forced to respond. They showed a decrease in drug choices at 15 mg/kg and a further decrease at 20 mg/kg, toward the random level of 50% drug choices. These results demonstrate the behavioral disruption produced by high doses of pentobarbital. The food-rewarded task has the advantage of allowing the animals to make neither choice if their response capability has been disrupted by the drug.

A progressive decrease in percentage of drug choice has generally been found in tests with pentobarbital doses lower than the training dose. These tests allow an estimation of the mean effective dose (ED_{50}), the dose that elicits the random response of 50% drug and 50% saline choice. Barry (1974) reported ED_{50} doses ranging from 3.5 to 6.5 mg/kg in three studies of rats trained to discriminate 10 mg/kg from the nondrug control condition. These findings suggest that for this drug the threshold between drug and nondrug choice occurs at approximately 50% of the training dose.

Trost and Ferraro (1974), in an elaborately designed food-rewarded task, trained monkeys to discriminate between two doses of pentobarbital (2 and 10 mg/kg). Tests at intermediate doses indicated that the approximate ED_{50} was below the midpoint of 6 mg/kg for each of the 3 monkeys (3, 3.5, and 5.5 mg/kg). The authors estimated the average ED_{50} at 4 mg/kg. This is only at 25% of the span of values from the lower dose (2 mg/kg) to the higher dose (10 mg/kg) used in training. The response to the higher dose therefore generalized to a remarkably low dose when the animals had been trained to discriminate this dose from a lower dose of the same drug. In the usual procedure of discriminating a high dose from the nondrug control condition, the nondrug condition may be more variable, differing in numerous ways from the high dose of a particular drug. The unidimensional discrimination between two doses of the same drug may be a useful technique for narrowing the range of the drug effects on which the discrimination is based, thereby increasing the specificity of the response. This is indicated by a steep dose–response curve for each of the three monkeys. The discrimination was learned on the basis of the quantitative differences between lower and higher doses rather than on the basis of the two specific doses used because the low-dose response was consistently made in tests with lower doses (.5 and 1 mg/kg) and the high-dose response in a test with a higher dose (12 mg/kg).

The drug effect as a discriminative signal apparently persists with a time course similar to other pharmacological actions. Krimmer (1974b) trained rats to discriminate pentobarbital (10 mg/kg, i.p.) from saline with food reinforcement or shock escape. The training sessions were always at 20 min after the drug or saline injection. The drug response was consistently selected in tests at 5 min after injection and usually even at the interval of 2.5 min. Tests at intervals longer than 20 min showed a gradual decrease to a level of about 50% drug response at 60 min after injection.

All studies using drugs as discriminative stimuli are inherently long-term experiments. They involve chronic drug treatment, which in the case of the barbiturates will produce physiological as well as behavioral tolerance. York and Winter (1975a,b) have assessed both the effects of tolerance on a barbital discrimination task and the effects of discriminative barbital training on spontaneous activity. York and Winter (1975a) established discriminative responding by giving food reward for bar pressing in one condition following injection of barbital or saline and withholding food reward in the other condition. They found that repeated daily hypnotic doses of barbital, administered for 8 days after acquisition of the discriminative response, produced a tolerance to the hypnotic action of barbital for all animals and attenuated the discriminative barbital response of animals not reinforced while in the barbital condition. However, the animals that received reinforcement while in the barbital condition continued to exhibit almost normal sensitivity for barbital, suggesting that the animal's sensitivity to the discriminable properties of a drug is influenced by the nature or strength of the reinforcer under that specific drug condition.

In another study, York and Winter (1975b) measured spontaneous activity of rats learning to discriminate barbital (80 mg/kg) from saline. As in the previous study, half the animals received food reward for bar pressing only following barbital, while the other half received food for bar pressing only after saline. The amount of spontaneous activity, measured in an activity chamber prior to the discrimination training, decreased when the rats were in the condition (drug or saline) associated with no food reinforcement. This phenomenon, which developed only after several months of training, indicates that the drug effect gradually became a conditional signal for the differential food and nonfood conditions. Earlier in the training, barbital increased spontaneous activity regardless of the drug-discrimination condition.

Various studies have tested the generality or specificity of the discriminative response by administering a novel drug condition in animals already trained to discriminate a drug from the control condition. Responses trained with pentobarbital generalize to other barbiturates, like secobarbital, amobarbital, barbital or phenobarbital (Overton, 1966); the nonbarbiturate hypnotics, chloral hydrate, paraldehyde, chlorolase, ethyl-carbamate, *t*-butanol, and meprobamate (Barry & Krimmer, 1972; Krimmer 1974; Overton, 1966); and the anxiolytic benzodiazepines, chlordiazepoxide and diazepam (Barry & Krimmer, 1972; Krimmer, 1974b; Overton, 1966). Drugs that do not elicit the pentobarbital-trained discriminative response include *d*-amphetamine, bemegride, gallamine, LSD, physostigmine, atropine, and β-erythroidine (Overton, 1966), chlorpromazine, (Barry & Krimmer, 1972; Overton, 1966); LSD and mescaline (Hirschhorn & Winter, 1975), tetrabenazine, hydroxyzine, and caffeine (Krimmer, 1974b).

Ambiguous results have been found when animals, trained to discriminate pentobarbital from the nondrug condition, are tested with ethyl alcohol as the novel condition. Overton (1966) reported that the drug response was elicited by ethyl alcohol (2400 mg/kg) when administered to animals trained to discriminate pentobarbital (20 mg/kg) from saline. Contrary to this generalization, Barry and Krimmer (1972) found that the saline response was elicited by alcohol (600 mg/kg) in animals trained to discriminate pentobarbital (10 mg/kg) from saline in a single lever-pressing task with food approach and shock avoidance learned under the alternative conditions. Moreover, in tests with higher alcohol doses (750–1200 mg/kg) the animals experienced behavioral toxicity indicated by nonresponding regardless of whether the approach response was associated with the drug or saline condition. Similarly, Krimmer (1974b) found that in animals trained to discriminate pentobarbital (10 mg/kg) from saline by pressing the correct one of two levers for food reward, tests with alcohol doses up to 1500 mg/kg elicited the saline response. Krimmer (1974b) also found that alcohol (1000–1500 mg/kg) did elicit the drug response in animals trained with pentobarbital (10 mg/kg) and saline to escape from shock by pressing the correct one of two levers; an ED_{50} of 790 mg/kg for alcohol was calculated for this task. The generalization from pentobarbital to alcohol, found in the shock-escape situations of Krimmer (1974b) and Overton (1966), suggests that a shock-escape task emphasizes the common properties of pentobarbital and alcohol.

The stimulus properties of pentobarbital and alcohol have been compared directly in a study by Krimmer and Barry (1973) in which animals were trained to discriminate between pentobarbital (10 mg/kg, i.p.) and alcohol (1000 mg/kg, i.p.) using a two-lever food-reward task, and in a study by Krimmer (1974) with the same dose of pentobarbital but a higher dose of alcohol (1500 mg/kg), using a two-lever shock-escape task. The discrimination between these similar drugs was learned by both groups, but the higher dose of alcohol appeared to be slightly more difficult to discriminate from pentobarbital. In both groups, the animals made the appropriate drug response when tested with higher doses of either training drug, indicating the differences in stimulus properties to be qualitative rather than quantitative.

The group trained to discriminate pentobarbital from the lower alcohol dose made the alcohol response when tested with various doses of chloral hydrate, paraldehyde, *t*-butanol, tetrabenazine, and chlorpromazine. Both groups made the pentobarbital response in tests with the benzodiazepines (chlordiazepoxide and diazepam).

Involvement of pharmacologically depressant actions in the discriminative pentobarbital signal is indicated by the observation that the CNS stimulant bemegride, in sufficient doses, can antagonize the discriminative pentobarbital response in rats (Overton, 1966). Pentylenetetrazol to a lesser extent antagonized the pentobarbital stimulus, while picrotoxin had little effect, and amphetamine none, on the pentobarbital stimulus (Krimmer, 1974a). Recent experiments by Johansson and Jarbe (1975) and Jarbe, Johansson and Henriksson (1975) have extended the work with pentobarbital discrimination to a new species, the gerbil. These authors have reported that bemegride and to a lesser extent pentylenetetrazol (but not amphetamine) will antagonize the discriminative properties of pentobarbital in gerbils (Johansson & Jarbe, 1975). Effective antagonism of the discriminative-stimulus properties of pentobarbital by bemegride was also apparent when gerbils failed to learn to discriminate the combined effect of these two drugs from saline (Johansson & Jarbe, 1975).

The depressant action apparently is a major component of the discriminative-stimulus complex of the barbiturates, but there still remains a high degree of stimulus specificity. This is evidenced by the failure of a discriminative pentobarbital response to generalize to Δ^9-THC (Jarbe *et al.*, 1975) or two dissociative anesthetics, ketamine and phencyclidine (Overton, 1975; Jarbe *et al.*, 1975).

To date, very little work has been done with neurotransmitters and purported transmitter substances to test for their involvement in drug discrimination, especially in the area of central sedatives. Rosecrans, Goodloe, Bennet, and Hirschhorn (1973), in a study dealing primarily with morphine, reported that treatment with *p*-CPA or α-MPT, which reduce brain levels of 5-hydroxytryptamine or norepinephrine, respectively, did not disrupt the discrimination of pentobarbital (15 mg/kg) from saline.

PURPOSES OF THE PRESENT STUDY

Pentobarbital was selected as the discriminative stimulus because of the extensive background information accumulated on this drug by ourselves and by others. Several features of this study are intended to provide unique and useful information.

Training and test sessions were begun within 15 sec after i.v. infusion. A chronically implanted cannula in the jugular vein of each animal enabled rapid infusion with minimal difficulty or disturbance. This unusual procedure tested attributes of the discriminative drug stimulus immediately after sudden onset of the drug effect. Remarkably rapid learning of the discrimination might be expected if the discriminative drug stimuli are similar to conventional sensory stimuli. With the more usual visual signals for discrimination learning, rapid adaptation (Woodworth & Schlosberg, 1954, p. 270) diminishes the effectiveness of the signal if the differential response occurs more than a few seconds after onset or other change of the signal.

This study was also designed so that reponse latencies both before shock onset (avoidances) and after onset (escapes) could be determined in both the drug and nondrug conditions. These data might reflect the depressant action of pentobarbital and indicate whether it changed during the course of discriminative learning.

A further novel feature of the study was that the discrimination consisted of differential active and passive shock-avoidance responses in a two-compartment box. This technique was designed to facilitate the learning because the responses are qualitatively different, easily trained, and do not require food deprivation.

Some aspects of these procedures and results have been reported in a recent article by Krimmer and Barry (1976). This chapter gives further detail and reports for the first time on the response latencies and on performance in the second and third trials of each session.

METHODS

Apparatus

A standard shuttle box (BRS/LVE model 164-04) with transparent Plexiglas side walls and ceiling, metal end walls, and a stainless steel grid floor was used throughout the study. The box, 45.4 cm long, 20.4 cm wide, and 19.7 cm high, was divided into two compartments, both 22.7 cm long, by a metal partition 4.3 cm high. One compartment was made opaque by attaching black paper to the outside of the Plexiglas walls and ceiling. This was intended to enhance the discriminability between the two compartments. Scrambled shock (200 V, ac) through a 150,000-Ω series resistor could be applied to one preselected compartment. Programming was through electromechanical equipment that also recorded avoidance and escape latencies.

Subjects and Maintenance

Male Wistar rats from Hilltop Lab Animals, Inc., Scottdale, Pennsylvania, were housed singly with food and water continuously available in the home cage. The room temperature was controlled at approximately 21° C, and the daily cycle of 12 hr lighted and 12 hr dark was automatically regulated. All testing was conducted during the lighted cycle of the day. The procedures for an initial group of four animals were replicated with 6 additional animals 2 months later. The data combined these 2 groups, omitting 1 animal that died after 26 training sessions.

Cannula Construction and Surgical Procedures

Cannulas were constructed from 13.5-cm segments of Silastic tubing (.64 mm inside diameter × 1.20 mm outside diameter, Dow-Corning Corp.). The end of the cannula, to be inserted in the vein, was cemented shut (Silastic cement). After the cement dried, clusters of small punctures through the Silastic tubing were made for a distance of approximately 4 mm above the sealed end, allowing infusion of fluids while eliminating back flow or the need for heparin (Krimmer & Barry, 1976). A narrow ring of Silastic cement, 3.5 cm from the sealed end, enlarged the cannula to approximately 1.5 mm outside diameter. A 3-mm collar of Silastic tubing (.70 mm inside diameter × 1.65 mm outside diameter) was cemented 4.5 cm from the sealed end. A piece of needle tubing (2 cm × 22 gauge) was cemented 1 cm into the open end of the cannula.

Approximately 24 hr before surgery, the animals were treated prophylactically with penicillin 200,000 units. Surgery was performed under pentobarbital (60 mg/kg i.p.) and followed procedures described by Krimmer and Barry (1976). The cannula was implanted chronically in the external jugular vein on the right side for half the animals and on the left side for the others. The cemented end of the cannula was inserted down the vein toward the heart as far as the collar permitted and a surgical tie made around the vein midway between the collar and cement enlargement. The open end of the cannula (with the needle tubing) was passed subcutaneously to the back of the neck region between the ears and anchored with the bare portion of the needle protruding through the skin.

Training and Testing

Discriminative response training in the shuttle box began 1 or 2 days after surgery. The animals were always placed in the transparent compartment facing away from the other compartment. The discriminative response was to avoid painful electric shock to the grid floor by an active response (being in the opaque compartment 5 sec later) under one condition (5 mg/kg pentobarbital or isotonic saline) or a passive response (being in the transparent compartment 5 sec later) under the alternative condition. The 10 animals were divided into 2 subgroups of 5 each in which the same drug condition was associated with the opposite active- or passive-avoidance responses. Injections were in a volume of .25 ml/kg for saline and all drug conditions except ethyl alcohol, which was in a 50% wt/vol solution so that the higher dose (500 mg/kg) required 1 ml/kg. All injections, including saline, were followed by a .15-ml flush of saline.

The injection procedure was to place the animal on a table while connecting a syringe through a short piece of cannula to the needle tubing protruding from the animal's scalp. The animal was disconnected and placed into the test box within 10 sec after drug injection and the saline flush. The weight of the animal on the floor activated a 5-sec timer, and the animal was removed at the end of 5 sec if it had made the appropriate avoidance response (active or passive). In the absence of an avoidance response, shock was applied to the occupied compartment until the animal moved to the other compartment.

The animal received further training in 2 additional trials, both at an interval of 10–15 sec after the preceding trial. During the intertrial interval the animal was retained in a nearby cage.

At least 4 hours later on the same day, a second session of 3 trials was given, divided equally between the pentobarbital and saline conditions for each animal. The sessions were conducted on 5 consecutive days per week. Tests under novel conditions were conducted exactly as training sessions except that only 1 trial was given, always during the first session of the day and without shock. At least 1 training session always intervened between successive tests. The animals were treated in a random sequence that was determined each day.

The drug and saline conditions in successive training sessions for the same animal were randomized except that there were 5 drug and 5 saline sessions in each group of 10 sessions and the saline or drug condition was always reversed after 2 consecutive sessions.

RESULTS

Successful acquisition of the discriminative response is indicated by the fact that each of the 9 animals reached Overton's (1969) criterion of 8 correct choices in 10 successive sessions on the first of 3 trials per day. The mean number of sessions to the start of the 10-trial criterion normalized by logarithmic transformation before calculations (Overton, 1975) was 5.7 sessions.

Figure 11.1 shows the average percentage correct responding for each of the 3 trials during the first 24 sessions and separately during the remaining 20 sessions for the drug and saline condition of the two subgroups (drug active, saline passive and drug passive, saline active). Performance on the first trial of each session was affected primarily by the animal's drug or saline condition. Averaging all 44 sessions, first-trial performance was appreciably better in the saline than drug condition, regardless of whether the response was active or passive. This difference was statistically reliable ($t = 9.36$, $df = 8$, $p < .05$). In the two upper graphs of Figure 11.1, the first-trial percentage correct was reliably higher for the drug-active than saline-passive response in Sessions 1–

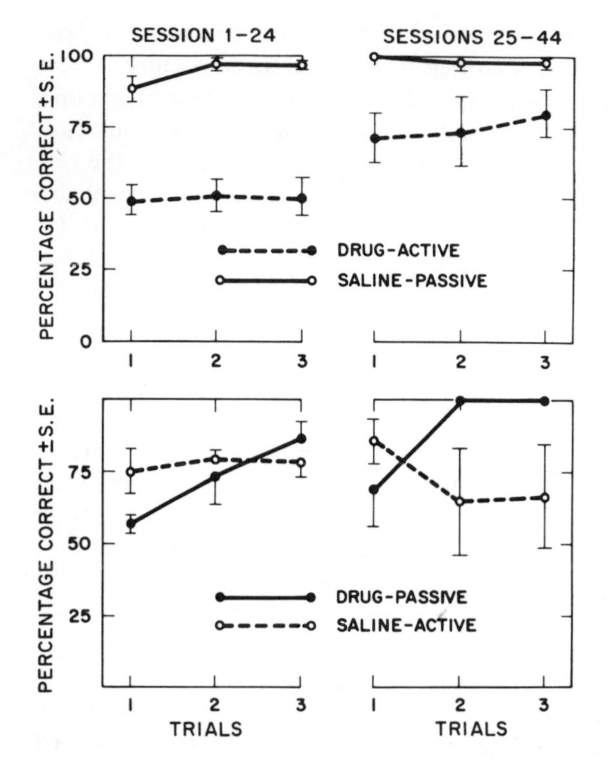

Figure 11.1. Average percentage correct (±*S.E.*) responding in drug sessions (filled circles) and saline sessions (open circles) for each of three trials, in Sessions 1–24 and 25–44. The upper graphs show the five drug-active, saline-passive animals; the lower graphs show the four drug-passive, saline-active animals.

24 ($t = 5.66$, $df = 4$, $p < .01$) and in Sessions 25–44 ($t = 3.33$, $df = 3$, $p < .05$).

On the second and third trials of each session, the type of response was more influential than the drug condition. Performance on these later trials of the same session improved substantially only when the passive response was required. This improvement was greater for the group required to make the passive response when in the drug condition, probably because the saline-passive group was close to 100% correct responses in the saline condition even on the first trial. In Sessions 25–44, the percentage correct saline-active responses actually declined after the first trial. When the 2 subgroups are combined, the average percentage correct response was higher under the drug than saline condition on Trials 2–3 in Sessions 25–44, contrary to the results for the first trial.

Each animal's latencies for completing the active response before shock onset (avoidances) and after shock onset (escapes) were averaged separately for the 3 trials in the same training session. These scores, averaged separately for Sessions 1–24 and 25–44, are graphically presented in Figure 11.2 for the drug-active and saline-active subgroups.

The drug-active subgroup completed the avoidance response more slowly. In sessions 1–24, this difference persisted throughout all three trials of the same session but was statistically significant only on the second trial ($t = 3.48$, $df = 7$, $p < .05$). In Sessions 25–44, the difference was much smaller and not statistically significant. The lower portion of Figure 11.2 shows that the speed of escape response was consistently slower by the drug-active subgroup only in Sessions 1–24. The difference was statistically significant on Trial 2 ($t = 3.88$, $df = 7$, $p < .01$). Both subgroups showed much longer escape latencies on the first trial of Sessions 1–24 than on any of the 3 trials of Sessions 25–44; the saline-active but not the drug-active subgroup showed greatly decreased escape latencies from Trial 1 to Trials 2–3 in Sessions 1–24.

Table 11.1 shows the average drug response made by the drug active, drug passive, and both subgroups combined in tests with novel drug conditions administered intravenously. Progressively increasing doses of pentobarbital (1.25–4.5 mg/kg) generally increased the percentage of drug responses. The maximum 100% drug response was at a dose (4.5 mg/kg) slightly below the training dose of 5 mg/kg. A disinhibitory effect of low doses of the drug is indicated by the substantially higher percentage of drug-active than drug-passive responses in the tests with 1.25, 2.5, and 3.75 mg/kg. The training conditions (0 and 5 mg/kg) did not show this preponderance of active over passive responses.

Table 11.1 also shows that chlordiazepoxide and, to a lesser degree, alcohol elicited increasing percentage drug responses in tests with progressively higher

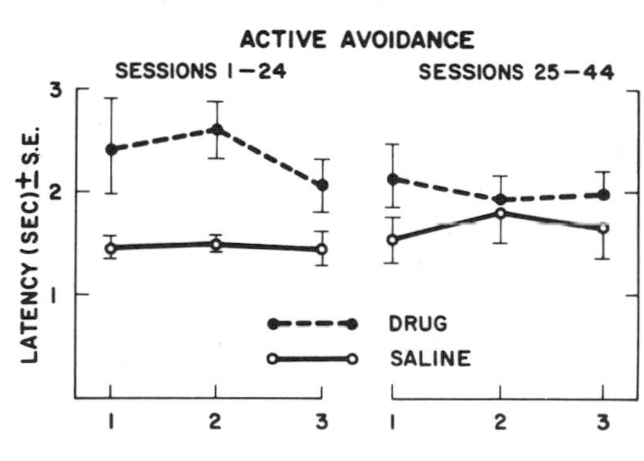

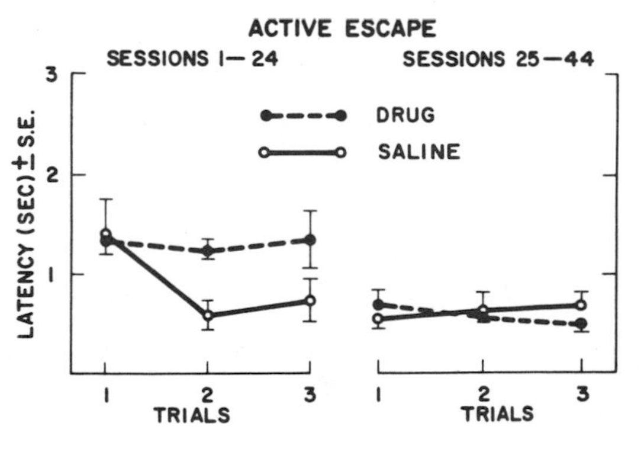

Figure 11.2. Average active avoidance latencies (±*S.E.*), upper graphs, and escape latencies (±*S.E.*), lower graphs, for each of three trials, in Sessions 1–24 and 25–44. Filled circles indicate the five drug-active animals, and open circles indicate the four saline-active animals.

Table 11.1
Mean Percentage Drug Response under the Training Conditions (0 and 5 mg/kg Pentobarbital) and under Novel Test Conditions .25 Min after Intravenous Infusion

	Drug active (N = 5)	Drug passive (N = 4)	Both groups (N = 9)
Pentobarbital (mg/kg)			
0	0	14	6
1.25	40	25	33
2.5	40	12	28
3.75	100	50	78
4.5	100	100	100
5	71	69	71
Chlordiazepoxide (mg/kg)			
1.25	100[a]	0[b]	40
2.5	60	75	67
5	100	75	89
Alcohol (mg/kg)			
125	40	25	33
500	60	50	56

[a] $N = 2$.
[b] $N = 3$.

doses of both drugs. Four of these five dosage conditions show a higher percentage of drug-active than drug-passive responses, indicating disinhibitory drug effects.

Barry (1974) has described a simplified method for calculating the dose of drug that elicits 50% drug response (ED_{50}). This method yields ED_{50} scores of 3 mg/kg for pentobarbital, 1.6 mg/kg for chlordiazepoxide, and 348 mg/kg for alcohol.

Table 11.2 shows the average drug response, again for the 2 subgroups separately and combined, when the training dose of pentobarbital (5 mg/kg) was tested at times longer than the usual interval of .25 min. The stimulus effect of pentobarbital was not consistently diminished below the training level as long as 7.5 min after infusion. At 15 min, the drug response was below 50% but remained higher than at .25 min after saline (Table 11.1). There was no

Table 11.2
Mean Percentage Drug Response at the Training Interval (.25 min) and at Longer Intervals after Intravenous Infusion of Pentobarbital (5 mg/kg)

Drug-test interval (min)	Drug active (N = 5)	Drug passive (N = 4)	Both groups (N = 9)
.25	71	69	71
1	80	75	78
2	60	75	67
7.5	80	50	67
15	20	50	33

consistent predominance of drug-active over drug-passive responses under these conditions.

In addition to the tests after intravenous infusion, shown in Tables 11.1 and 11.2, the animals were given 2 tests at 5 min after i.p. injection. The saline injection elicited 40% drug-active and 25% drug-passive responses (33% for both groups combined). A high dose of pentobarbital (10 mg/kg) elicited 80% drug-active and 75% drug-passive responses (78% for both groups combined).

DISCUSSION

Numerous studies have shown that barbiturates are discriminable from the control condition, but with the prevalently used i.p. route, a substantial time interval, usually 5 min or longer, is necessary for discriminable drug effects (Krimmer, 1974b). The procedures reported in this chapter were developed to investigate the discriminable attributes of pentobarbital immediately after i.v. administration and to compare this immediate effect with the discriminable attributes after i.p. administration. The results, therefore, have methodological value as well as theoretical significance.

The data demonstrate that the i.v. route of administration and short-onset interval are effective conditions for discriminative learning with pentobarbital in the rat. Subsequent tests with the training drug indicate that the discriminable properties of pentobarbital are qualitatively similar for the i.v. and i.p. routes but, as expected, differ quantitatively. The qualitative similarities between the i.v. and i.p. routes suggest that the advantages of lower doses and shorter time intervals, using the i.v. route of administration, are outweighed by the disadvantages of the deleterious effects of the surgical intervention and nonfunctional cannulas on the longevity of the study.

The results of this study, however, have several important theoretical implications. The apparent lack of any major advantage for the short onset period supports a suggestion by Overton (1974) that drug-discrimination learning differs from the more usual discrimination learning, in which the response generally occurs within a few seconds after the onset or change of a visual, auditory, or other sensory signal. The critical difference may be that drugs produce more pervasive, generalized changes in the CNS.

Overton (1973) reviewed evidence that the discriminable properties of drugs may be the same as the attributes that lead to self-administration. Presumably, these are the reinforcing effects of drugs that are typically effective only with immediate, sudden onset following intravenous infusion. Contrary to these effects, the discriminable pentobarbital signal at the short interval after i.v. infusion was qualitatively similar to the signal at the longer interval after i.p. injection and apparently persisted throughout the duration of the pharmacological action of the drug.

The present study used a novel procedure for discriminative training. Differential active and passive shock-avoidance responses were learned in a two-compartment box. Previous methods have used locomotor choice responses or an operant choice between two levers. The types of responses have usually been equivalent under both drug conditions, while only the direction of the response is different (Barry, 1974). The advantages and disadvantages of training qualitatively different responses rather than equivalent alternatives have been discussed by Kubena and Barry (1969a,b). They used differential approach and avoidance responses in an operant conditioning lever-pressing situation.

The present technique allowed rapid training of the differential active- and passive-avoidance responses. There was no need to train a lever-pressing response or to maintain the animals on a food-deprivation schedule. Although the incentive (shock avoidance) was the same under both the drug and saline conditions, the qualitatively different types of response (active and passive) may have facilitated the discrimination learning.

The lower percentage of correct responses under the drug than saline condition on the first trial of the session, shown in Figure 11.1, may indicate a behaviorally toxic effect of pentobarbital at 15 sec after i.v. infusion. This is suggested by the optimal discrimination at a dose of 4.5 mg/kg, slightly lower than the training dose of 5 mg/kg (Table 11.1). Comparison between the two counterbalanced subgroups (drug active, saline passive and drug passive, saline active), shown in Figure 11.1, indicates that the drug had a greater detrimental effect when an active- rather than passive-avoidance response was required. This would be expected with an excessively high dose of the hypnotic-sedative agent.

In the second and third trials, a general preference for passive rather than active avoidance was manifested (Figure 11.1). This may be because of the painful shocks received in both compartments. The tendency for a passive freezing or crouching response was apparently countered in the first trial, probably because of the short interval after the handling and perhaps because of the physical sensations associated with the intravenous saline or drug infusion. The second and third trials were at longer intervals. A strong preference for the passive-avoidance response has been found in animals trained with similar procedures at intervals of 10–20 min after i.p. injection, in some of the same animals in a subsequent stage (Krimmer & Barry, 1976), and also in a prior unpublished experiment.

Active avoidance and escape latencies, shown in Figure 11.2, indicate that in Sessions 1–24 the behavioral impairment caused by pentobarbital persisted throughout the 3 trials. After the first trial, escape latencies decreased greatly in the saline but not pentobarbital condition. During the later stage (Sessions 25–44), the similar latencies under the drug and saline conditions suggest that tolerance developed to the disruptive effects of pentobarbital. This is probably attributable to a behavioral compensatory response in this situation rather than a general weaker response to the drug because the percentage of correct discriminative responses under the drug was higher in sessions 25–44 than in Sessions 1–24 (Figure 11.1).

Combining the 2 subgroups, Table 11.1 shows consistent dosage functions and Table 11.2 shows consistent time functions. These results agree with corresponding tests in animals trained at 10–20 minutes after i.p. injection (Barry, 1974; Krimmer & Barry, 1976). In the present study, with training at a short time after i.v. infusion, the drug response was elicited by lower doses, and the duration of the drug effect was shorter. These are the differences to be expected from the pharmacological effects of the i.v. compared with the i.p. route of administration.

Another aspect of the data shown in Table 11.1 is a disinhibitory effect of the low doses of pentobarbital and of the doses of chlordiazepoxide and alcohol that were tested. This is in accordance with the effects of pentobarbital, chlordiazepoxide, and several other sedative drugs reported by Barry and Krimmer (1972) in rats trained by differential food-approach and shock-avoidance responses under the discriminative pentobarbital (10 mg/kg, i.p.) and saline conditions. A discrepancy is the tendency for a disinhibitory effect of alcohol in the present experiment. The locomotor response may have been more compatible with this effect of alcohol than the lever-pressing response used by Barry and Krimmer (1972).

REFERENCES

Ando, K. The discriminative control of operant behavior by intravenous administration of drugs in rats. *Psychopharmacologia,* 1975, *45,* 47–50.

Barry, H., III. Classification of drugs according to their discriminable effects in rats. *Federation Proc.,* 1974, *33,* 1814–1824.

Barry, H., III, & Krimmer, E. C. Pentobarbital effects perceived by rats as resembling several other depressants but not alcohol. *Proc. 80th Annual Convention, Amer. Psychol. Assoc.,* 1972, *7,* 849–850.

Hirschhorn, I. D., & Winter, J. C. Differences in the stimulus properties of barbital and hallucinogens. *Pharmacol. Biochem. Behav.,* 1975, *3,* 343–347.

Jarbe, T. U. C., Johansson, J. O., & Henriksson, B. G. Δ^9-Tetrahydrocannabinol and pentobarbital as discriminative cues in the Mongolian gerbil (*Meriones unguiculatus*). *Pharmacol. Biochem. Behav.,* 1975, *3,* 403–410.

Johansson, J. O., & Jarbe, T. U. C. Antagonism of pentobarbital induced discrimination in the gerbil. *Psychopharmacologia,* 1975, *41,* 225–228.

Krimmer, E. C. Selective antagonism of the discriminable properties of pentobarbital by several stimulants. *Federation Proc.,* 1974, *33,* 550. (a)

Krimmer, E. C. Drugs as discriminative stimuli. *Disser. Abstr. Internat.,* 1974, *35,* 4572-B. (b)

Krimmer, E. C., & Barry, H., III. Differential stimulus characteristics of alcohol and pentobarbital in rats. *Proc. 81st Annual Convention, Amer. Psychol. Assoc.,* 1973, *8,* 1005–1006.

Krimmer, E. C., & Barry, H., III. Discriminative pentobarbital stimulus in rats immediately after intravenous administration. *European J. Pharmacol.,* 1976, *38,* 321–327.

Kubena, R. K., & Barry, H., III. Two procedures for training differential responses in alcohol and nondrug conditions. *J. Pharm. Sci.,* 1969, *58,* 99–101. (a)

Kubena, R. K., & Barry, H., III. Generalization by rats of alcohol and atropine stimulus characteristics to other drugs. *Psychopharmacologia,* 1969, *15,* 196–206. (b)

Overton, D. A. Discriminative behavior based on the presence or absence of drug effects. *Am. Psychologist,* 1961, *16,* 453–454.

Overton, D. A. State-dependent or "dissociated" learning produced with pentobarbital. *J. Comp. Physiol. Psychol.,* 1964, *57,* 3–12.

Overton, D. A. State-dependent learning produced by depressant and atropine-like drugs. *Psychopharmacologia,* 1966, *10,* 6–31.

Overton, D. A. Visual cues and shock sensitivity in the control of T-maze choice by drug conditions. *J. Comp. Physiol. Psychol.,* 1968, *66,* 216–219.

Overton, D. A. Control of T-maze choice by nicotinic, antinicotinic, and antimuscarinic drugs. *Proc. 77th Annual Convention,* Amer. Psychol. Assoc., 1969, *4,* 869–870.

Overton, D. A. State-dependent learning produced by addicting drugs. In S. Fisher & A. M. Freedman (Eds.), *Opiate addiction: Origins and treatment.* Washington, D.C.: V. H. Winston, 1973. Pp. 61–75.

Overton, D. A. Experimental methods for the study of state-dependent learning. *Federation Proc.,* 1974, *33,* 1800–1813.

Overton, D. A. A comparison of the discriminable CNS effects of ketamine, phencyclidine and pentobarbital. *Arch. Intern. Pharmacodyn.,* 1975, *215,* 180–189.

Rosecrans, J. A., Goodloe, J. R., Bennett, G. J., & Hirschhorn, I. D. Morphine as a discriminative cue: Effects of amine depletors and naloxone. *European J. Pharmacol.,* 1973, *21,* 252–256.

Trost, J. G., & Ferraro, D. P. Discrimination and generalization of drug stimuli in monkeys. In J. M. Singh & H. Lal (Eds.), *Drug addiction.* Vol. 3. *Neurobiology and influences on behavior.* New York: Stratton Intercontinental Medical Book Corp., 1974. Pp. 223–239.

Woodworth, R. S., & Schlosberg, H. *Experimental Psychology* (rev. ed.), New York: Holt, 1954.

York, J. L., & Winter, J. C. Assessment of tolerance to barbital by means of drug discrimination procedures. *Psychopharmacologia,* 1975, *42,* 283–287. (a)

York, J. L., & Winter, J. C. Long-term effects of barbital on spontaneous activity of rats trained to use the drug as a discriminative stimulus. *Psychopharmacologia,* 1975, *42,* 47–50. (b)

II

Research Methods and New Techniques

12

Experimental Design and Data Analysis in Studies of Drug Discrimination: Some General Considerations

Ronald L. Hayes
Neurobiological and Anesthesiological Branch
National Institute of Dental Research

Two general experimental approaches characterize pharmacological evaluations of the discriminative cue properties of drugs: (*1*) the dose–response assessment of the ability of a compound to exercise discriminative control over behavior; (*2*) the determination of whether or not the qualitative aspects of the drug-produced or interoceptive cue fall within a known pharmacological class; that is, does the behavior under stimulus control of the drug generalize to cues that previously have been shown to be common to one class of drugs (e.g., barbiturates) but to no other? The purposes of these studies could include the definition of the mechanism or mechanisms that mediate the psychoactive effects of the drug. Data from these efforts could also advance our understanding of the physiological determinants of consciousness and its alteration and thereby provide insights into processes that underlie certain psychopathologies. A more immediately pragmatic goal would be the implementation of improved screening techniques useful in the development of new therapeutic agents or the evaluation of their potential abuse liability.

While the specific goals of these studies as well as details of their execution may vary, an essential feature of all such experiments is an appraisal of the discriminative control of behavior by drugs. Thus, a discussion of their design and analysis is largely reducible to a set of considerations generally applicable to all these procedures. There is no intent, then, to provide here a broad knowledge of experimental design, and the assumption is made that the reader has some background in the evaluation of behavioral experiments. Rather, this chapter will briefly review some representative types of operant behavioral paradigms found in the literature, followed by an elaboration of problems common to the design and analysis of any experiment to study the properties of drug-produced discriminative stimuli.

DESIGNS USED IN DRUG DISCRIMINATION STUDIES

One of the simplest and most commonly used tasks in drug discrimination studies is the 2-bar operant task (Hirschhorn, Hayes, & Rosecrans, 1975). Typically, rats are taught to discriminate between the presence and absence of a drug by the individual presentation of each of these 2 conditions in a double-alternation sequence; that is, 2 days of 15-min sessions of positive-reinforcement availability on only 1 bar paired with drug administrations, followed by 2 days of positive reinforcement on only the opposite bar paired with the omission of drug administration. Early in training, animals are continuously reinforced, but the schedule is gradually made leaner by the introduction of stricter fixed-ratio or variable-interval contingencies. Simultaneously, an unreinforced period is introduced at the beginning of the session, and this is slowly extended to a duration of about 2 min. Responding during the initial unreinforced period, both in the presence and absence of the drug state, is obviously unconfounded with feedback to the animal about the appropriateness of its responses. Therefore, these responses constitute a basic datum by which the stimulus control of the drug can be evaluated.

Aside from two-bar tasks employing only positive reinforcers, analogous procedures employing aversive motivators in a Sidman avoidance task have been reported (Shannon & Hoffman, 1975). Rats can learn to press one bar with sufficient frequency to avoid unsignaled electric shock in a drug state and to press the opposite bar to avoid shock in the absence of the drug.

Another common behavioral task employs the T-maze in which rats are trained to escape from shock (Overton, 1964). On each trial, a rat is dropped into the maze with the shock already on, and the rat must choose the one goal box in which the grid floor is unshocked. Multiple trials are given daily under either a drug or undrugged condition, and each state is always associated with a given correct response; for example, a right turn, or choice of black goal box, is required under one drug state, and a left turn, or white box, is correct in the other state. Stimulus conditions are usually presented on alternate days, and only choices on the first trial of each day of training are used to evaluate the stimulus control of the drug. Similar procedures have been successfully used with food (Barry, Koepfer, & Lutch, 1965) or shock avoidance (Bindra & Reichert, 1966) are reinforcers.

A third category of drug-discrimination tasks is characterized by the pairing of different reinforcement schedules with different drug states (Harris & Balster, 1971). In the simplest form of this design, 1 food-reinforced bar-press schedule (e.g., DRL 20) is paired exclusively with drug administration, and a second schedule (e.g., FR 50) effective on the same bar is associated only with saline injections. Drug conditions are alternated randomly on successive days of 1-hr sessions. Since each schedule results in a characteristic response patterning even in extinction, tests of responding in each drug state but in the absence of reinforcement are used to assess stimulus control. Data are presented in the form of cumulative response records.

A fourth category, perhaps more correctly subsumed under the third, pairs an active shuttle-box avoidance requirement with one drug state and a passive-avoidance contingency with the other (Holmgren, 1964). In these tests, the animal is required in one drug state to alternate between the two sides of a box at a frequency sufficient to avoid unsignaled shock (active avoidance). In the other state, this alternation results in shock presentation, but the failure to alternate is unpunished (passive avoidance).

The preceding outline of operant behavioral procedures used in discrimination is not intended to be definitive. Indeed, if the reader recognizes that within certain limitations drug-produced or interoceptive stimuli are procedurally interchangeable with exteroceptive or environmental cues in discrimination tasks, it becomes obvious that there are a great number of possible paradigms. Parameters associated with exteroceptive cues such as duration and intensity are generally more manipulatable than drug-associated cues, and these considerations will impose the primary constraints on experimental designs.

In selecting a design, the experimenter should first ask which task most directly and simply addresses the issue being studied. If one wishes to assess only the potential ability of a compound to exert stimulus control over a task, a procedure incorporating different behaviors, reinforcement schedules, and/or motivational conditions in a single study is obviously confounded by these variations. Thus, control procedures and interpretation of data in such a study are unnecessarily complex. Of course, one would ideally like to examine the behavioral generality of a phenomenon across different experimental situations. It is important, however, to appreciate that an experimenter is asking a qualitatively and significantly different question when employing an aversive rather than an appetitive reinforcer, and this issue is even more critical when such variations as these occur within the same experiment. Other important criteria that dictate the choice of designs should be economy in training and run-time. More important still is the directness and simplicity with which discrimination data can be objectively quantified, made easily interpretable, and presented in a format that allows ready comparison with data from other discrimination tasks.

EVALUATION OF DISCRIMINATION PERFORMANCE

Regardless of the design employed, the problem of evaluating the degree of stimulus control exerted by a given treatment may be reduced to two simple and related questions: Did the animal detect or fail to detect the relevant cue? Are there possible nonsensory (e.g., attentional or motivational) variables influencing the animal's discrimination performance? While this chapter does not pretend to offer an entirely unequivocal resolution to even these questions, what is proposed here is a uniquely rigorous effort to deal with these issues. In the following discussion, two independent, quantifiable variables will be presented. One, D′, may be viewed as an index of the appropriateness of the animal's responding and, at asymptotic levels of training, related exclusively to the physical discriminability of the stimulus and the ability of the animal's sensorium to detect it. The other, response bias (RB), represents the possible nonsensory contribution to the animal's discrimination behavior and reflects generalized predispositions by the subject to respond in one mode or another. That is, RB simply quantifies a phenomenon familiar to all students of animal learning—the tendency of many subjects to respond preferentially by turning left instead of right in a maze or by pressing the left bar rather than the right one in an operant chamber, independent of whether or not the response is correct and will be reinforced. However, as I shall point out later, not all changes in RB need be the result of strictly behavioral changes in response bias. In the specialized context discussed here, RB can also be mediated by pharmacologically induced changes in the discriminative stimulus.

Many readers will recognize similarities between the analyses outlined here and methods associated with the Theory of signal Detection (TSD) (Swets, Tanner, & Birdsall, 1961). The signal-detection model attempts to dichotomize variables that contribute to the detection of a stimulus. One variable, d′, is said to be related only to the physical discriminability of the stimulus or signal and the sensitivity of the organism's sensory apparatus. The other variable, β, is related exclusively to nonsensory variables and is affected by such things as changes in the reward associated with the detection of a stimulus. Signal Detection Theory, then, attempts to account quantitatively for the well-known phenomenon that nonsensory variables can influence whether or not a subject reports the presence of a stimulus. The following discussion is the result of a conscious attempt to gain some of the rigorous inferential advantages associated with the theory without having to comply with the details of its procedural requirments. The formal benefits of the TSD are impressive, however, and further efforts to make discrimination studies consistent with its methodology should be encouraged.

As suggested earlier, an animal can indicate that it detects or fails to detect a stimulus, and the stimulus can actually be present or absent. A factorial combination of these possibilities yields four outcomes: (*1*) Animal responds as if stimulus were present, and it is present (a "hit"); (*2*) animal responds as if stimulus were not present, and it is present (a "miss"); (*3*) animal responds as if stimulus were present, and it is not present (a "false alarm"); (*4*) animal responds as if stimulus were not present, and it is not present (a "correct rejection"). A brief consideration of the either–or nature of these four possible outcomes will reveal that only two are independent events. Stated differently, the occurrence of hits and false alarms eliminates the possibility of the occurrence of misses and correct rejections. Therefore, knowledge of the frequency of occurrence of the former two terms yields that same information for the latter two terms. Thus, it is unnecessary to incorporate frequency data on misses and correct rejections into any analyses.

Figure 12.1 presents a generalized schematic for the evaluation of discriminative responding that is consistent with the considerations presented in the previous paragraph. The ordinate plots the percentage of responses on which the animal gave a stimulation-appropriate response, that is, responded on the bar always paired with the drug state, thereby indicating it "detected" the drug-produced stimuli. Line A plots that percentage for those occasions on which drug stimulation actually was present (hits). Line B plots that

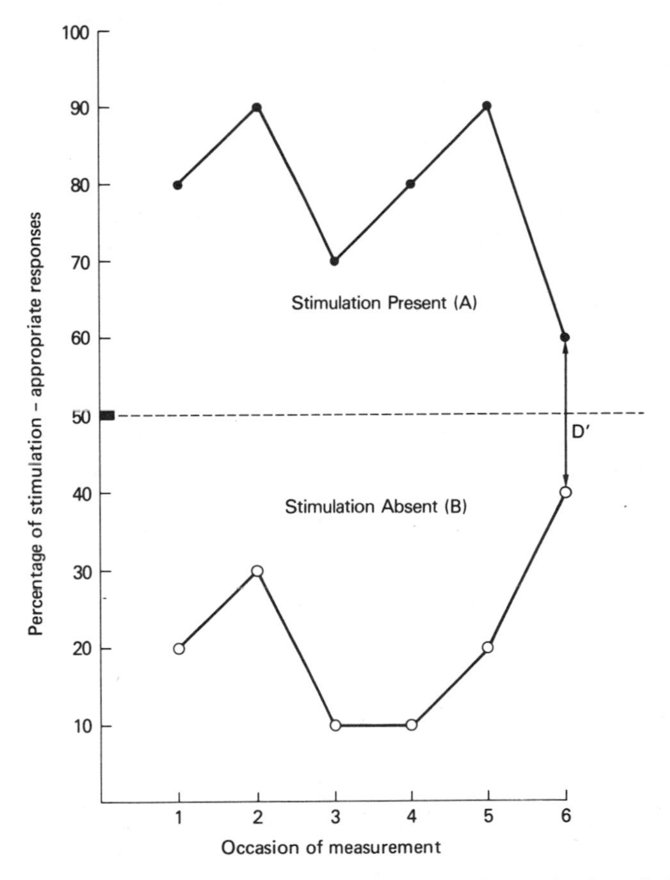

Figure 12.1 Generalized data presentation schematic for evaluation of discriminative responding. (See text for details.)

percentage for those occasions on which stimulation was in fact not present (false alarms). It should be emphasized that hit and false-alarm percentages are computed separately. That is, hits are expressed as a percentage of the total number of responses made on only those trials on which the drug-produced stimulus was actually presented. Similarly, false-alarm percentages are computed from response totals observed only on trials on which stimulation was absent.

All designs employed in drug-discrimination studies employ binary tasks potentially analyzable by the procedures outlined above. Two-bar operant tasks would have the responses made on the drug reinforced bar, expressed independently as percentages of the total number of responses occurring in each condition, plotted separately for both drug and nondrug treatments. Responses made on the drug-reinforced bar would constitute hits when made in the presence of a drug and false alarms when made in the absence of the drug. T-maze data, while restricted to percentages of responses for grouped data only, could be similarly plotted. Other tasks would require more detailed criteria for deciding what constitutes the presence or absence of a stimulation-appropriate response, but such criteria should not be difficult to develop. The need for such criteria, however, is a more general disadvantage to any objective analysis of their data and complicates the use of the simplest inferential statistics.

The properties of data plotted as shown in Figure 12.1 allow analysis of outcomes bearing upon the physical detectability of the drug-produced stimuli independent of the possible influence of nonsensory variables. Since line A plots hits or correct responses, and line B plots false alarms or incorrect responses, the distance between them in percentage points, D′, is an index of the appropriateness of discriminative behavior. Since D′ is the percentage of hits from which erroneous responses have been subtracted, the value may simply be viewed as an "adjusted" quantification of the animal's ability to detect the presence or absence of stimulation. Thus, discrimination behavior shown on Occasion 1 of Figure 12.1 is superior to that on Occasion 6, which has fewer hits and more false alarms. The D′ value presented here is monotonically related to d′ employed by formal application of TSD. It is important to note that a statistically reliable difference between hit and false-alarm percentages is the single criterion for inferring stimulus control by a drug rather than any reference to a 50% chance level of responding. Thus, it is possible to observe reliable discriminative responding even when both hit and false-alarm rates are greater than or less than 50%. This situation would reflect a significant contribution of RB to the animal's performance, as the following discussion makes clear.

Possible nonsensory contributions to discrimination performance are seen on Occasions 1–3 of Figure 12.1, where hit and false-alarm percentages covary without affecting values of D′. Simultaneous, equal, and unidirectional shifts in these measures suggest a change in RB unaccompanied by a change in the ability of the animals to detect the relevant stimuli. Recall that RB reflects generalized predispositions by the subject to respond in one mode or another, and contingent upon the behavioral paradigm employed, the value simply quantifies such phenomena as a bar-press bias or a preference for turning left in a maze. Note that Figure 12.1 illustrates a change from an RB favoring the bar paired with drug administration on Occasion 2 to an RB favoring the bar not associated with the drug on Occasion 3. What the graph shows is a simultaneous and equal reduction in both hit and false-alarm percentages from Occasion 2 to Occasion 3. While it may not be immediately apparent that this shift can reflect a change in the subject's predisposition to respond on a given bar, the following discussion elaborates on this point. For example, assume there were 100 responses sampled in each of the four data points plotted on Occasions 2 and 3. On Occasion 2, 90 of those 100 responses made in the presence of a drug were made on the correct bar (A condition). If the left bar was the appropriate one for stimulation, then there were 10 responses made on the right bar. In the absence of a drug on Occasion 2 (B condition), 30 responses were made on the stimulation-appropriate left bar, thereby producing a 30% false-alarm rate; 70 responses, then, were made on the right bar. Thus, for the A and B conditions on Occasion 2, there was a total of 120 responses made on the left bar. A similar computation for Occasion 3 will reveal that only 80 responses were made on the left bar. Thus, on Occasion 2, 60% of all 200 responses were made on the left bar, while only 40% were made on that bar on Occasion 3. Yet the D′ value is the same for both occasions.

An example of the possible genesis of the RB shown on Occasion 3 would be the consistent pairing of an aversive drug state with responding on one bar. This could conceivably result in a predisposition of the animal to respond on the opposite bar. To the extent such a bias represents a learning phenomenon, its onset would be expected to be a somewhat gradual approximation of a typical learning curve. Moreover, this statistical control for RB makes unnecessary operational controls such as counterbalancing bar assignments associated with stimulus conditions. It should be noted that it is also possible to have simultaneous, unidirectionai, but unequal shifts in hit and false-alarm rates. These changes by definition would reflect a change both in bar bias and discrimination performance.

While there are several possible quantifications of RB, one of the most direct would simply reflect a comparison of variations about the 50% level of responding associated with discrimination performance on each occasion of measurement for each stimulus condition. For example, such variation is +30% for the stimulation-present condition on Occasion 1 of Figure 12.1 [e.g., $(+80) - (+50) = +30$] and −30% for the stimulation absent condition [e.g., $(+20) - (+50) = -30$]. A comparison of these 2 values [$(+30) + (-30) = 0$] indicates the complete absence of a response bias. Similar computations yield response-bias values of +20% and −20% for Occasions 2 and 3, respectively. An RB value of +20% reflects an RB that favors the report of detection of the drug stimulus, while an RB of −20% quantifies a bias that favors the report of no drug stimulation.

Discrimination studies commonly evaluate generalization of cues associated with the compound used in training to cues produced by a test drug. Typically, the test compound is simply substituted for the training drug and the animals tested in the absence of any reinforcement. Complete generalization would be inferred if the false-alarm rate, that is, responding on the drug-appropriate bar in the presence of the test compound, did not differ significantly from the hit rate observed in the presence of the training drug. Partial or incomplete generalization would be characterized by a false-alarm rate in the presence of the test compound significantly greater than the false-alarm rate observed in the absence of the training drug but significantly less than the hit rate recorded in the presence of the training drug.

Whenever practical, test compounds should be administered in conjunction with training compounds to assess the possibility of interactions between them that affect discriminative responding. Possible interactions could be a potentiation or an antagonism of the interoceptive cue by the test compound. Potentiation would be evidenced by a significantly greater hit rate in the presence of both compounds than in the presence of the training drug alone. A significant reduction in hit rate would constitute antagonism. The exposure of these relationships obviously requires that training procedures produce hit rates that reflect a less than optimal level of performance, thereby preventing ceiling effects from masking decrements or increments in discrimination behavior. Careful dose–response assessments of a compound's ability to exercise discriminative control would make possible an evaluation and subsequent use of submaximal performance levels. If a test compound produced both potentiation and generalization, the plot of such changes would look like a change in RB. However, the transient, drug-associated nature of the phenomenon would help rule out a nonpharmacological explanation. If a test compound simply reduced D′ by producing both a partial antagonism and incomplete generalization, it would suggest that the compound had an interoceptive effect but one qualitatively different from that produced by the training drug.

Aside from explicitly examining generalization, researchers often evaluate the effects on discrimination performance of such agents as amine-synthesis inhibitors. These manipulations commonly represent an effort to define the role of specific neurotransmitters in the production of interoceptive cues. For example, if depletion of central nervous system levels of a transmitter abolished drug-produced discriminative behavior, one may infer the amine is required to produce the interoceptive cue. These agents also should be evaluated in the presence and absence of training drugs. A significant but approximately equal reduction in both hit and false-alarm rates produced by a test compound (no change in D′) requires a quite different explanation than a reduction in hit rate alone (reduced value of D′). That either outcome was associated transiently and exclusively with drug administration, however, would argue against mediation by a nonpharmacological process like learning.

In summary, this chapter has presented a conceptual approach to the evaluation of discrimination studies that has its foundation in the TSD (Swets *et al.*, 1961). In addition, while several experimental designs were presented, two-bar operant tasks were offered as the most amenable to this analytical approach. These tasks sample relatively large numbers of responses and represent a straightforward operationalization of questions related to discriminability and detection. They can also employ appetitive behaviors unconfounded by the stress produced by aversive motivators. Moreover, because they require execution of a similar operant in both stimulus conditions, these responses should be similarly affected by certain relevant variables. Thus, discrimination performance in these tasks should be robust to nonspecific facilitatory or inhibitory effects produced by pharmacological manipulations or variations in deprivation or amount of reinforcement.

REFERENCES

Barry, H., Koepfer, E., & Lutch, J. Learning to discriminate between alcohol and a nondrug condition. *Psychol. Rep.*, 1965, *16*, 1072.

Bindra, D., & Reichert, H. Dissociation of movement initiation without dissociation of response choice. *Psychonomic Sci.*, 1966, *4*, 95.

Harris, R. T., & Balster, R. L. An analysis of the function of drugs in the stimulus control of operant behavior. In T. Thompson & R. Pickens (Eds.), *Stimulus properties of drugs*. New York: Appleton, 1971. p. 111.

Hirschhorn, I. D., Hayes, R. L., & Rosecrans, J. A. Discriminative control of behavior by electrical stimulation of the dorsal raphe nucleus: Generalization to lysergic acid diethylamide (LSD). *Brain Res.*, 1975, *86*, 134.

Holmgren, B. Conditional avoidance reflex under pentobarbital. *Bol. Inst. Estud. Med. Biol.* (Mexico), 1964, *22*, 21.

Overton, D. A. State-dependent or "dissociated" learning produced with pentobarbital. *J. Comp. Physiol. Psychol.*, 1964, *57*, 3.

Shannon, H. E., & Holtzman, S. G. A pharmacologic analysis of the discriminative effects of morphine in the rat. *Problems of Drug Dependence Meeting*, 1975, *37*, 698.

Swets, J. A., Tanner, W. P., & Birdsall, T. G. Decision processes in perception. *Psychol. Rev.*, 1961, *68*, 301.

13

Statistical and Methodological Considerations in Drug–Stimulus Discrimination Learning

Harold Zenick and Joel E. Greene
Department of Psychology
New Mexico Highlands University

INTRODUCTION

Any book purporting to serve as an advanced text and reference source should accomplish at least two goals: One, the reader should be informed as to "what has been done." This objective is most often met, albeit one has to sort through the various biases before drawing his own conclusions. The second objective is that the reader should be able to clearly discern *how* "what has been done" was done. This information is especially valuable for the researcher attempting an initial experimental venture. Furthermore, this objective should certainly be stressed in meeting the text utilization of this book.

Information provided should range from how the animals were handled to how the data were handled. It is, in fact, the latter consideration to which the present chapter will be devoted. Although there have been several monographs and book chapters in which drug-stimulus discrimination learning has been discussed (Overton, 1971; Overton & Winter, 1974; Thompson & Pickens, 1971), none of these sources has really addressed the problem of data analysis; it certainly can be a problem in any experimental design that involves repeated measures with a small number of subjects. In fact, after more than a decade of research in this area, only one author has offered a critical analysis of methods and data interpretation in the field of drug-stimulus discrimination learning (Overton, 1974).

This chapter reviews studies concerned with drug-stimulus discrimination learning and the appropriateness of the statistics applied in data analysis. The advantages (of which there were few) and the pitfalls (of which there were many) are discussed. Some of the design constraints in pursuing research in this area are noted. The reader is provided with a variety of statistical alternatives that may be used based on experimental design and task employed. In addition, some statistical checks on design manipulations (e.g., counterbalancing vs. single alternation) and some novel applications in this area are described.

Although we reviewed the majority of studies published in drug-stimulus discrimination learning, we have concentrated on the major authors in the area. The main reason for our choice was to benefit the new investigator entering the field, for it is these authors who will provide the literature base for his experiments. This chapter by no means invalidates the results of prior research; in most instances, errors in statistical application have only increased the conservativeness of the test. Thus, the statistical differences that have been found serve to magnify the results and are a tribute to the strength of the drug-stimulus discrimination methodology. Awareness of experimental constraints, however, become more important as designs become more sophisticated. Consideration of these factors in initial basic design research can provide the researcher with the proper set to approach more complex problems. In these instances, he can avoid errors that will sacrifice power and jeopardize the discovery of new relationships.

MAJOR PHASES IN DRUG-STIMULUS DISCRIMINATION STUDIES

Establishment of Drug-No-Drug Discrimination

The initial phase determines whether the drug (D) in question produces interoceptive cues that are different from the no-drug (ND) state. Pilot work should answer several questions, including level of training dosage, order of treatment administration, intertrial and intersession interval, trial length, and the injection-test interval. The latter variable is somewhat difficult to assess. A review of the literature on the drug in question may be of little help, since all studies note the interval selected but few comment on rationale. In examining central nervous system (CNS) depressants, loss of righting reflex may serve as a time marker (Overton, 1966, 1971); however, this physiological index may not accurately reflect maximal brain concentration of the drug, and it is of no use in examining CNS stimulants or hallucinogens. Another approach, and one certainly more precise, is to establish curves of brain uptake over time. Surprisingly, few labs have employed this technique (Hirschhorn & Rosecrans, 1974b). Even such an exact approach, however, may encounter difficulties because the behavioral effects of some drugs do not coincide with maximal brain concentrations (Harley-Mason, Laird, & Smythies, 1958). Such inconsistencies may be attributed to metabolites being the active agent rather than the parent compound. In spite of this shortcoming, brain-uptake analysis is the preferred method. Such data will not only indicate the appropriate injection-test interval but also will reflect the adequate interval between test sessions to eliminate the possibility of carryover.

If for some reason (availability of facilities, expertise, etc.) brain analyses are not feasible, there is an alternative approach to assessing the temporal parameters of the drug cue. The experimenter may establish a D–ND discrimination, given a specific injection-test interval. Then subjects are tested at varying injection-test intervals, and the alterations in level of discriminability recorded (Browne & Ho, 1975a,b; Hirschhorn & Rosecrans, 1974b; Kuhn, Appel, & Greenberg, 1974). The optimal interval is reflected by the highest degree of discriminability witnessed. A major weakness of this approach is that the length of the interval may be a direct, inverse function of training dosage employed. Clearly, however, this approach does provide some alternative to analytic procedures.

Given that these variables are determined, D–ND training is initiated. The majority of studies run the animal until some criterion is attained, before proceeding to the next experimental stage. Considerations involved in a trials-to-criterion approach are detailed later in the chapter.

Establishment of Dose-Response Generalization Curves and Generalization of Test Drugs

The next phase of experimentation establishes the ED_{50} for the training drug. This is especially relevant for equating dosages of different drugs to examine the degree of generalization between them. The problem is that the slope of the generalization gradient will be influenced by the dosage of the training drug (Overton, 1974). Too high a dosage may sharply steepen the gradient, whereas too low a dosage may impede acquisition of the D–ND discrimination. A recent, ingenious study by Greenberg, Kuhn, and Appel (1975) serves to reinforce this point. After establishing a 95% level of discrimination between 80 μg/kg of LSD versus saline, drug responding to 10 μg/kg of LSD generalized only 30% of the time. However, additional training at this dosage not only increased responding to 83% but also extended the slope of the generalization gradient such that 5 μg/kg and 2.5 μg/kg of LSD also produced a majority of drug responses (89 and 58%, respectively).

Test-drug generalization follows a similar format in that various dosages of either agonists or antagonists are administered and the degree of drug responding assessed. Alterations of the approach may involve (*1*) administration of a blocking agent prior to drug injection or (*2*) administration of metabolites or structural derivatives followed by assessment of the degree of drug generalization. It is primarily this phase of experimentation that elucidates mechanisms of drug action, in this respect, drug stimulus generalization techniques have been proved to be as sensitive a gauge as any pharmacological procedure.

MAJOR TASKS EMPLOYED IN DRUG-STIMULUS DISCRIMINATION STUDIES

T-Maze Position Discrimination Tasks

The design of the apparatus primarily conforms to the basic T-maze structure. An interesting modification of this design, however, is the use of a three-compartment box; the animal is placed in the middle compartment and required to exit to one of the sides (Rosecrans, Goodloe, Bennett, & Hirschhorn, 1973; Schechter, 1973a,b, 1974; Schechter & Rosecrans, 1972a, 1973).

The procedures involved in this task are basic. Essentially, one side is designated the D side, the other, the ND side. Position preference is controlled by dividing subjects randomly into two equal groups, with one group required to turn left at the choice point under a condition and the other group required to go right under the same condition. "Massed trials within sessions" training procedures are used almost exclusively. Training techniques vary, however, in that some laboratories may employ a free-choice, self-correcting procedure, whereas others use forced choice. These procedures are discussed further in the next section. Perhaps the major limiting factor in this design is that data analysis is limited to recording the choice on only the first trial within a session, since first-trial fate determines choices on remaining trials. One might say that the subject operates under a "Win, stay—lose, shift" philosophy.

Reinforcing stimuli have primarily been escape from continuous shock (e.g., Schechter & Rosecrans, 1972a) or water (e.g., Henriksson & Jarbe, 1972; Jarbe & Henriksson, 1973, 1974). The water-escape maze offers an excellent alternative if the drug produces anorexia or elevated pain thresholds. The task may also employ some combination of reinforcers such that the animal may receive positive reinforcement for a correct response or punishment (shock) for entering the incorrect compartment (e.g., Schechter & Rosecrans, 1971a,b, 1972a,b).

Bar-Press Tasks

The apparatus employed usually is an operant chamber equipped with two levers, with one designated the drug lever. Again, appropriate controls for position preference need be exercised. Although a variety of schedules have been employed to maintain the behavior, it seems that some variable-interval (VI) component is desirable. Extinction periods (neither bar delivers reinforcement) are interposed between training sessions or at the start of a training session and the number of responses on each bar recorded. Training under a VI schedule serves to maintain response rate during extinction. Furthermore, since neither bar reinforces, the "stay or shift" behavior encountered in the T-maze task is avoided. Thus, several responses/animal are recorded to provide an adequate data base, obviating the problem in the T-maze task of being able to utilize only first-response/animal/test session as the dependent variable.

An interesting modification of this task is the use of only a single bar as the manipulandum (Barry, Steenberg, Manian, & Buckley, 1974; Kubena & Barry, 1972; Winter, 1973, 1974). In this instance, one treatment serves as S^+, the stimulus in whose presence responses are reinforced; the other treatment is the S^{Δ}, the stimulus in whose presence responses on the same bar are punished by delivery of shock. Extinction periods are imposed as test sessions. Controls are run to eliminate the possibility of a bias in the interaction between drug state and nature of the reinforcing stimulus. This approach seems highly precise in estimating the stimulus-discriminative properties of drugs.

DESIGN CONSIDERATIONS IN DRUG-STIMULUS DISCRIMINATION STUDIES

Forced Constraints in Design Construction

This type of research has a number of features that may make one type of design more attractive than another. In the stimulus-discrimination tasks, each subject is trained to two habits, and both of these habits should be stabilized at relatively equal strengths before test trials are run. This means that a considerable amount of time per subject may have to be devoted to establish these habits.

The number of training sessions varies somewhat, depending on specific D–ND conditions, number of trials per session, motivating and/or reinforcing conditions, and strain of animals. For example, most studies using the T-maze task would be covered by a range of 10–20 sessions with from 4–10 trials per session. Each session usually uses relatively massed trials and lasts from 10–20 min. Bar-press tasks may require from 10–40 sessions, with each session varying from 10–30 min.

Related to the above is that the number of trials per session is limited to time of the maximum effect and may be influenced by effects of the drug and dosage. There is a tendency, then, as a consequence of both these features, to use the same subject in a variety of test-drug and dosage conditions. Therefore, the designs used generally do not use independent, random groups but attempt to rely on various types of sequencing approaches using the same group of animals for a variety of test conditions. Although this may be experimentally justified, it does pose problems in selecting the statistical analyses that are to be employed.

Because of the amount of time involved and the relatively high cost of subjects, the tendency is to use small groups. To the extent that the data show considerable variation, to that extent more subjects are needed if the statistical tests are to be powerful enough to detect differences. Fortunately, in much of the research conducted, the effects are fairly large and, as near as can be determined, fairly uniform within a group of subjects. Under such conditions small group size is not the problem it might otherwise be. As the discriminations become finer and finer, however, the problem is likely to become more important.

In spite of these constraints, we believe that many researchers have spread their numbers too thin. The average number of test-drug comparisons per study ranges from 4 to 8. If one multiplies these figures by an average of 3 levels/drug and about 2 trials/animal/level, this range increases dramatically, and these figures are only an average. For example, a study by Huang and Ho (1974b) employed 20 animals in establishing the D–ND discrimination. Then various subgroup combinations of 5 animals were used in comparing 15 other drugs with the training drug. If one adds in the levels of each drug, this number rises to 33 comparisons.

The statistical problem that arises is that the comparisons involving the same subjects should allow for this fact. The subset of 5 animals given a test-drug condition could be compared with the remaining 15 subjects on the training-drug condition. This would be an independent groups' comparison. The 3 to 1 ratio between the 2 groups should be viewed as requiring caution in interpreting the results. Suppose now a comparison was being made between two dosages of the same test drug. If the two subsets of five animals are different, an independent groups' comparison can be made, though the group size is too small to expect much power regardless of statistics employed. What if the two subsets share animals in common? Then there is no legitimate statistical analysis that allows use of the data from all of the subjects.

A more common design is to train animals in a D–ND discrimination and then use the same animals in all test conditions. Usually training-trial sessions are interposed between test sessions in order to maintain the original discrimination. In this design, all comparisons must be made in such a way as to allow for the fact that the data are from the same subjects.

Unfortunately, these studies have ignored the influence of prolonged exposures to behavioral cues in the apparatus. If the researcher insists on such repeated testing, then he should attempt to assess whether or not continued exposure differentially influences the degree of generalization in progressive test comparisons. It would seem that total reliance on randomization is not advisable in studies using protracted time and test trials with small sample size. Another danger inherent in these methods is the continual retraining/maintaining of the discrimination between test sessions. Such procedures may confound overtraining and performance, with the potential for this interaction increasing as the number of comparisons increases. For example, Schechter and Rosecrans (1972d) trained seven animals to criterion. These animals were then administered 12 drug-dosage conditions, each repeated twice. Test trials were conducted twice a week; retraining, 3 times a week. In short, testing and retraining continued over a 12-week period. This pattern was followed in most of the studies conducted in their laboratory. In contrast, Jarbe and Henriksson (1973) used 68 animals divided into 10 groups. Original training and testing took 18 days. Although test comparisons were based on groups of 6 and 8, the statistical analyses are defensible, and the problems of overtraining are avoided. A series of papers by Iwahara and colleagues have indicated that overtraining may in fact "override" existing drug state (Iwahara & Noguchi, 1972, 1974).

There are no set rules governing the number of comparisons; however, statistical analysis may allow the investigator to assess the influence of continued testing. This assessment may then determine how many comparisons can be conducted with a single group of animals.

Current Pitfalls in Data Analysis

One of the major problems with the research reviewed is that many studies provide inadequate descriptive data. Common measures reported involve frequencies and/or percentages of subjects that respond in a certain way, the latency of escape, and number of correct responses. In general, the use of percentages should be discouraged unless the number on which the percentages are based is somewhere between 50 and 100. Otherwise, it is more desirable simply to report the frequencies. The use of percentages with groups of less than 20, for example, means that a single subject is worth 5 or more percentage points.

Jarbe, Johansson, and Henriksson (1975, p. 406) present a table in which percentage of first-trial choices into a designated arm is shown. The group size ranges from 12 to 78. As a consequence, the various percentages are not exactly comparable, and portions of the apparent differences are attributable to the differences in the number of subjects in each group. Though perhaps not serious in itself, the practice may give the unwary reader an inaccurate impression.

Another fallacy occurs in descriptive data using percentages based on number of subjects times number of trials. This results in an inflated number, and although the percentage is based on a sufficient number from the point of view mentioned previously, it is faulty because each subject contributes more than a single trial to the total figure. This has the consequence that the elements of the total number lack independence. In such cases, inferential statistics such as Chi-square or the binomial are not appropriate. For examples of this practice, see Schechter and Rosecrans (1971a, 1972d). A good general rule is that the descriptive presentation should correspond to the test of significance. For example, if the statistic tests the differences between means, then means should be presented.

Another difficulty with some of the studies is the failure to indicate exactly how the statistics have been applied to the data, so that the reader is not able to reconstruct the analysis. For example, only to state the p value of a Chi-square gives no idea as to how/what particular chi-square formula was employed, how a contingency table (if used) was set up, or what was the basis for expected values. Reporting that the analysis of variance was used without giving a breakdown of degrees of freedom or reiterating the design under test makes it impossible to determine exactly what was done. The same is true of uses of the sign test and some of the other nonparametric techniques. Without some indication of how the data were organized for the purposes of applying a particular inferential technique, it is impossible to evaluate the appropriateness of the applications.

One of the glaring neglects in these studies is the failure to present any data on variability. Yet inspection of the data may reveal that the main effect was on variability rather than on the measure of central tendency reported. The studies using mazes take measures on which the calculation of standard deviations is meaningful only during the training of the discrimination. Many fail to report descriptive information on this phase. Jarbe *et al.* (1975), however, report heterogeneity of variance during training. In another study by Jarbe and Henriksson (1973), the training data suggest that the drug groups are more variable than the no-drug groups. To determine the significance of differences between variabilities, a number of statistical tests could be used. The *F*-test can be used if only two independent groups are to be compared. Overall tests comparing more than two groups have been developed and may be found in most standard statistics texts (Myers, 1972; Winer, 1971). Although the latter tests may lack power, there may still be value in using them for this purpose. If the comparisons are made on dependent measures, for example D–ND training trials for the same group, there is a *t*-test formula that may be useful (McNemar, 1969, p. 282).

Legitimacy of Trials to Criterion Assessment

For the most part, the training of the D–ND discrimination is done to a certain number of trials-to-criterion, the typical one being 8 of 10 trials-to-criterion. The purpose of such a procedure is to equate the performances. Certain characteristics are inherent in this type of procedure that should be explained. It equates the performances for the two habits involved. The measure, as such, is correlated positively with measures of error as well. However, as a measure, it fails to equate for the number of reinforcements as well as failing to equate for the number of nonreinforcements. Consequently, the experimenter is not certain whether an animal is making the discrimination on the basis of avoiding a side under certain cues or approaching a side under certain cues. This factor is important if the above differences in determinants of the behavior interact with the drug states. This would be particularly true, for example, if a certain drug were more effective in terms of avoiding the presence of certain cues, as opposed to approaching in the presence of certain cues. It also means that although the intent of trials-to-criterion is to equate groups as well as equate habits, it does so only in the sense of being able to observe a particular kind of performance. Unless records of the acquisition of the habits are retained for the various groups, it is possible that the groups could be different in actual basis for the particular habit.

Some of the studies use equal trials-to-criterion with a forced-choice procedure during training. In this case, the measures are reduced to the number of sessions to reach criterion. Such a procedure controls for the number of reinforcemens versus nonreinforcements and may prove to be, in the long run, the more desirable approach for establishing the habit. There is one distinct disadvantage to the forced-choice approach, however. The investigator cannot asses the influence of the drug on the learning habit per se through an error analysis of trials for each subject in each training session. Such information may be of value in examining difference in trials-to-criterion or the degree of discriminability attained by each animal. Ironically, of all the studies employing free choice, only one has reported an error analysis (Overton, 1975).

Another consequence of using trials-to-criterion is that it may result in a loss of *cases* because of a failure to reach criterion. In describing the results, it is important to account for the losses of cases to avoid the possibility that the failure to reach criterion is also associated with a particular experimental condition. Two problems are involved here. First, it is possible to interpret the loss of case as simply an accidental or chance loss that occurs to an average subject, in contrast to a loss resulting from an animal's inability to perform. In the first case, the lost animal may be replaced, although this is not commonly done. Failing to do so, however, should not unduly affect the data. In the second instance, simply to analyze the data may result in a bias in the group involved because the data are based on animals able to meet the criterion. Even replacement of the subjects is questionable. The problem has no ready solution except to develop procedures that minimize the loss of numbers. Second, if loss of cases results in a changing number in some groups, it may be necessary to modify the statistical analysis accordingly. For example, the analysis of variance (ANOVA) for more than a one-way analysis requires correction if there are disproportionate numbers from group to group. Even in the case of contingency tables, especially in two- and three-way contingency tables, unequal numbers may pose difficulty in interpreting the results. Care must be taken, therefore, to evaluate the best course of action if cases are lost. None of the studies reviewed replaced subjects, yet ranges from 1 out of 14 to 2 out of 3 animals failing to reach criterion have been reported. A loss of 2–3 animals of 10 is typical. The extent of the bias created is difficult, if not impossible, to assess (cf., Schechter & Rosecrans, 1972a; Schechter, 1973b, 1974).

STATISTICAL APPLICATION—THE T-MAZE TASKS

Use of *t* and Chi-Square

Relatively few of the reports employing the T-maze task use the *t*-test. Part of the reason is that only one or two transfer test trials are used, thus giving no basis for even ordinality in the measures. A few studies, however, have employed such tests. Jarbe and Henriksson (1973) used the one-tail test in order to assess the significance of difference over a series of trials in transfer of tests of the change and no-change groups. It is interesting to note that in this study a two-way ANOVA could have provided the basic analysis of the data, although it still would have been necessary to employ the separate t-tests. When many t-tests are used, allowance must be made for more than one comparison on the same data. A discussion of the consequences of this and the correction for it appears elsewhere in this paper.

The most common statistic employed in the evaluation of transfer data is the chi-square (χ^2). Unfortunately, in the majority of the reports reviewed, the description of how χ^2 is applied is ambiguous, and one is not certain that it is correct.

In spite of this, the usage of χ^2 and some of the weaknesses of its application can be explained. Table 13.1 presents the standard format for indicating the degree of D–ND discriminability. There is an $N = 8$: with a total of 80 trials, the animals make only 15% error in the D and ND states. One of the difficulties in determining the appropriateness of the χ^2 statistic is that there is no statement as to what the expected values should be. Is it to be assumed that the expected values are 50%, that is, if neither state exercises an influence, should the animal's behavior be random? Or should the assumption, which is qualified by the statement of significance at the bottom of Table 13.1, be that the expected here is what the animal does in the appropriate state? That is, if one wants to know whether or not the number of drug responses under saline is different from the number of drug responses under the drug, the drug percentage of responding is used as the expected value. In this instance, 85% is the expected value. A χ^2 is run on these data, and it is found that the probability of the difference from the drug score, being attributable to chance,

Table 13.1
Application of Chi-Square to Drug–Stimulus Discrimination Data

Drug	Dose (mg/kg)	N	Number of trials	Drug responses (%)
Training drug	2	8	80	85[a]
Saline	—	8	80	15[b]
Test drug	4	8	16	10[b]
	8	8	16	40[b]
	16	8	16	80

[a] Probability of difference from saline score being due to chance; $p < .001$, chi-square test.

[b] Probability of difference from training drug (2 mg/kg) being due to chance; $p < .001$, chi-square test. If test drug has pharmacological properties similar to training drug, then HO: percentage of drug responding should be equal.

$$\chi^2 = \frac{\Sigma(O - E)^2}{E} = \frac{(1.6 - 13.6)^c}{13.6} + \frac{(13.4 - 2.4)}{2.4} = 61.01^b$$

[c] Expected values derived by taking 85% of 16.

Table 13.2
Use of Exact Probability with Dichotomous Data from a T-Maze Task in a Drug-Stimulus Discrimination Experiment

Training drug	Dose (mg/kg)	*N*	Number of animals[a] reaching criterion
Training drug	4	8	8
Test drug	10	8	4[b]
	20	8	6

[a] Criterion is reflected by the expected value for drug responding established for the training drug.
[b] Significantly fewer animals reached criterion at this level of the test drug ($p \leq .03$).

is less than $p \leq .001$. This application is quite inaccurate: There is complete lack of independence in the two cells providing the df, that is, the animals that contribute to correct responding, are the same animals that contribute to incorrect responding. Furthermore, the experimenter has made several observations on each subject and is counting them as independent. Here the experimenter has yielded to the "inflated number fallacy." The true number in this experiment is 8, not 80, and the χ^2 should be calculated on the expected frequencies with regard to that number.

Misuse of χ^2 may be further examined by looking at the comparisons made between the test drug and its dosages and those of the training drug. In this instance, the expected values can be more readily specified since it is apparent that the null hypothesis (H_o) is that if the test drug and its dosages possess pharmacological properties similar to the training drug, then the test drug should yield similar response percentages. As a result, the experimenter may take the percentage of responding to the training drug, determine the expected percentage of responding, and compare this with the actual percentage of responding to the test drug. This formula is illustrated in the lower portion of Table 13.1. Again, the same criticisms occur. The two cells contributing the frequencies are not independent, and the experimenter is, in fact, working with an inflated number.

The alternative statistic that may be employed is determining exact probabilities by use of the binomial distribution. Table 13.2 shows how this may be done. Say that during training 8 animals reached the 80% criterion. In examining the test drugs, the experimenter uses the actual number employed to determine how many of the animals under the test dosage reach the probability level of .80. For example, with a test-drug dosage of 10 mg/kg, 4 of 8 animals reached the .80 criterion. Examination of the binomial table will indicate that a significant number of subjects did not make the drug response 80% of the time, $p \leq .03$. On the other hand, at 20 mg/kg of the test drug, only 2 animals failed to respond at the probability of .80. This indicates that this dosage of test drug yields responding that is not different from that seen with the training dosage. The exact probability application here does not suffer from the problems encountered with χ^2.

In using contingency tables to describe the results of test drugs, a distinction must be made between comparisons made on independent groups and those based on the same group. The former case is illustrated in Figure 13.1. The usual χ^2 from 2×2 tables may be calculated, provided N equals number of subjects and the expected frequencies are greater than 5—some prefer 10. Should the expected frequencies be too small, Fisher's exact probability for 2×2 tables may be used.

Many studies use the same animals to compare test drugs. In such instances, the contingency table should be set up as shown in Figure 13.2. Again, the usual χ^2 formula may be applied provided that N equals number of subjects—not the number of measures—and the expected frequencies are acceptably large.

Frequently, two or more trials are conducted with the same test drugs. This practice leads to the temptation to use trials times subjects as N. Comment has already been made on this fallacy. What is necessary is to arrive at a single classification for each subject. Granted, this will be largely an arbitrary decision on the part of the experimenter, but there seems to be no way to avoid the issue. Whatever method is used, it should be adequately described so that the reader may evaluate it. Suppose, for example, that two trials are given with the test drug. If an animal goes to either the D or ND side on both trials, the classification is made accordingly. But what should be done if the animal chooses the D side on one trial and the ND side on the other? One possibility is to make the classification at random. If several subjects are involved, the assignments might be equalized across categories, for example, one-half of the subjects assigned to the D side and one-half to the ND side. This tends to make the test of significance conservative.

If more investigators reported the trials-to-criterion necessary to establish a discrimination, ANOVA would probably be more widely used. As it is, only two studies using the T-maze calculated ANOVA (Hill, Jones, & Bell, 1971; Jarbe *et al.*, 1975). Such data would be instructive, because different drugs could well require different amounts of discrimination training. In such

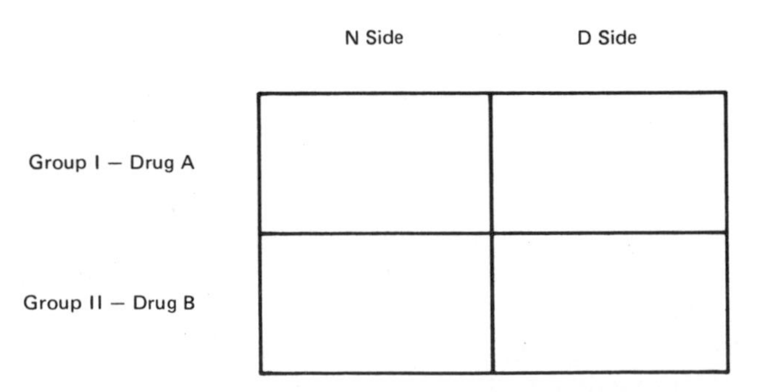

Figure 13.1. χ^2 Contingency table for independent group.

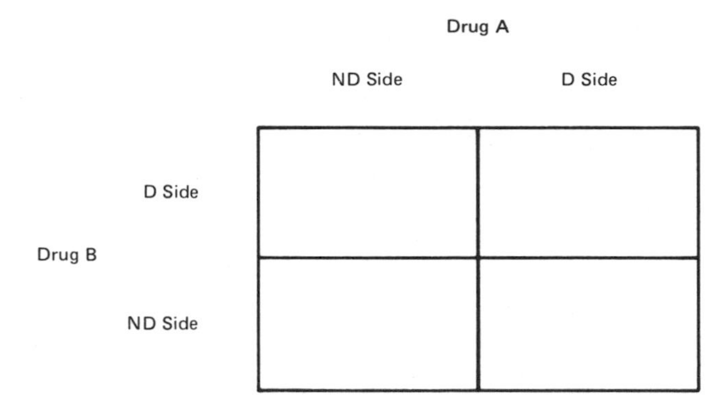

Figure 13.2. χ^2 Contingency table for related samples.

instances, it might be well to extend the research, using each drug both as an original discriminative drug and as a test drug.

The alternatives for the analysis of test trials are limited to those already suggested unless measures having ordinal characteristics are taken.

Nonparametric Alternatives

Jarbe and Henriksson (1973) studied state-dependent learning using a water T-maze. The design employed independent groups; the Mann–Whitney test was used to compare the groups. Unfortunately, descriptive information, means and standard deviations, are lacking, so one can only surmise why the ANOVA and/or *t*-tests were not calculated. The analysis in this study is marred by trial-by-trial comparisons made with the intent of concluding on which trials differences between groups cease to occur. This was done by reporting *p* values for each comparison. There are two objections to this procedure: (*1*) Multiple comparisons for the same groups require corrections of the *p* values; (*2*) *p* values should not be interpreted as reflecting the importance or extent of effect of a variable.

Jarbe and Henriksson (1974) used a sign test to evaluate transfer drug effects. The description of its exact application is sketchy, but it probably involved comparing the test-drug response with training-drug response. Because transfer results were accepted only if the preceding and following training trials were correct, the signs used could only be = or –. Animals that failed the above criterion were retrained and transfer tests rerun. Therefore, *N* would not change in the comparison, as would normally be the case when + and – signs were counted with ties excluded. Two aspects of this analysis might be deliberated. As described, all combinations of D and ND training conditions may precede and follow the transfer drug trial according to random assignment. The problem is whether or not a bias could occur depending on the specific training conditions that sandwich in the test trial. The second problem is whether or not H_0 should be 50:50 for this particular test. Rejection of this hypothesis may not necessarily mean that positive transfer or equivalence to training drug has occurred. What would be the situation if H_0 were 90:10 or some other extreme short of 100:00? Discussion of some of the points involved may be found in Rozeboom (1960) and Wilson and Miller (1964). An additional uneasiness is created by the procedure of adding training trials when the criterion is not met. There is no indication of how many times this was required.

Jarbe *et al.*, (1975) used a Kruskal–Wallis one-way analysis because of heterogeneity of variance. This is one of the few studies that evaluated independent training groups on the trials-to-criterion measure. The analysis seems to have been well chosen. Comment has been made elsewhere on the appropriateness of describing the trials-to-criterion required during training of the discrimination.

STATISTICAL APPLICATIONS—BAR-PRESS TASKS

Cue-State Saliency

With the initial D–ND comparison, the situation is somewhat different from that involving the T-maze task. In some instances, the animal is, in fact, run to a criterion, that is, the animal is assumed to have learned the initial discrimination when it presses the appropriate bar at least 80% of the time (Morrison & Stephenson, 1969; Schechter & Rosecrans, 1972e). Statistical comparisons are the same as comparing trials-to-criterion performance on the T-maze task; since the animal has attained a criterion, it serves little purpose to test for significant difference between D–ND discrimination.

Some studies, rather than running the animal to criterion, run the animal until its performance is stable. As a rule, this means that over days the animal's performance maintains a baseline between 65 and 80% (Bueno & Carlini, 1972; Hirschhorn & Rosecrans, 1974c). In this case, it is important to know whether the animal, when drugged, stabilizes at a higher percentage of correct responses than random performance. This comparison is completely one-sided, however, since it fails to consider the degree of control exercised by the ND state. The comparison does not answer the important question of the equivalency of each state to impact cues to the animal. For example, one study showed equivalent performance in both states (65%) when the drug level was 200 μg/kg of nicotine. However, in another group of subjects, receiving 400 μg/kg, drug responding was about 80%; in the ND state, stable performance was only 65% correct (Hirschhorn & Rosecrans, 1974b). This nonequivalency has also been noted by York and Winter (1975). The differences in cue-state equivalency may be influenced by the type of drug as well as dosage. This situation is related to, but not identical with, the problem of overinclusiveness described by Overton (1974) which seems more restricted to T-maze tasks. Overton's point is that, during test sessions, the animal must indicate whether or not the test drug is similar to either the D or ND state. The bar-press task, with its opportunity for multiple responding, avoids this problem, for random responding will be observed if this is the case. However, a bias may cause the animal to favor one state over the other simply because the degree of D–ND cue saliency was not equated during training. The problem of cue-state equivalency also applies to trials-to-criterion measures because the animal may reach criterion under one state more rapidly than under another. Prolonged training to reach criterion under both states introduces problems such as differences in total number of reinforcements for each state, as well as other variables discussed previously. The animal's performance in the ND state should be examined as closely as D performance. Analyses should compare the percentage level of correct responding within states in addition to the standard comparison contrasting degree of drug responding across states; obviously, the analyses answer different questions.

Two-Sample Comparisons

These statistics compare the initial D–ND training and, second, examine the generalization curve for the training drug. Surprisingly enough, many studies report no statistics on the training drug-test drug comparison, although this is the third phase of most of the experiments (Bueno & Carlini, 1972; Greenberg *et al.*, 1975; Hirschhorn & Rosecrans, 1974b; Huang & Ho, 1974a,c).

The parametric statistic of choice in making D–ND or D–D comparisons is the *t*-test, which may be run on the difference in response rates on the appropriate lever to determine whether the D–ND discrimination has been established at any one point in time (Barry, 1968; Hirschhorn & Rosecrans, 1974a,b; Huang & Ho., 1974b). However, the *t*-test should not be applied to multiple two-sample comparisons between the training drug and various test drugs without appropriate correction in *p* values since, as the number of comparisons increases, Type I error, falsely rejecting the null hypothesis, increases. The degree to which this error increases may be estimated for independent groups by the formula $Pj = 1 - (1 - \alpha)^k$, where Pj is the joint probability, α is the significance level, and k is the number of independent experiments. For example, in a study examining 4 test drugs, the assumed α is .05, whereas α would actually be $Pj = 1 - (1 - .05)^4 = .18$. That is, the probability of rejecting the null hypothesis has been increased from 5 to 18%. Since most of the studies in this area do not employ independent groups, an exact estimate of the joint probability level cannot be made. However, the moral still applies: Running multiple two-sample comparisons is not a satisfactory technique of statistical analysis.

A further complication of multiple comparisons concerns the reference against which to judge. Ryan (1960) has pointed out that in conducting multiple comparisons, the experimenter is actually faced with the existence of three main rates of error as presented by the following statements:

(*a*) Error rate per comparison equals number of comparisons incorrectly called significant; (*b*) error rate per experiment = number of comparisons incorrectly called significant total number of experiments; (*c*) error rate per experiment = number of experiments containing erroneous statements of significance the total number of experiments.

All these rates are equal when only a single comparison is made or the "complete" null hypothesis is true, that is, all samples come from a single population. When these assumptions are not met, however, Ryan (1959) urges that the experiment be used as the unit in computing error rates rather than the individual comparison. A method for adjusting significance levels to control for the error rate of the whole set of comparisons is presented (Ryan, 1960). For a more complete discussion of error rate, the reader is referred to Miller (1966) and Petrinovich and Hardyck (1969).

The existence of different error rates creates a problem, since there are no hard-and-fast rules as to which should be measured in a given experimental design. Furthermore, the type of multiple-test comparison employed may assess one type of error rate but at the cost of increasing the conservativeness or liberalness of the test. Winer (1971) has presented a table that reflects the differences in treatment totals required for .01 level of significance for different tests. The conclusion is that the application of different methods will yield widely disparate results, with the *t*-test being too liberal and the Scheffé method too conservative. Thus, there is no clear-cut path to choose. However, the experimenter can increase the correctness of his choice by good insight into the questions he wishes to answer and the types of error he will tolerate.

In assessing the degree of D–ND discriminability, Cameron and Appel (1973) applied the one-way ANOVA to their data. The authors qualified this application with the statement "The validity of the test perhaps can be questioned, however, because of the small number of subjects" (two/group). The problem is not that small sample size affects the validity but rather the power; specifically, in this instance, it increases the probability of committing Type II error, retaining a false null hypothesis, thus making the test extremely conservative. This may, in fact, explain Cameron and Appel's failure to find significance or only marginal significance in several of the comparisons made in the five experiments comprising this study. Perhaps the greater dilemma that is magnified with small number is that a single aberrant subject may determine the significance of the experimental manipulation, depending on the group to which it is assigned.

Although the ANOVA as used by Cameron and Appel (1973) essentially reduces to a *t*-test, the ANOVA is valuable for assessing the variability contributed by other design manipuations. For example, Winter (1973, 1974) employed a 2 $\times$ 2 repeated measures ANOVA in assessing degree of discriminability. In this case, since he used a single manipulation (see "bar press tasks"), he needed not only to assess differential responding, S^D versus S^Δ but also to examine the influence of the possible combinations involved in assigning treatments to conditions. (That is, S^D may be either drug or vehicle, the same being true for S^Δ.) Results indicated that only the $S^D - S^\Delta$ stimulus conditions exercise a significant influence on rate of responding.

The ANOVA also allows the experimenter to examine directly the differences in degree as well as rate at which the discrimination of various drugs is acquired. For example, Jones, Hill, and Harris (1974), in employing a groups x trials repeated ANOVA followed by individual comparisons, showed that a *d*-amphetamine-versus-saline discrimination was acquired more quickly than the *l*-amphetamine-versus-saline discrimination, although the latter eventually reached a comparable level. The third compound, *para*-hydroxyamphetamine, failed to acquire discrimination control. Kuhn *et al.* (1974) have also used ANOVA to assess the trials (days, sessions) effect. However, the degrees of freedom presented in this study seem to be in error, although this conclusion results in part from an inadequate description of application.

The χ^2 statistic has been employed in some of the bar-press studies (Kuhn *et al.*, 1974; Morrison & Stephenson, 1969; Schechter & Rosecrans, 1972e). However, the errors of usage discussed in the section of T-maze statistics are encountered in these instances as well.

Another method of comparing D–ND or D–D performance is the Wilcoxon's paired signed-ranks test. Among the studies that have used the Wilcoxon, some have applied it inappropriately; in others, the description of application is insufficient to judge whether or not it has been applied correctly (Hirschhorn & Rosecrans, 1974c). For example, in Figure 13.3, if the experimenter were to use the Wilcoxon test, he would need to consider the D–ND performance across all sessions for each subject, as would be true of the *t*-test. Incorrect application would consist of deriving a difference score within each session collapsed across subjects. The Wilcoxon is an extremely powerful test in that it is sensitive to both the direction and magnitude of a difference. It might be noted in passing that the Wilcoxon often, rather than the *t*-test, has been used for the D–ND discrimination. In fact, one group that had employed the *t*-test in earlier studies (Hirschhorn & Rosecrans, 1974,a,b) used the Wilcoxon for the same assessment in a later study (Hirschhorn & Rosecrans, 1974c). One can only assume that the paired *t*-test was not used because of failure to meet some assumption of *t*. This speculation is left to the reader since the information was not provided by the authors.

The Mann–Whitney U is another ranking test for bar-press data. Browne, Harris, and Ho, (1974) and Browne and Ho (1975a,b) offer rare examples of reports employing this test and giving the rationale, that is, lack of homogeneity of variance. Unfortunately, some experiments using this statistic have either failed to meet the independent-groups requirement (Browne & Ho, 1975a,b; Hirschhorn & Winter, 1971; Ho & Huang, 1975) or have met it only when the D–ND responding was compared for the initial test session (Browne

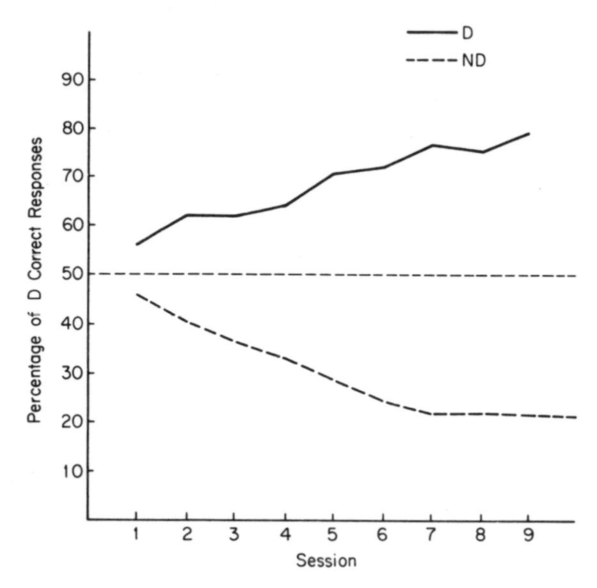

Figure 13.3. Acquisition curve reflecting typical drug–no drug (D–ND) discrimination learning on a bar-press task.

et al., 1974). In the latter study, with continued test sessions, the same subjects vacillated between D–ND states such that group performances over time were related and additional comparisons beyond the first test session are suspect.

Multiple-Sample Comparison

The appropriate statistics may be examined for comparing test drugs versus training drug or for examining the training drug's generalization curve. When three or more samples or conditions are compared in an experiment, it is desirable to use a statistical test that will indicate whether or not there is an overall difference among the conditions before one picks any pair to test for a significant difference between them. For example, if comparisons between several dosage levels of the same drug are desired, an overall test should be performed first. If the number of dosages equals 5, for example, the number of comparisons two at a time is 10. The experimenter now has 10 opportunities rather than one to reject the null hypothesis and has increased Type I error from 5% to 40% (Siegel, 1956). (See previous discussion of multiple comparisons).

If the same subjects are involved, a test comparing repeated measures is needed. The parametric statistic of choice would be the ANOVA with one factor repeated. However, the heterogeneity of variance across dosages or groups may rule against this application (Jarbe *et al.*, 1975). Two alternative nonparametric statistics may be employed as overall tests of significance: the Friedman (1937) ANOVA for *k* related samples or the Kruskal–Wallis for independent samples. Two studics havc employed the latter statistic (Jarbe *et al.*, 1975; Overton, 1966); none has employed the Friedman. Table 13.3 shows how this statistic would be employed. The table presents each subjects percentage of drug responses at the training dose of 20 mg/kg, as well as responding at other levels of the drug. The initial step is to rank each subjects performance across the different dosages. If dosage has no effect, then the rank for a subject at any 1 dosage should be random and the sum of the column ranks should be essentially equivalent. Thus H_0 is that the subjects ranks are independent of the conditions and the mean ranks of the various columns are equal. Following this overall comparison, if there is a significant difference, it is then appropriate to use multiple comparisons to examine these differences more closely. The power of the Friedman can be compared with the *F*-test. In a comparison reported by Friedman (1937), the Friedman ANOVA rejected H_0 26 times at the .05 level, with the *F*-test rejecting H_0 24 times at the .05 level; consequently, the 2 tests appear similar in their power. The Friedman seems acceptable as a repeated-measures examination of test drugs versus the training drug or to establish a generalization curve for the training drug.

ADDITIONAL CONSIDERATIONS IN DRUG-STIMULUS DISCRIMINATION DESIGN

Manipulation Checks

D–ND Sequencing

The manipulation check of the D–ND ordering is important, especially since some authors note that accuracy improves when the drug state is held constant over several sessions and that accuracy decreases with changing drug states (Kuhn *et al.*, 1974). The techniques, ranging from the most frequent to least frequent, are single alternation, randomization, double alternation, and counterbalancing. Whereas single alternation was employed predominantly in early studies, double alternation now seems to be more popular, especially in the bar-press tasks. In these, the responses is acquired on a single alternation pattern to reduce the probability of the bars' developing cue properties. Then discrimination training is maintained on a double-alternation schedule. Each treatment session preceded equally often by a session with the same and the opposite treatment.

Only one experimenter employed a statistical analysis to check the effect of the preceding session on a subsequent test (Winter, 1973, 1974). The comparison is the amount of drug responding in test periods following a drug training session versus the amount of responding following a no-drug training

Table 13.3
Application of Friedman's ANOVA

Subject	Dosage (mg/kg) 5	10	20[a]	25	30
1	20[b]	50	70	75	75
2	22	45	72	70	80
3	27	48	68	70	69
4	15	42	80	78	80
	Ranked scores, dosage (mg/kg)				
1	1	2	3	4.5	4.5
2	2	1	4	3	5
3	1	2	3	5	4
4	1	2	4.5	3	4.5
	5	7	14.5	15.5	18

[a] Training dosage.
[b] Percentage drug responses/total responses.
$$\chi r^2 = \frac{12}{Nk(k+1)}[5^2 + 7^2 + 15.5^2 + 18^2] - 3N(k+1) = 51.83^c.$$
[c] $p \leq .001$.

session. Either the paired *t*-test or Wilcoxon's signed-ranks test is appropriate for analysis. A semiexperimental check on sequencing was imposed by Bueno and Carlini (1972) who, after training animals with a single alternation pattern, interposed consecutive, "same" treatment sessions. That discriminability was unaltered served as evidence that the treatment, not pattern of administration, was the response-control agent.

The Strength of Relationship between Independent and Dependent Variables

Most experimenters should have some idea of the overall strength of the association between independent and dependent variables that a significant finding actually represents. The investigator should not rely only on statistical significance as evidence for a strong statistical association. The statement of significance only notes that a relationship exists, not the strength of the relationship. Hays (1973) provides a statistic, Omega² (ω^2), for assessing the strength of an association. This formula can be modified to indicate the sample size needed, given the ω^2 and alpha level desired by the experimenter. Only one study (Jarbe *et al.*, 1975) assessed the strength of the relation between independent and dependent variables using ω^2. They reported the strength of relationship between dosage and sessions to criterion to be 39%, a very healthy association. With increasing design complexity, it is even more important to attempt some assessement that will reflect the contribution of the experimenter's manipulation.

Assessment of Tolerance

A few studies have ingeniously used the D–ND discriminability in assessing the development of tolerance to a drug (Bueno & Carlini, 1972; Hirschhorn & Rosecrans, 1974a; York & Winter, 1975). This technique has advantages over standard procedures for assessing tolerance in that the discrimination is established at smaller dosages that do not incapacitate the animal. Thus, tolerance may be examined at a dosage range that is more relevant to the drug abuser. Two of the studies employed accompanying physiological index to assess the development of tolerance. Bueno and Carlini (1972) measured changes in rope-climbing ability with chronic marihuana administration before and during the establishment of a D–ND discrimination. York and Winter (1975) use the development of tolerance to the hypnotic action of barbital as an index of the influence on a previously learned D–ND discrimination. Although no accompanyind index was employed by Hirschhorn and Rosecrans (1974a), they followed York and Winter's scheme (1975) of assessing the effects of tolerance on an already trained discrimination habit. Furthermore, a check that tolerance had occurred was conducted at the conclusion of the chronic administration phase. In this instance, morphine-tolerant rats exhibited signs of narcotic withdrawal when treated with naloxone.

In two of the studies (Bueno & Carlini, 1972; Hirschhorn & Rosecrans, 1974a), the subjects maintained a significant D–ND discrimination, although the degree was reduced, indicating the development of limited tolerance. The York and Winter study (1975) produced conflicting results. No impairment was observed in animals when the drug served as S^D (saline = S^Δ). Yet tolerance was evident in animals when the discrimination was reversed (S^D = saline; S^Δ = drug). It might be noted, however, that the latter group had not shown as high a degree of discriminability compared to their counterparts before chronic administration.

The paired *t*-test has been employed in assessing changes in discriminability attributed to tolerance. This application is accurate if only a single point in time is examined (York & Winter, 1975). However, if assessment occurs periodically during the development of tolerance (Hirschhorn & Rosecrans, 1974a), a repeated measures ANOVA is recommended in assessing the interaction between length of exposure and degree of discriminability.

On the whole, these studies seem to indicate that the drug-stimulus discrimination design may be used not only to assess the influence of tolerance on this task but that, this task is an excellent means for assessing whether or not tolerance develops to various drugs. Certainly, any technique is welcome that sheds additional light on the tolerance controversy surrounding certain popular drugs of use and abuse.

REFERENCES

Barry, III, H. Prolonged measurements of discrimination between alcohol and nondrug states. *J. Comp. Physiol. Psychol.*, 1968, *65*, 349–352.

Barry, III, H., Steenberg, L., Manian, A., & Buckley, J. P. Effects of chlorpromazine and three metabolites on behavioral responses in rats. *Psychopharmacologia*, 1974, *34*, 351–350.

Browne, R. G., Harris, R. T., & Ho, B. T. Stimulus properties of mescaline and *n*-methylated derivatives: Difference in peripheral and direct central administration. *Psychopharmacologia*, 1974, *39*, 43–56.

Browne, R. G., & Ho, B. T. Discriminative stimulus properties of mescaline: Mescaline or metabolite? *Pharmacol. Biochem. Behav.*, 1975, *3*(*1*), 424–428. (a)

Browne, R. G., & Ho, B. T. Role of serotonin in the discriminative stimulus properties of mescaline. *Pharmacol. Biochem. Behav.* 1975, *3*, 429–435. (b)

Bueno, O. F. A., & Carlini, E. A. Dissociation of learning in marihuana tolerant rats. *Psychopharmacologia*, 1972, *25*, 49–56.

Cameron, O. G., & Appel, J. B. A behavioral and pharmacological analysis of some discriminable properties of d-LSD in rats. *Psychopharmacologia*, 1973, *33*, 117–134.

Friedman, M. The use of ranks to avoid the assumptions of normality implicit in the analysis of variance. *J. Am. Stat. Assoc.*, 1937, *32*, 675–701.

Greenberg, I., Kuhn, D. M., & Appel, J. B. Behaviorally induced sensitivity to the discriminable properties of LSD. *Psychopharmacologia*, 1975, *43*, 229–232.

Harley-Mason, J., Laird, A. H., & Smythies, J. R. Delayed clinical reactions to mescaline. *Confin. Neurol.*, 1958, *13*, 152–155.

Hays, W. L. *Statistics for the social sciences.* New York: Holt, 1973.

Henriksson, B. G., & Jarbe, T. Δ^9-Tetrahydrocannobinol used as discriminative stimulus for rats in positive learning in a T-shaped water maze. *Psychon. Sci.*, 1972, *27*, 25–26.

Hill, H. E., Jones, B. E., & Bell, E. C. State dependent control of discrimination by morphine and pentobarbital. *Psychopharmacologia*, 1971, *22*, 305–313.

Hirschhorn, I. D., & Rosecrans, J. A. Morphine and Δ^9-tetrahydrocannobinol: Tolerance to the stimulus effects. *Psychopharmacologia*, 1974, *36*, 243–253. (a)

Hirschhorn, I. D., & Rosecrans, J. A. Studies on the time course and the effect of cholinergic and adrenergic receptor blockers on the stimulus effect of nicotine. *Psychopharmacologia*, 1974, *40*, 109–120. (b)

Hirschhorn, I. D., & Rosecrans, J. A. A comparison of the stimulus effects of morphine and lysergic acid diethylamide (LSD). *Pharmacol. Biochem. Behav.*, 1974, *2*, 361–366. (c)

Hirschhorn, I. D., & Winter, J. C. Mescaline and lysergic acid diethylamide (LSD) as discriminative stimuli. *Psychopharmacologia*, 1971, *22*, 64–71.

Ho, B. T., & Huang, J. T. Role of dopamine in d-amphetamine-induced discriminative responding. *Pharmacol. Biochem. Behav.*, 1975, *3*, 1085–1092.

Huang, J. T., & Ho, B. T. The effect of pretreatment with iproniazid on the behavioral activities of β-phenylethylamine in rats. *Psychopharmacologia*, 1974, *35*, 77–81. (a)

Huang, J. T., & Ho, B. T. Discriminative stimulus properties of d-amphetamine and related compounds in rats. *Pharmacol. Biochem. Behav.*, 1974, *2*, 669–673. (b)

Huang, J. T., & Ho, B. T. Effects of nikethamide, picrotoxin and strychnine on 'amphetamine-state'. *European J. Pharmacol.*, 1974, *29*, 175–178. (c)

Iwahara, S., & Noguchi, S. Drug state dependence as a function of overtraining in rats. *Japanese Psychol. Res.*, *14*, 141–144.

Iwahara, S., & Noguchi, S. Effects of overtraining upon drug-state dependency in discrimination learning in white rats. *Japanese Psychol. Res.*, 1974, *16*, 59–64.

Jarbe, T. U. C., & Henriksson, B. G. Effects of Δ^8-THC, and Δ^9-THC on the acquisition of a discriminative positional habit in rats. *Psychopharmacologia*, 1973, *31*, 321–332.

Jarbe, T. U. C., & Henriksson, B. G. Discriminative response control produced with hashish tetrahydrocannabinols and other drugs. *Psychopharmacologia*, 1974, *40*, 1–16.

Jarbe, T. U. C., Johansson, J. O., & Henriksson, B. G. Δ^9-Tetrahydrocannabinol and pentobarbital as discriminative cues in the mongolian gerbil (meriones unguiculates). *Pharmacol. Biochem. Behav.*, 1975, *3*, 403–410.

Jones, C. M., Hill, H. F., & Harris, R. T. Discriminative response control by d-amphetamine and related compounds. *Psychopharmacologia*, 1974, *36*, 347–356.

Kubena, R. K., & Barry, H. Stimulus characteristics of marihuana components. *Nature*, 1972, *235*, 397–398.

Kuhn, D. M., Appel, J. B., & Greenberg, I. An analysis of some discriminative properties of d-amphetamine. *Psychopharmacologia*, 1974, *39*, 57–66.

McNemar, Q. *Psychological statistics*. New York: Wiley, 1969.

Miller, Jr., R. G. *Simultaneous statistical inference*. New York: McGraw-Hill, 1966.

Morrison, C. F., & Stephenson, J. A. Nicotine injections as the conditioned stimulus in discrimination learning. *Psychopharmacologia*, 1969, *15*, 351–360.

Myers, J. L. *Fundamentals of experimental design*. Boston: Allyn & Bacon, 1972.

Overton, D. A. State-dependent learning produced by depressant and atropine-like drugs. *Psychopharmacologia*, 1966, *10*, 6–31.

Overton, D. A. State-dependent or dissociated learning produced with pentobarbital (revised with commentary). In J. A. Harvey (Ed.), *Behavioral analysis of drug action*. Glenview, Illinois: Scott Foresman, 1971. Pp. 56–83.

Overton, D. A. Experimental methods for the study of state-dependent learning. *Federation Proc.*, 1974, *33*, 1800–1813.

Overton, D. A. A comparison of the discriminable CNS effects of ketamine, phencyclidine, and pentobarbital. *Arch. Intern. Pharmacodyn.*, 1975, *215*, 180–189.

Overton, D. A., & Winter, J. C. Discriminable properties of drugs and state dependent learning. *Federation Proc.*, 1974, *33*, 1785–1836.

Petrinovich, L., & Hardyck, C. D. Error rates for multiple comparison methods: Some evidence concerning the frequency of erroneous conclusions. *Psychol. Bull.*, 1969, *71*, 43–51.

Rosecrans, J. A., Goodloe, M. H., Bennett, G. J., & Hirschhorn, I. D. Morphine as a discriminative cue: Effects of amine depletors and naloxone. *European J. Pharmacol.*, 1973, *21*, 252–256.

Rozeboom, W. W. The fallacy of the null-hypothesis significance test. *Psychol. Bull.*, 1960, *57*, 416–428.

Ryan, T. A. Multiple comparisons in psychological research. *Psychol. Bull.*, 1959, *56*, 26–47.

Ryan, T. A. Significance tests for multiple comparisons of proportions, variances, and other statistics. *Psychol. Bull.*, 1960, *57*, 318–328.

Schechter, M. D. Ethanol as a discriminative cue: Reduction following depletion of brain serotonin. *European J. Pharmacol.*, 1973, *24*, 278–281. (a)

Schechter, M. D. Transfer of state-dependent control of discriminative behavior between subcutaneously and intraventricularly administered nicotine and saline. *Psychopharmacologia*, 1973, *32*, 327–335. (b)

Schechter, M. D. Effect of propranolol, d-amphetamine and caffeine on ethanol as a discriminative cue. *European J. Pharmacol.*, 1974, *29*, 52–57.

Schechter, M. D., & Rosecrans, J. A. CNS effects of nicotine as the discriminative stimulus for the rat in a T-maze. *Life Sci.* 1971, *10*, 821–832. (a)

Schechter, M. D., & Rosecrans, J. A. Behavioral evidence for two types of cholinergic receptors in the CNS. *European J. Pharmacol.*, 1971, *15*, 375–378. (b)

Schechter, M. D., & Rosecrans, J. A. Effect of mecamylamine on discrimination between nicotine and arecoline-produced cues. *European J. Pharmacol.*, 1972, *17*, 179–182. (a)

Schechter, M. D., & Rosecrans, J. A. Nicotine as a discriminative stimulus in rats depleted of norepinephrine or 5-hydroxytryptamine. *Psychopharmacologia*, 1972, *24*, 417–429. (b)

Schechter, M. D., & Rosecrans, J. A. Lysergic acid diethylamide (LSD) as a discriminative cue: drugs with similar stimulus properties. *Psychopharmacologia*, 1972, *26*, 313–316. (c)

Schechter, M. D., & Rosecrans, J. A. Nicotine as a discriminative cue in rats: Inability of related drugs to produce a nicotine-like cueing effect. *Psychopharmacologia*, *27*, 379–387. (d)

Schechter, M. D., & Rosecrans, J. A. Atropine antagonism of arecoline-cued behavior in the rat. *Life Sci.*, 1972, *11*, 517–523. (e)

Schechter, M. D., & Rosecrans, J. A. d-Amphetamine as a discriminative cue: Drugs with similar stimulus properties. *European J. Pharmacol.* 1973, *21*, 212–216.

Siegel, S. *Nonparametric statistics*. New York: McGraw-Hill, 1956.

Thompson, T., & Pickens, R. (Eds.) *Stimulus properties of drugs*. New York: Plenum Press, 1971.

Wilson, W. R., & Miller, H. A note on the inconclusiveness of accepting the null hypothesis. *Psychol. Rev.* 1964, *71*, 238–242.

Winer, B. J. *Statistical principles in experimental design*. New York: McGraw-Hill, 1971.

Winter, J. C. A comparison of the stimulus properties of mescaline and 2,3,4-trimethoxyphenylethylamine. *J. Pharmacol. Exp. Therap.*, 1973, *185*, 101–107.

Winter, J. C. The effects of 3,4-dimethoxyphenylethylamine in rats trained with mescaline as a discriminative stimulus. *J. Pharmacol. Exp. Therap.* 1974, *189*, 741–747.

York, J. L., & Winter, J. C. Assessment of tolerance to barbital by means of drug discrimination procedures. *Psychopharmacologia*, 1975, *42*, 283–287.

14

A Functional Analysis of the Discriminative Stimulus Properties of Amphetamine and Pentobarbital

Daniel W. Richards, III
Division of Social Sciences
Houston Community College

A functional analysis of a drug's discriminative-stimulus properties examines the variables affecting the ways a drug acquires and maintains control over a given aspect of behavior. The analysis addresses itself to the problem of possible interactions between a drug's discriminative properties and its unconditioned behavioral effects, as well as the other discriminative stimuli of the training environment. An attempt is made to fit a drug's discriminative properties into a functional explanation of behavioral adaptation. This analysis considers a drug treatment as a stimulus and is not dependent on an understanding of neural mechanisms or biochemical events mediating the transduction of chemical compounds to a psychological stimulus.

The role of response-reinforcement contingencies in drug discrimination has not been examined systematically. Most investigators agree that discriminative control maintained by external stimuli may be altered significantly by changing the reinforcement contingencies of the training environment (Nervin, *1973*). It is reasonable to wonder whether the same relationship exists in situations in which drugs acquire discriminative control of behavior. Or, asked differently, can the response-reinforcement relationships, behavioral contingencies, used in training environments of drug-discrimination tasks affect the cueing properties of a given drug?

In *The Stimulus Properties of Drugs,* Catania (1971) discusses the importance of behavioral contingencies in understanding the functional properties of discriminative control.

> We therefore can proceed in one of two ways when we set about to study the discriminative effects of drugs. On the one hand, we can examine the functional properties of discriminative control; on the other, we can search for the mechanism of action or, in other words, for the drug receptor. In the first case, we establish a stimulus–control relationship between drugs and behavior and then, by administering different dosages of the drugs or by administering different drugs, we explore the functional properties of this control. Thresholds, the dimensions along which drugs are generalized, and interactions with behavioral contingencies such as the schedule of reinforcement or the temporal parameters of stimulus presentation [p. 153].

This paper describes a series of experiments to investigate the functional properties of amphetamine and pentobarbital discriminative control.

Hearst, Korensko, and Poppen (1964) reported that in a line–tilt discrimination task, training with a variable interval (VI) schedule generated a "peaked" generalization gradient, while training under a differential reinforcement of low (DRL) response-rate schedule resulted in a relatively flat gradient. These authors suggest that the difference between the gradients of each schedule condition was the result of the animals' response patterns adapted to meet the reinforcement demands of each schedule. The two schedule contingencies are differentially dependent on different proportions of internal versus external cues of the controlling stimulus complex. Thus, in tasks in which the discriminative cue originates from an external source such as the line–tilt discrimination task, behavior controlled by DRL contingencies is less sensitive to changes occurring with an external discriminative stimulus than when the task is under VI control. Haber and Kalish (1963) reported the same schedule effect on generalization gradients using a wavelength continuum.

Kramer and Rilling (1970) explained the shift in the internal versus external stimulus complex produced by the DRL contingencies to be the result of mediating behaviors required by animals to space their lever responses by a temporal interval between two successive lever responses. These activities by their consistent nature serve as discriminative stimului for subsequent behavior (Ferster & Skinner, 1957). The mediating behaviors take the form of a responses chain, in which each response acts as a discriminative (or eliciting) stimulus for the next response in the chain. By adopting such a behavioral strategy, it is possible for the animal to accomplish the crucial feature of the DRL schedule, to separate lever responses by a temporal interval.

The present investigation attempts to reverse the Hearst *et al.* (1964) paradigm by comparing the acquisition and generalization of discriminative control under DRL and VI schedules, using a drug-discrimination task in which the discriminative cues originate in the animals' internal environment. To ensure that differences in discriminative control were caused by training under two different behavioral contingencies and not by a property of the drug, two very different drugs, amphetamine and sodium pentobarbital, were employed as the two drug-cue conditions.

EXPERIMENT 1: DRL-VI SCHEDULE EFFECTS ON ACQUISITION

Methods and Procedures

Subjects were 36 adult male albino (Sprague–Dawley) rats, divided into six equal groups. Three groups were given discrimination training under DRL contingencies, and three groups were trained under VI contingencies. All training and testing sessions occurred in five standard two-lever operant chambers (Scientific Prototype, Model 100). Behavioral contingencies, response-reinforcement totals, and cumulative recorders for each chamber were controlled by a central solid-state bank of programming equipment.

The DRL rate schedule reinforces only those lever responses that are separated from previous ones by a certain interresponse time. The animals were shaped to perform the two-lever multiple DRL-15-sec, DRL-15-sec schedule by gradually increasing the interresponse timer across the pretraining sessions. The operative reinforcement lever was alternated daily throughout the pretraining sessions to ensure the animals' equal experience on both response levers. Premature or incorrect lever responses reset the interresponse interval timer.

The variable-interval schedule reinforces responses on the correct lever at intervals that average a certain time. The animals were shaped to perform the 2-lever multiple VI-60-sec, VI-60-sec schedule by increasing the interval means across the pretraining sessions. The operative reinforcement lever alternated daily throughout pretraining. Incorrect and premature-correct lever responses were not reinforced.

The basic paradigm is diagrammed in Tables 14.1 and 14.2. Three different variations of the A–B lever–lever cue conditions were used under each of the two schedule conditions. The three A–B cue variations were: (*a*) 1 mg/kg of *d*-amphetamine versus saline; (*b*) 15 mg/kg of pentobarbital versus saline; and (*c*) 1 mg/kg of *d*-amphetamine versus 15 mg/kg of pentobarbital.

Table 14.1
Training Experimental Procedures and Behavioral Contingencies

	Injection cue	Reward condition	Schedule	Testing (no reward)
Group 1	Drug A	Lever A	DRL	$\frac{\text{Lever A responses}}{\text{Total response}}$
	Drug B	Lever B	DRL	$\frac{\text{Lever B responses}}{\text{Total response}}$
Group 2	Drug A	Lever A	VI	$\frac{\text{Lever A response}}{\text{Total response}}$
	Drug B	Lever B	VI	$\frac{\text{Lever B response}}{\text{Total response}}$

Table 14.2
Drug Discrimination Paradigm

Schedule	Response-reinforcement contingency
Differential reinforcement of low response rates (DRL—15 sec)	Appropriate lever responses separated by given temporal interval are rewarded Typical response pattern $R_L \rightarrow$ Collateral response sequence or activity $\rightarrow R_L$
Variable interval (VI—60 sec)	Appropriate lever responses are reinforced at intervals that average 60 sec. Typical response pattern—steady consistent response rates without collateral behaviors

All animals received an equal amount of preliminary training to shape stable performance levels before discrimination training was started. Forty 20-min sessions constituted drug-discrimination training. An A or B drug injection was given 15 min before each training session to designate the corresponding A–B lever condition operative during the session. Before every fifth training session, the animals were given a 10-min test probe followed by 20 min of discrimination training. During the test, reinforcement was not available regardless of lever response. Lever responses for each test probe were recorded at the end of each test and the reward mechanism turned on for the normal discrimination session that followed. The A–B drug–lever cue conditions for the 4 training sessions separating the acquisition test probes were arranged in one of the following sequences for the 10 tests occurring throughout acquisition: ABAB, AABB, BABA, BAAB, ABBA. The order was balanced both within and between groups to minimize order effects due to extraneous sources.

Results and Discussion

The total number of responses on each lever in the 10 acquisition tests was combined to form 5 acquisition test blocks. Each test block was composed of 2 observations of each animal, each under the 2 training drug-cue conditions (A and B).

Sessions with 10 or fewer responses were defined as incidents of behavioral toxicity. The number of such toxic sessions was recorded for each group and is listed in Table 14.3.

Note that the incidence of toxic drug effects occurred only in groups trained under the DRL contingency. Since the total response density was substantially lower under DRL contingencies than under VI contingencies, the apparent susceptibility of the DRL condition to toxic effects was simply the result of the differences between the absolute baseline response levels. Behavioral toxicity was defined by any session with 10 or fewer total responses. Animals treated with pentobarbital reduced their total number of responses under both conditions. Amphetamine-treated animals increased their total number of responses under VI contingencies, but decreased response totals under the DRL contingencies.

The toxic effects of amphetamine and pentobarbital were manifested in two different ways. The amphetamine-treated animals ceased responding because of drug-induced stereotypy. In contrast, the pentobarbital-treated animals ceased responding because the animals became anesthetized and stopped all movements temporarily.

In both cases, each drug's behavioral toxicity was overcome as all animals developed behavioral tolerance. In the case of amphetamine, this is the first time that behavioral tolerance has been reported for amphetamine-induced stereotypy.

The data for each of the five acquisition tests were arranged into a Schedule X Groups X Blocks repeated-measures analysis of variance model. Preliminary tests of homogeneity of variance, using noninteracting group means for the toxic sessions (Winer, 1971, p. 281–283) yielded a nonsignificant *F*-Max test.

A significant schedule effect, $F(1,60) = 17.8$, indicated that the training-reinforcement contingency was an important variable in the drug's functional properties. Furthermore, comparisons of the drug training groups under each schedule indicated that the schedule effect was present for amphetamine and pentobarbital regardless of the drug's being paired with saline or the other drug condition. Figure 14.1 shows summaries of the analysis-of-variance data for the respective groups and schedule conditions.

EXPERIMENT 2: SCHEDULE EFFECTS DURING GENERALIZATION

An experiment was undertaken to determine whether or not the schedule effect observed in Experiment 1 was restricted to the initial phases of learning, or whether it also held for well-learned tasks. Schuster and Balster (1977)

Table 14.3
Incidence of Behavioral Toxicity during Acquisition Training

		Acquisition tests				
Schedule	Task	1	2	3	4	5
DRL	P.S. anesthesia	2	—	—	1	—
	A.S. stereotypy	4	2	2	2	—
	A.P. stereotypy	3	—	—	—	—
	anesthesia	1	—	—	1	—
VI	P.S. anesthesia	—	—	—	—	—
	A.S. stereotypy	—	—	—	—	—
	A.P. stereotypy	—	—	—	—	—
	anesthesia	—	—	—	—	—

Behaviorally toxic effects = any test with <10 total responses.

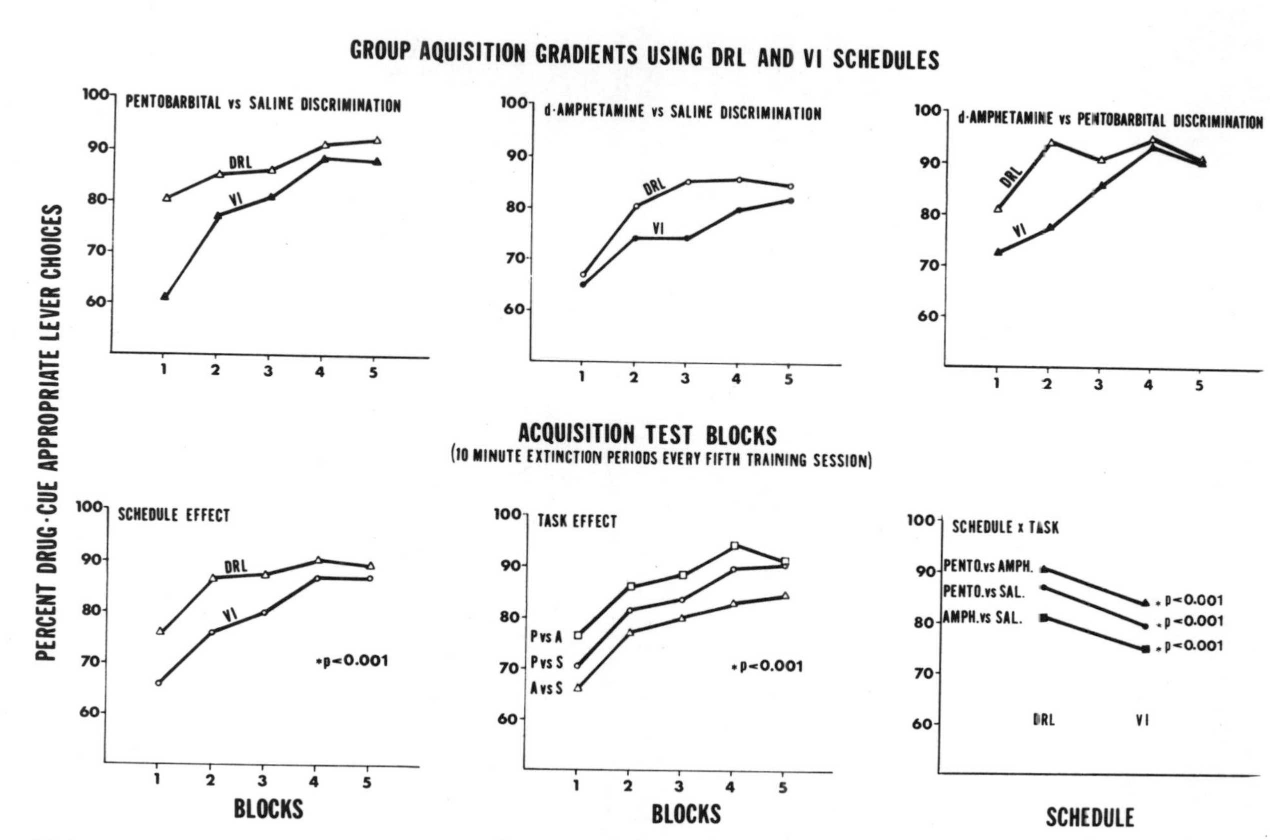

Figure 14.1 A comparison of group discrimination ratios (cue lever responses/total responses) for animals trained under DRL and VI schedules of reinforcement.

noted important parameters of drug discriminability to include: (*a*) rate of acquisition; (*b*) degree of response control at asymptote (percentage of drug-cue-appropriate lever choices); and (*c*) *lever-choice* responses to novel training drug dosages in generalization tests.

Methods and Procedures

After acquisition procedures were completed, all subjects were given 12 separate 10-min generalization tests occurring every fifth session. Four 20-min retraining sessions, including two exposures of each group's A and B drug lever conditions, always occurred between each generalization test session. After each 10-min test, the animal was placed in its home cage. The order of test administration was determined by constructing two 6 (subjects in each group) × 6 (test dosages) Latin squares and administering to each subject the test dosage on the order dictated by the 2 squares. Each test was separated from the next one by 4 retraining sessions.

Results and Discussion

Test results for each drug condition were arranged into a Schedule × Tests analysis of variance model. Three separate analyses indicated that effects due to schedule contingencies are not significant enough to be distinguished in generalization. The proportion of lever-choice responses for the three drug training groups under each schedule condition are shown in Figure 14.2. The absence of any systematic difference in lever-choice responses to novel testing conditions suggests that the schedule effect is restricted to the acuqisition phases of discrimination training.

Presumably, the important difference between the two schedules of reinforcement is the type of response patterns adopted by the animals under each condition. Since a variable-interval reinforcement schedule rewards lever responses at differing intervals, it is to the animal's advantage to press the appropriate lever at consistent, moderately high response rates. Under DRL contingencies that require the animal to separate successive lever responses with a temporal interval, however, a variety of collateral, stereotyped behavioral chains are needed to fill the pause between successive responses (Bruner & Renusky, 1961; Dews & Morse, 1958; Hodos, Ross, & Brady, 1962; Holz, Agrin, & Ulrich, 1963; Kapostins, 1963; Laties & Weiss, 1962; Segal & Holloway, 1963; Wilson & Keller, 1953). This collateral behavior is believed to serve as mediating behavior in the sense conveyed by Ferster and Skinner's (1957) definition: "Behavior occurring between two instances of the response being studied . . . which is used by the organism as a controlling stimulus in subsequent behavior [p. 729]." The specific collateral behavior is functionally related to the efficiency of subjects in spacing their responses (Laties, Weiss, Clark, & Reynolds, 1965), so that it produces a set of internal cues for the dis-

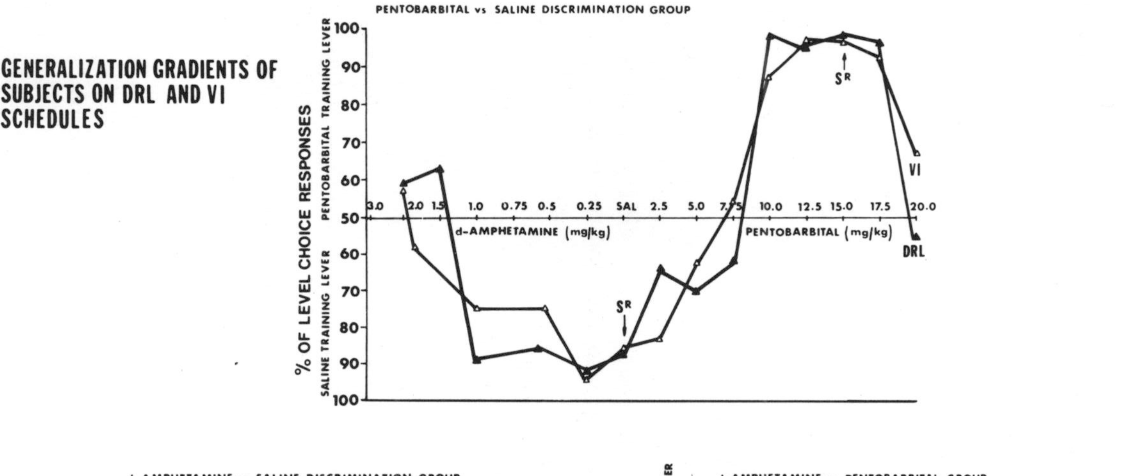

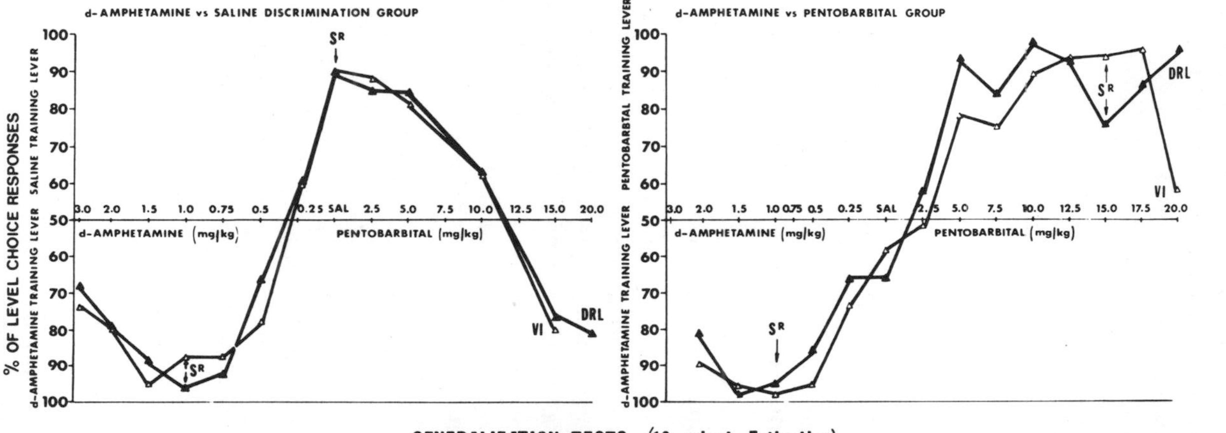

Figure 14.2 A comparison of discrimination ratios (novel drug lever/total responses) of animals given the same drug-discrimination training under DRL and VI contingencies to novel drug-cue conditions.

criminating stimuli. These then enable the animal to respond appropriately after the minimum waiting time.

The difference between response patterns of each schedule becomes important in considering the difference in rate of acquisition observed in this study of the animals being trained under DRL contingencies. The attention already being devoted to the interoceptive cues of DRL timing enables animals more readily to associate different drug conditions with corresponding changes in reinforcement. This explanation does not suggest that the magnitude of drug response is any different for the two schedule conditions; the degree of discriminative control and the response of animals to novel dosages and drug conditions were not found to be significantly different for the two conditions. Apparently the most significant factor distinguishing the two schedule conditions is the manipulation of the animal's attention. In tasks that associate internal discriminative events with appropriate lever-choice responses, such as drug discrimination, DRL contingencies produce a shift in attention that facilitates acquisition of discriminative responding.

In summary, we observed that manipulation of the reinforcement schedule during drug-discrimination training significantly improved the rate in which animals acquired discriminative control of responding for both amphetamine and pentobarbital. This was found with either drug's being discriminated from saline or from the other drug. However, other measures of discriminability—degree of stimulus control (percentage of drug-cue-appropriate responses) at asymptote, generalization gradients to novel dosages and drug conditions—were not significantly different for the two schedules. This suggests that the important difference between the two conditions was the schedule-related shift in internal versus external cueing. The greater dependency of DRL contingencies on interoceptive cueing facilitates the acquisition of drug-discriminative response control for both the stimulant and the tranquilizer.

FUNCTIONAL ANALYSIS MODEL

Adaptive Significance of Drug Discriminative Control

In a strict sense, behavioral adaptation refers to modifications in behavior to meet reinforcement demands of the environment. In most drug-discrimination studies, it is assumed implicitly that changes in lever-choice responses result in corresponding increments in reinforcement-appropriate responses. It is true that responses must occur on the drug-appropriate lever in order to be rewarded. But it is also true that not every correct lever response is reinforced. This is especially important when the schedule demands are very restrictive, as in the case of the DRL situation. Under the timing requirements of DRL contingencies, we find tremendous differences in reinforcement-appropriate responses even though animals may make a proportionately equal number of drug-appropriate lever-choices. To better understand the relationship between lever-choice response-control and reinforcement-appropriate responses, we shall examine the data from experiment 1.

Drug versus Nondrug DRL Discrimination

Drug versus nondrug discrimination under DRL contingencies could be viewed as both lever-discrimination and schedule-performance tasks. In the case of lever discrimination, the proportion of lever-choice responses under each cue condition is the common response measure. The response measure indicating schedule performance is the number of reinforcement-appropriate responses under each drug condition. The fact that each animal performs the same schedule on each lever and each animal is tested under both conditions makes it possible for each animal to serve as its own control. Lever-choice control and schedule performance can be assessed, therefore, by comparing each subject to itself under the two training-drug conditions. Figure 3 shows the development of lever-choice discriminative control and corresponding number of reinforcement-appropriate lever responses for each of the two DRL drug-discrimination groups across the five acquisition test blocks. In both the amphetamine and pentobarbital-discrimination groups, the animals performed at a significantly higher level of cue-appropriate lever choices under the drug condition than under saline. It is also true, however, that both groups made more reinforcement-appropriate responses under a saline condition than under their respective paired-drug condition. This was an unexpected finding. It indicates that even though an animal may choose the appropriate lever more often under the drug-cue condition, it actually adapts better to the environmental demands of the discrimination task without a drug cue. Drug-discriminative response control might be accompanied by a penalty, that is, by the unconditioned disruptive effects of the drug to schedule performance on the appropriate lever.

MEASURES OF RESPONSE TOPOGRAPHY IN DRUG DISCRIMINATION

A useful model for seeking explanations of the adaptive significance of drug-discriminative control must account for the possible interaction between the acquired discriminative properties and the unconditioned behavioral effects of the drug. Ideally, such a model should incorporate response measures that would simultaneously indicate values of each of these two drug effects individually.

The model proposed here attempts to account for both changes in lever-choice behavior and changes in response patterning as related to schedule performance. The possible responses emitted in DRL drug-discrimination studies may be differentiated into the three separate response classes diagrammed in Figure 14.4. Alpha (α) responses, which are correct lever

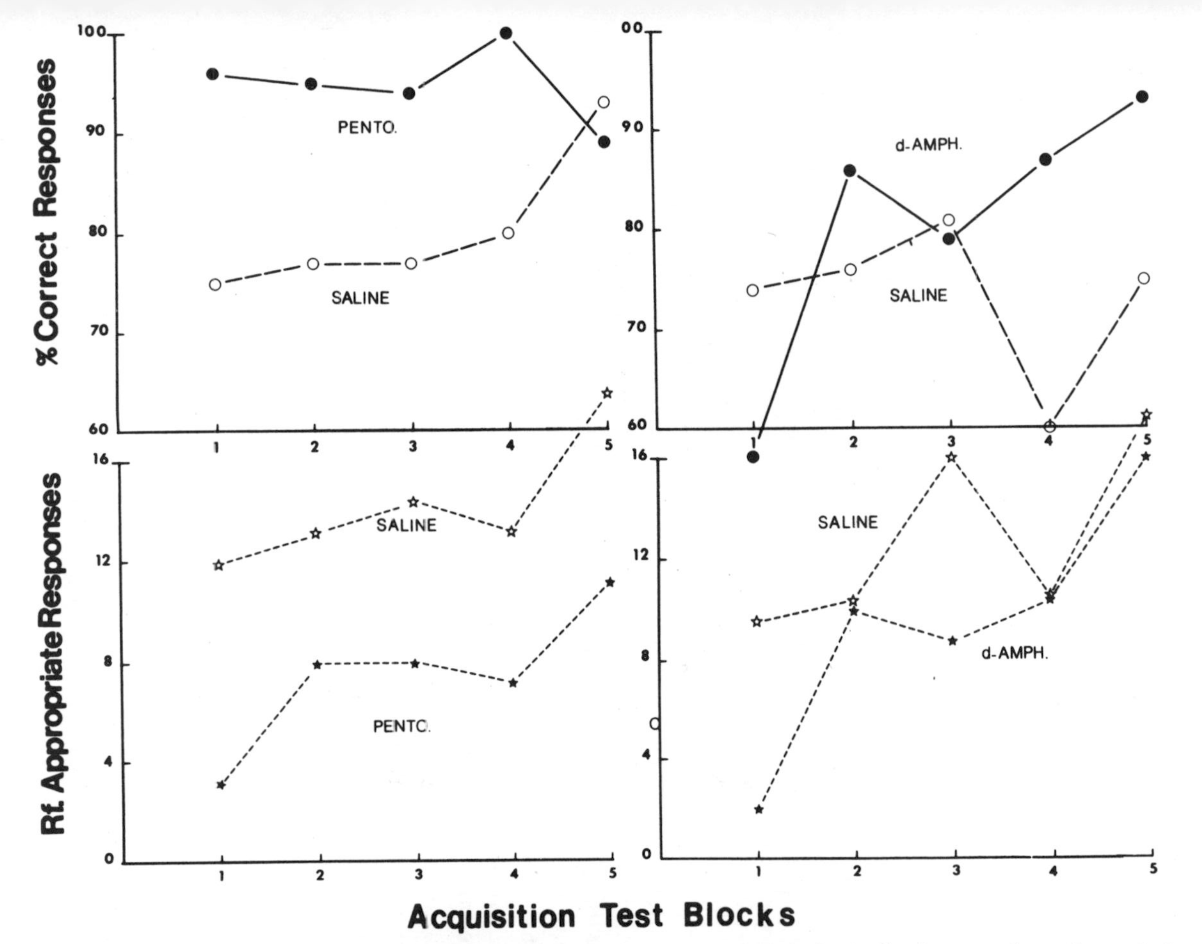

Figure 14.3 A comparison of behavioral adaptation and discriminative response control of animals given amphetamine and phentobarbital discrimination training from the nondrug (saline) condit on.

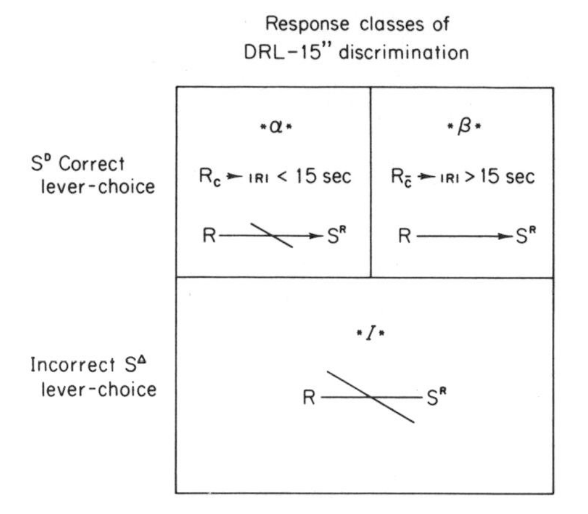

Figure 14.4 Response-class categories of DRL response topography model. Alpha (α) responses are premature, appropriate cue-lever responses. Beta (β) responses are appropriate cue-lever responses separated from previous responses by minimum pause of 15 sec or more. Inappropriate (*I*) responses are responses on the noncue lever.

responses are separated from previous response by an interval less than 15 sec; beta (β) responses, which are correct lever responses occurring at least 15 sec after the previous response and are the only ones reinforced during normal training conditions; and I responses, which are incorrect or inappropriate lever-choices. The latter occur on the inappropriate lever at undetermined intervals after previous responses.

Measures for the three response classes are obtained by dividing the total responses into separate proportions, which then are the response-probability measures of each class. The model is sensitive to changes in response pattern as well as changes between levers, and it lends itself to examination of two functional properties of drug-discriminative control, using this model. Those two ways are to consider effects produced by discrimination training and by generalization testing.

Effects of Drug-Discrimination Training

Drug-discrimination training affects behavior in two important ways. The first is by the development of behavioral tolerance. When drugs are repeatedly administered to animals performing the behavioral demands of a reinforcement schedule, and there is a diminution of drug-induced disruptive schedule effects over trials, it is refered to as behavioral tolerance (Schuster, Dockens, & Woods, 1966).

The second major behavioral mechanism affected by discrimination training is the acquisition of discriminative-response control brought about by the repeated associations of alternating drug treatments with simultaneous alternations in reinforced lever-choice responses. Figure 14.5 shows the effects of dis-

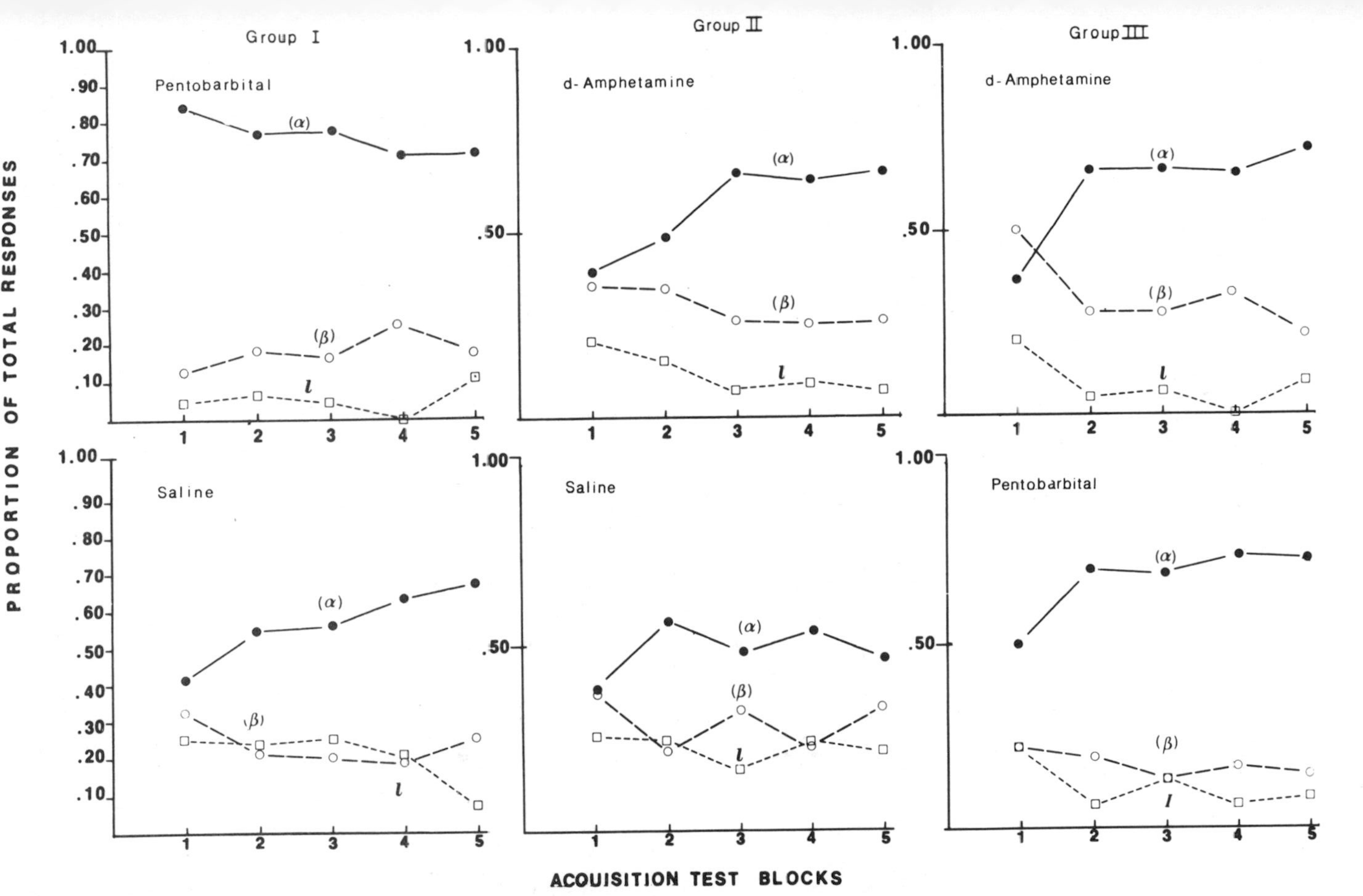

Figure 14.5 Response class probability gradients in drug-discrimination acquisition: correct α, β; and incorrect I. Comparison of response-class probabilities of DRL drug-discrimination group means across five test blocks of acquisition training.

crimination training. Group I, trained to discriminate pentobarbital from saline, evidenced two different sets of response-class probabilities under the drug and nondrug conditions. Under the pentobarbital condition, the proportion of incorrect lever responses (I) remained at about the same low level throughout all phases of training; this indicates that pentobarbital developed a high degree of stimulus control over the lever-choice responses in these animals almost immediately. The first few tests under the drug condition produced abnormally high alpha and low beta response-probability scores compared to the values obtained from the same animals under the nondrug condition. The alterations of these alpha and beta values under the drug condition indicate that the drug disrupted the animal's normal response pattern. With continued training, the alpha and beta levels adjusted so that by the fifth acquisition block these values approximated the corresponding levels reached under the saline condition. The figure shows that the animals developed behavioral tolerance to schedule-disruption effects even though the drug continued to cue appropriate lever-choice responses.

Group II subjects were trained to discriminate amphetamine from saline. As discussed previously, these subjects were affected more than any other group by the incidence of behavioral toxicity (Table 14.3) during the initial phases of training. The response-class probabilities of this group on the first tests were very similar under drug and nondrug conditions. The toxic effects of amphetamine were manifest by the tendency of these animals to perseverate in their execution of the mediating behaviors of the collateral response chain demanded by the DRL schedule. Amphetamine administration actually improved the efficiency of these animals to respond appropriately to the schedule demands. About 50% of these animals correctly delayed their lever responses according to the required interresponse time. The total number of responses for the entire group was so low, however, that four of the six subjects were judged to be toxic, an effect eventually overcome when they developed a higher level of alpha responses and corresponding lower levels of beta responses. This change in response-class probabilities suggests that these animals increased their lever responses and decreased their interresponse interval mediating behaviors. By doing this, they developed behavioral tolerance to the toxic effects of amphetamine.

Note that the direction of change in alpha and beta response classes is different across training for amphetamine and pentobarbital. In the case of pentobarbital, the animals developed behavioral tolerance to its schedule-disruptive effects reflected by the reduction of alpha and increase of beta response probabilities. In contrast, behavioral tolerance to amphetamine developed by the increase of alpha and decrease of beta levels.

Group III subjects were trained to discriminate pentobarbital from amphetamine. This group showed a level of *I* responses throughout training lower than either of the two other groups, indicating that the two drugs are more discriminable from each other than either drug is from the nondrug condition. This group was less impaired by toxic drug effects than either of the other two groups.

Effects of Generalization Testing

An important consideration of the functional analysis model is the effects produced by manipulations of the quantitative or qualitative dimensions of the discriminative stimulus.

In part, the drug pair combinations selected for this study were based on the "mixed method" recently articulated by Overton (1974). This method makes it possible to compare the discriminability of two drugs in terms of qualitative and quantitative dimensions. To analyze discriminability in a quantitative dimension is akin to stimulus-intensity gradients and can be obtained by using novel test dosages of the training stimulus (Shuster & Balster, 1976). A qualitative dimension of discriminability is a more complex problem and involves the phenomenon of "transfer test overinclusiveness" (Overton 1974). Overton has suggested that when animals given discrimination training with one drug are given a second drug they will, under certain conditions, yield results showing the two drugs to be identical when in fact they are different. Apparently the weakness of cross-drug generalization arises from the fact that the animal has no way to respond when the test drug is like neither of the training drug conditions. The cue drug combinations employed in experiments described in this paper allow the identification of errors of overinclusion by having already demonstrated in acquisition the discriminability dimension being tested in generalization. Stated differently, acquisition training has demonstrated that amphetamine and pentobarbital are both discriminably different from saline as well as from each other. Therefore, if an animal responds differently, it must be evidence that transfer-test overinclusion has taken place. Using this model, it should be possible to identify overinclusiveness when it occurs and, in that way, better understand its role in drug discrimination.

The response-class gradients for the three training groups across the generalization test dosages are graphed in Figure 14.6. The gradients obtained for Group I (P vs. S) subjects indicate that these animals successfully covaried their lever-choice behavior between the 2 training conditions (5 and 15 mg/kg of pentobarbital) but failed to change levers only alpha and beta response probabilities when tested with novel amphetamine dosages. That is, they showed "transfer-test overinclusion" to all of the amphetamine test dosages.

The gradients of Group II (A vs. S) again indicate that these animals were able to shift the majority of their responses to the other lever when they were tested with intervening amphetamine dosages. These animals also responded to dosages of less than 12.5 mg/kg of pentobarbital with saline-appropriate lever

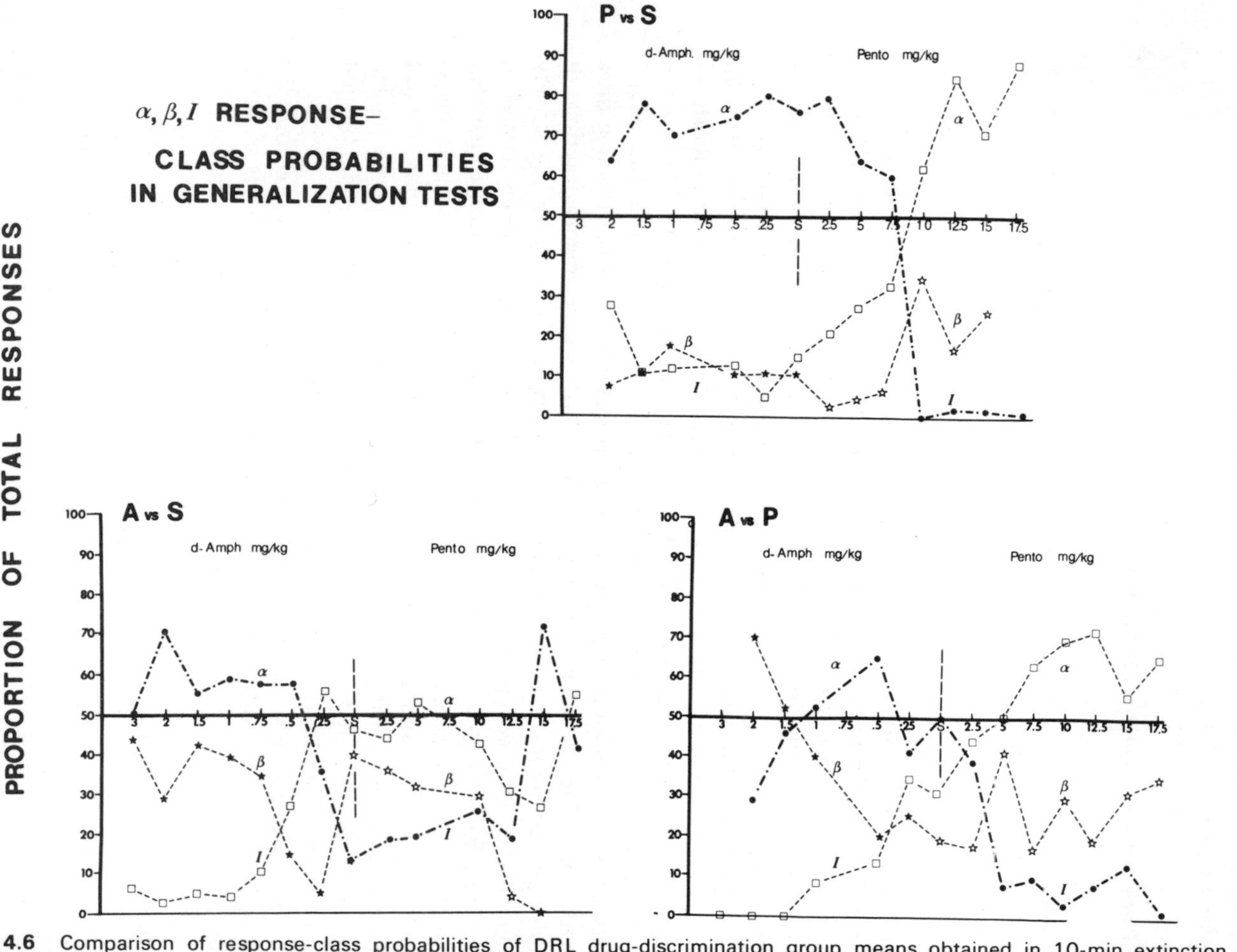

Figure 14.6 Comparison of response-class probabilities of DRL drug-discrimination group means obtained in 10-min extinction tests following novel drug-cue pretreatments. Correct: α and β, incorrect: *I*.

choices. Thus, both groups responded overinclusively to novel drug treatments that extended beyond the training continuum.

Group III (A vs. P) presented a situation in which the novel drug condition lies, theoretically, between the two training dosages. In this group, there were no errors of transfer-test overinclusiveness because the animals responded with approximately random lever choices to saline. These results indicate that animals will respond to discriminably different drug treatments with random behavior when the test drug is different from the two training conditions. But tests with novel drug treatments that fall to either side of the training drugs are highly susceptible to errors of overinclusion.

Analysis of Amphetamine-Induced Stereotypy and Discriminative Control

In addition to its ability to identify errors of overinclusive responding. The model provides alpha and beta gradients across incremental amphetamine dosages. These reflect the alterations in response topography that accompany the gradual development of stereotyped behavior. In animals trained to discriminate amphetamine, the drug-induced stereotypy became toxic behaviorally at dosages of 3 mg/kg or higher.

The word "stereotypy" has been used in psychopharmacology in connection with amphetamine-induced hyperactive behaviors (Fog, 1972; Randrup & Munkvad, 1967). The biochemical and neural mechanisms underlying this syndrome have been studied extensively in animals (Fog, Randrup, & Pakkenberg, 1967; Munkvad, Pakkenberg, & Randrup, 1968; Costall, Naylor, & Olley, 1972). Types of behavioral acts have been studied extensively (Ellinwood, Sudilovoky, & Nelson, 1972), as have the various species susceptible to this syndrome (Randrup & Muukvad, 1967). Very little is known, however, about the role stereotypy has in normal ongoing behavior. It is true that an important pharmacological relationship exists between amphetamine-induced stereotyped behavior in animals and the psychotic status associated with the psychopathology of schizophrenia in man (Snyder, 1972; Snyder, Banerjee, Yamanura, & Greenberg, 1974). Empirical support for this relationship comes from the striking fact that neuroleptic drugs found to have strong antipsychotic effects clinically antagonize or prevent amphetamine-induced stereotypy in animals (Fog, 1972; Snyder, Taylor, Coyle, & Meyerhoff, 1970).

As we consider discriminative control in this model, it is also possible to determine the changes occurring to the response topography under various dosages of amphetamine since both measures are reported simultaneously. What is interesting in drug generalization is the apparent generalization of stereotyped behavior reflected by the changes in alpha and beta response probabilities under increasing amphetamine dosages. Results from Group II (A vs. S, Figure 14.6) and Group III (A vs. P, Figure 14.6) showed that the major effects observed in tests with 1 mg/kg or more of *d*-amphetamine was altered response topography, that is, changes in the animals alpha and beta values and not their lever choices. Both groups of animals became increasingly efficient in performing the schedule demands. Beta value increased and alpha value decreased until they closely approximated each other. At about 3 mg/kg, the behavior of the animals in these two groups stopped because of toxic stereotypy. The behavioral blockage invariably occurred because the animals were away from the response lever, being engaged in some sort of repetitious act that confined them to a small portion of the chamber away from the response lever. The increase in drug dosage seemed to facilitate the performance of the interresponse interval mediating behaviors demand by the timing requirements of the DRL schedule up to a point. Beyond that point, however, the animals' activity became aimless and never returned to the response lever. The change in response topography was also evident in the cumulative records of these animals.

Figure 14.7 compares 3 cumulative records of one subject. The animal was trained to discriminate amphetamine from saline. The cumulative records are from 10-min generalization tests conducted under extinction conditions. The top record shows the animal's response to a test with 1 mg/kg of *d*-amphetamine. Note that all of the responses are restricted to the drug-appropriate lever. The gradual slope and regular occurrence of reinforcement slash indicator in the cumulated record is typical of the topography generated by a DRL-15-sec schedule. This response pattern and corresponding gradient, however, are dramatically changed with higher amphetamine dosages. The

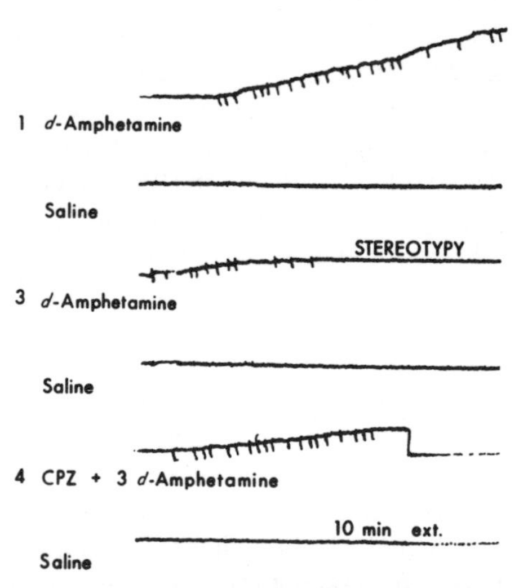

Figure 14.7 Typical amphetamine versus saline subject. Comparison of three cummulative test records of one rat given amphetamine- versus saline-discrimination training.

record just below the 1 mg/kg test is the results of the same animal's being tested with 3 mg/kg *d*-amphetamine. Notice the responses all occur on the drug-appropriate lever (marks on the five pen line are smudge marks and not lever responses), but the response topography is very different. The flatter slope of the drug-appropriate response line indicates the animal is responding much less than before. Notice the interval between successive reinforcement-appropriate slashes contains fewer responses and in some instances is longer than before. In fact, the major change in the animal's behavior observed in these 2 test records is the dramatic increase in away from the lever time in the test session with 3 mg/kg *d*-amphetamine. The animal's time away from the lever was consumed by repetitious performance of some component of its collateral response chain. The tendency for the animal to maintain the behavior seemed to increase as the test session progressed. However, the last slash mark on the line was in fact the last response even though the animal was left in the environmental chamber for an additional 2-hr observation period. The changes in this animal's behavior from the test with 1 mg/kg of *d*-amphetamine, in which the alpha value was much higher than the beta value to a point where the two values approximated each other as in the test with 3 mg/kg of *d*-amphetamine, was accompanied by an increasing and ultimate dominance of stereotyped behavior. The behaviorally toxic effect of the 3 mg/kg dosage is the blockage of ongoing schedule performance by the repetitious performance of a single act. This suggests that the drug's cueing properties may be different from the behavioral toxic effects brought about by amphetamine-induced stereotypy. To test the possible differentiation of drug effects of amphetamine discriminative properties from stereotyped behavior, an attempt was made to antagonize amphetamine stereotypic toxicity without impairing amphetamine discriminative control. To do this, the same subject was pretreated with 4 mg/kg of chlorpromazine (CPZ) 45 min before being tested with 3 mg/kg of *d*-amphetamine. The results of the test with CPZ pretreatment are shown in Figure 14.7. Note the absence of any extended interreinforcement-appropriate slash intervals. This indicates that the stereotypical behavioral blockade was successfully antagonized by CPZ pretreatment. Notice also that all responses occurred on the drug-appropriate lever; discriminative properties were unaffected. This implies that the basis of amphetamine-induced stereotypy can possibly be differentiated chemically from amphetamine's discriminative stimulus properties.

ACKNOWLEDGMENTS

The experiments were conducted while the author was a postdoctoral fellow in the Section of Neuropsychopharmacology, Texas Research Institute of Mental Sciences, Houston, Texas.

The author gratefully acknowledges the assistance of Carol Meyers, Brenda Connelly, and Jim Kendall of the Texas Research Institute of Mental Sciences, and Rusty Kelley of Texas Institute of Rehabilitation and Research, Houston, in the collection and analysis of the experimental data.

REFERENCES

Bruner, A., & Renusky, S. H. Collateral behavior in humans. *J. Exp. Anal. Behav.*, 1961, *4*, 349–350.

Catania, A. C. Discriminative stimulus functions of drugs: Interpretations. In T. Thompson & R. Pickens (Eds.), *The stimulus properties of drugs.* New York: Appleton, 1971. Pp. 149–155.

Costall, B., Naylor, R. J., & Olley, J. E. The substantia nigra and stereotyped behavior, *Eur. J. Pharmacol. 18*, 95–106, 1972.

Dews, P. B., & Morse, W. H. Some observations on an operant in human subjects and its modification by dextroamphetamine. *J. Exp. Anal. Behav.*, 1958, *1*, 359–364.

Ellinwood, E. H., Sudilovsky, A., & Nelson, L. Behavioral analysis of chronic amphetamine intoxication. *Biol. Psychiat.*, 1972, *4*, 215–230.

Ferster, C. B., & Skinner, B. F. *Schedules of Reinforcement.* New York: Appleton, 1957. P. 37.

Fog, R. On stereotypy and catalepsy: Studies on the effect of amphetamines and neuroleptics in rats. *Acta Neurol. Scand.*, 1972, *Suppl. 50, 48*, 10–63.

Fog, R. L., Randrup, A., & Pakkenberg, H. Aminergic mechanisms in corpus striatum and amphetamine induced stereotyped behavior, *Psychopharmacologia*, 1967, *11*, 179–183.

Haber, A., & Kalish, H. I., Prediction of discrimination from generalization after variations in schedule of reinforcement. *Science*, 1963, *142*, 412–413.

Hearst, E. Korensko, M. B., & Poppen, R. Stimulus generalization and the response-reinforcement contingency. *J. Exp. Anal. Behav.*, 1964, *7*, 369–380.

Hodos, W., Ross, G. S., & Brady, J. V. Complex response patterns during temporally spaced responding. *J. Exp. Anal. Behav.*, 1962, *5*, 473–479.

Holz, W. C., Azrin, N. H., & Ulrich, R. E., Punishment of temporally spaced responding. *J. Exp. Anal. Behav.*, 1963, *6*, 115–122.

Kapostins, E. E. The effects of DR schedules on some characteristics of word utterance. *J. Exp. Anal. Behav.*, 1963, *6*, 281–290.

Kramer, T. J., & Rilling, M., Differential reinforcement of low rates: A selective critique. *Psychol. Bull.*, *74*, 225–254, 1970.

Laties, V. G., & Weiss, B. Effects of alcohol on timing behavior, *J. Comp. Physiol. Psychol.*, 1962, *55*, 85–91.

Laties, V. G., Weiss, B., Clark, R. L., & Reynolds, M. D. Overt "mediating" behavior during temporally spaced responding. *J. Exp. Anal. Behav.*, 1965, *8*, 107–116.

Munkvad, I., Pakkenberg, H., & Randrup, A. Aminergic systems in basal ganglia associated with stereotyped hyperactive behavior and catalepsy. *Brain Behav. Evol.*, 1968, *1*, 89–100.

Nervin, J. A. Stimulus control. In J. A. Nervin & G. S. Reynolds (Eds.), *The study of behavior learning, motivation, emotion, and instinct.* Glenview, Illinois: Scott, Foresman, 1973. Pp. 115–148.

Overton, D. A., Experimental methods for the study of state-dependent learning. *Federation Proc.*, 1974, *33*, 1800–1813.

Randrup, A., & Munkvad, I. Stereotyped activities produced by amphetamine in several animal species and man. *Psychopharmacologia*, 1967, *11*, 300–310.

Schuster, C. R., & Balster, R. L. Discriminative stimulus properties of drugs. In T. Thompson & P. B. Dews (Eds)., *Advances in behavioral pharmacology.* Vol. 1. New York: Academic Press, 1977.

Schuster, C. R., Dockens, W. S., & Woods, J. H. Behavioral variables affecting the development of amphetamine tolerance. *Psychopharmacologia*, 1966, *9*, 170–182.

Segal, E. F., & Holloway, S. M. Timing behavior in rats with water drinking as a mediation. *Science*, 1963, *140*, 888–889.

Snyder, S. H. Catecholamines in the brain as mediators of amphetamine psychosis. *Arch. Gen. Psychiat.*, 1972, *27*, 169–179.

Snyder, S. H., Banerjee, S. P., Yamamura, H. I., & Greenberg, D. Drugs, neurotransmitters, and schizophrenia. *Science*, 1974, *184*, 1243–1253.

Snyder, S. H., Taylor, K. M., Coyle, J. T., & Meyerhoff, J. L., The role of brain dopamine in behavioral regulation and the actions of psychotropic drugs. *Amer. J. Psychiat.*, 1970, *127,* 117–125.

Wilson, M. P., & Keller, F. S. On the selective reinforcement of spaced responses. *J. Comp. Physiol, Psychol.*, 1953, *46,* 190–193.

Winer, B. J. *Statistical principles in experimental designs.* New York: McGraw-Hill, 1971.

15

Discriminative Control of Behavior by Electrical Stimulation of the Brain: A New Neuropharmacological Research Strategy

Ronald L. Hayes
Neurobiological and Anesthesiological Branch
National Institute of Dental Research

David J. Mayer
Department of Physiology
Virginia Commonwealth University

One of the most intriguing issues in the study of psychoactive agents is the precise definition of the central nervous system (CNS) process or processes that mediate the discriminable properties of drug stimuli. While this question is of intrinsic concern to anyone interested in pharmacological actions, the issue touches on more general problem of describing the possible structures and determinants of consciousness. Questions about the nature of conscious experience and the varieties of its affective, perceptual, and cognitive components have a universal immediacy that transcend traditional academic divisions. Different speculations have spawned whole schools of philosophies, and formal studies of the content of consciousness comprised some of the earliest experimental research in psychology. Aside from the significant philosophical and heuristic value of an examination of consciousness, there are more pragmatic benefits potentially derived from these studies. Many psychopathological syndromes, including drug abuse, are often described and/or explained by use of terms whose referents are found exclusively in conscious experience. Consequently, elucidation of the mechanisms underlying aspects of consciousness could advance significantly our appreciation of those processes that contribute to these pathological states.

The use of psychoactive agents as discriminable stimuli has the potential to reduce the uncomfortably amorphous phenomenon of consciousness to an objectively quantifiable and uniquely manipulatable set of behaviors. It is true that the study of the mechanisms underlying the psychoactive effects of drugs may proceed successfully, and practical benefits may be derived in the absence of any formal invocation of conscious experience. Indeed, the very real philosophical problems attending the scientific use of the term are sufficient reasons for caution. It is both instructional and appropriate, however, to point out the larger context in which these inquiries occur.

This chapter introduces combinations of behavioral and neuropharmacological techniques that should allow singularly direct evaluations of the sites and mechanisms of actions of drug-produced stimulus effects mediated by the CNS. These techniques require only the synthesis of existing physiological and behavioral technologies and allow powerful inferences about drug actions not possible by use of other procedures. The primary rationale of these paradigms is to capitalize on the properties of direct electrical or pharmacological intervention at restricted brain loci. Brain stimulation and microinjection of chemical agents obviously have their direct effects mediated by more limited anatomical sites than do systemic applications of drugs. In general, the consequences of brain stimulation may also be expected to be short-lived as compared to drug-produced effects. Moreover, the use of brain stimulation makes possible the application of sophisticated neurophysiological techniques to characterize more fully the neural substrate participating in the stimulus effects of these different treatments (Deutch, 1964).

The following sections first outline the general behavioral paradigms followed by discussions of the decisions possible in the context of given experimental manipulations. Finally, two specific experiments are reviewed in detail, including analyses of data outcomes and possible technical refinements.

PROCEDURES FOR EVALUATING STIMULUS CONTROL OF BEHAVIOR BY ELECTRICAL STIMULATION OF THE BRAIN

The interoceptive cue effects of brain stimulation may be considered procedurally interchangeable with those produced by systemic administrations of psychoactive drugs. Thus, any experiment designed to evaluate the discriminative properties of drugs is appropriate for similar analyses of the stimulus properties of brain stimulation. Consideration is given here to two variants of positively reinforced two-bar operant tasks performed by rats in Skinner boxes housed in sound- and light-attenuating environmental chambers. Two-bar tasks are well suited to the experiments outlined here because, as compared to other behavioral procedures, they yield easily quantifiable data bearing directly on the detectability of the stimulus. Also, subjects emit the same operant under all treatment conditions rather than differentially withholding responses or responding at different optimal rates in different stimulus conditions. Thus, discrimination performance in each stimulus condition is equally affected by nonspecific facilitory or inhibitory effects of various manipulations, including simple deviations in the degree of food deprivation.

In both types of 2-bar tasks, the rat is trained to press 1 bar in the presence of brain stimulation and the other in the absence of such stimulation. In the first or "single-trial" variant, each of the 2 conditions is commonly presented individually in a double-alternation sequence, that is, 2 days of 15-min sessions of stimulation paired with reinforcement availability on only one bar followed by 2 days of nonstimulation paired with reinforcement availability on only the opposite bar. Early in training, animals are continuously reinforced (FR 1), but the schedule is gradually increased to a FR-3. Simultaneously, an unreinforced period is introduced at the beginning of the session, and this is slowly extended to a duration of 2 min.

Responding during the initial unreinforced period, both in the presence and absence of brain stimulation, is obviously unconfounded with feedback to the animal on the appropriateness of its responses. Therefore, these responses constitute a basic datum by which the stimulus control of brain stimulation can be evaluated. Figure 15.1 presents a commonly used graphic representation of discrimination performance. The top line (A) plots the average percentage of total responses made on the stimulation-reinforced bar during the 2-min periods in which the animals were stimulated. The bottom line (B) plots the same percentage for responses made in the absence of brain stimulation. Thus, the top and bottom lines present data that describe the appropriateness of the animals' responding along the two independent response dimensions defined by the presence or absence of brain stimulation. Since the top line reflects correct responding, and the bottom line reflects incorrect responding, the distance between them (D′) is an index of the degree of stimulus control.

We developed the second or "multiple trial" variant of the two-bar task (Hayes, 1975) to take advantage of the fact that unlike interoceptive cues produced by drugs, stimuli associated with brain stimulation are temporally punctate, not seeming to long outlast that stimulation. Thus, rather than exposing the animal to only a single stimulus condition on any given daily session, one can alternate rapidly within each session between conditions associated with the presence or absence of brain stimulation. Consequently, while the behavioral requirements outlined below are straightforward, accurate administration of these contingencies obviously requires extensive use of programmable logic equipment.

Rats are given 80 daily trials separated by a 20-sec intertrial interval (ITI), defined by the darkening of the environmental chamber and the absence of any treatment or reinforcement for a bar press. Unless made during the "observation interval" discussed below, bar presses during the last 10 sec of the ITI increase the duration of the ITI an additional 10 sec. This is done to suppress responding during the ITI.

At the end of the ITI, the house light comes on, and the animal is either

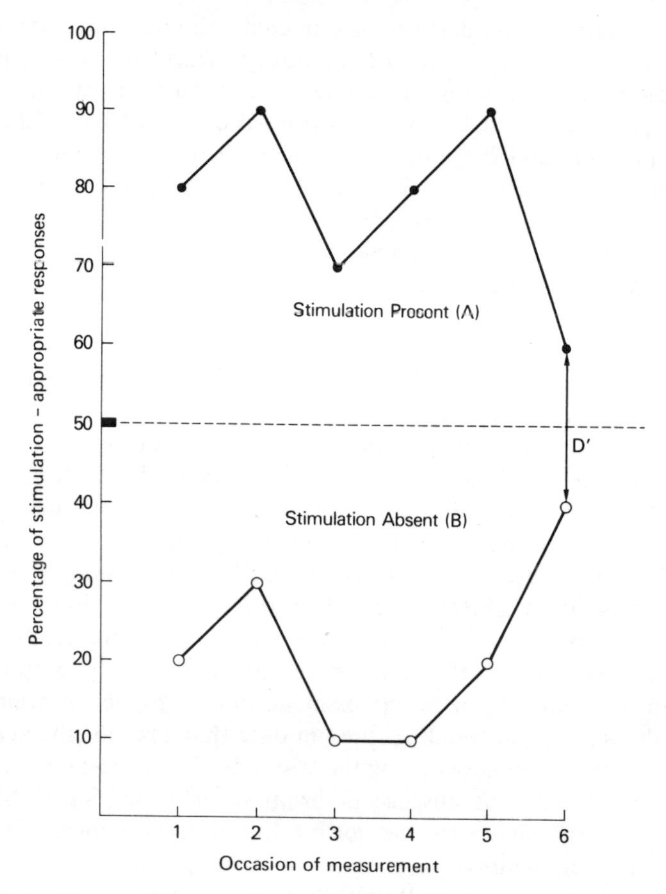

Figure 15.1 Generalized data presentation schematic for evaluation of discriminative responding.

stimulated (S) or not stimulated (N), and reinforcement is made available under a FR-1 schedule to the left or right bar, conditional on the presence or absence of brain stimulation.

Reinforcement on the left bar is always paired with brain stimulation (S trials), and reinforcement on the right bar always occurs in the absence of stimulation (N trials). The probability of an occurrence of an S or N trial is always equal on any given daily session. Training begins with longer trial durations, but within three daily sessions trial durations can be reduced to a final value of 15 sec. In the case of an S trial, brain stimulation begins .5 sec before the end of the ITI. This is done to ensure that the animal has an "observation interval" prior to initiation of a trial during which the animal can attend to the relevant stimuli. Bar presses during this interval extend the ITI for an additional .5 sec.

The "first choice" bar press responses after the initiation of each S or N trial, obviously made in the absence of any feedback to the animal regarding the appropriateness of its behavior, constitutes the primary datum of the experiment. As has been illustrated previously in Figure 15.1, these responses can be saliently characterized by a two-line graph. One line (A) reflects the percentage of correct first-choice responses made in the presence of stimulation, and the other (B) represents the percentage of incorrect first-choice responses made in the absence of stimulation. Again, D′ quantifies the degree of stimulus control by brain stimulation.

The multiple-trial procedure possesses certain advantages over the single-trial design. First, the approximately 10-day training time observed for the multiple-trial task is at least one-third of that experienced with the use of single trials. The procedure also gives explicit recognition to the operational requirements of the Theory of Signal Detection (TSD), primarily because the paradigm samples behavior in a binary choice, discrete trial task. (An extended discussion of TSD is beyond the scope of this section, and more detailed information may be found in Swets, Tanner, and Birdsall, 1961.) It is important, however, to note that the rigorous treatment given by TSD to sensory and motivational variables argues strongly that serious consideration be given to incorporating TSD whenever possible into data analyses of discrimination studies.

INTERPRETATION OF BEHAVIORAL DATA DERIVED FROM STUDIES OF STIMULUS CONTROL OF BRAIN STIMULATION

Recall that D′ quantifies the degree of stimulus control. The quantification of discrimination behavior by this value has an additional property of exposing changes in the predisposition of the animal to respond to a given bar. Note that the D′ value shown in Figure 15.1 does not change across the first three occasions of measurement. As compared to Occasion 1, however, performance on Occasion 2 reflects an increased predisposition to respond on the bar associated with stimulation. Similarly, performance on Occasion 3 shows a decreased predisposition to respond on that bar. Thus, it can be seen that equidistant and unidirectional shift of both lines about the 50% chance level of responding result from changes in response bias unaccompanied by changes in stimulus control. Of course, there can also be behavioral changes resulting from alterations in both response bias and stimulus control, and this situation would result in unidirectional but unequal shifts of both lines. Chapter 12 gives a more detailed treatment of all of these issues.

The quantification of response bias makes unnecessary such other controls for this variable as counterbalancing bar and stimulus conditions across animals. Moreover, it allows some inferences about the possible motivational significance of manipulations. For example, one might expect that the consistent pairing of aversive brain stimulation with responding on a given bar would result in a predisposition of the animal to respond on the opposite bar, which is never paired with such stimulation. In general, response bias changes of this type would be expected to reflect learning processes and therefore display a gradual onset.

One of the more basic questions one might ask about stimulus control is whether it generalizes to another stimulus condition. Typically, a group of animals would be trained to discriminate the presence or absence of brain stimulation, and their discrimination responding under both these conditions would be evaluated in the presence of a third manipulation, such as systemic administration of a psychoactive drug. To infer generalization, it is necessary that discriminative responding in the presence of the drug and in the absence of stimulation approximate responding in the presence of stimulation alone. Simply stated, administration of the drug must result in the animals' pressing the bar paired with brain stimulation even when no stimulation is present.

Stimulus effects of the drug in the presence of brain stimulation should also be studied. The drug could potentiate the discriminable cue effects of brain stimulation, resulting in animals' more frequently pressing the correct bar in the presence of stimulation. In order to maximize the possibility of seeing such an effect, pretreatment values of brain stimulation should be reduced sufficiently to produce suboptimal discrimination behavior. While the graphic plot of the simultaneous appearance of both generalization and potentiation would look like a change in response bias, a rapid onset of this change should be helpful in characterizing it as a sensory rather than motivational phenomenon. A drug could also produce a pharmacological blockade of the cue produced by electrical brain stimulation. This would be evidenced by a reduction in the percentage of correct responses in the presence of stimulation with no concomitant effect on discriminative responding in the absence of stimulation.

It is possible that the introduction of the test stimulus in generalization studies could simply increase erroneous responding and reduce D′. This would be consistent with the interpretation that the test stimulus was dissimilar to the cue effects associated with both the presence and the absence of brain stimulation. Of course, the possibility of behavioral toxicity would have to be eliminated, which is usually possible by analyses of such variables as response rates across treatment conditions.

Several observations suggest that instances of complete generalization between brain stimulation and systemic drug administration may not be frequent. Brain stimulation, although activating only a restricted amount of neural tissue, may be presumed to produce a number of stimulus effects, some or all of which may gain stimulus control over the behavior of any given subject. Systemic drug administration would affect an even greater neural substrate and should be expected, therefore, to produce an even greater number of potential cues. This multiplicity of cues theoretically reduces the probability that the same subset of cues will gain stimulus control over the behavior of a group of subjects in both conditions and thereby produce generalization.

Even the presence of complete generalization does not eliminate the possibility of the presence of additional cues distinguishing the two stimulus conditions but not utilized by the subjects. By always testing for generalization in the absence of reinforcement, the learning of a refined discrimination employing these cues can be avoided. This is routinely done in the single-trial procedure, animals never being exposed to the 13-min training sessions on test days. Animals trained by the multiple-trial procedure can also be reliably tested during 2-min extinction periods identical to those used in the single-trial task (Hayes, 1975).

EXAMPLES OF EXPERIMENTS EMPLOYING BRAIN STIMULATION AS A CUE FOR DISCRIMINATIVE RESPONDING

Brain stimulation used in the two studies outlined here was always delivered by a constant current stimulator through a chronically implanted bipolar electrode constructed of twisted stainless-steel wire, 200 μm in diameter and insulated with Teflon except at the cut cross-sections of their tips. The stimulation consisted of 20 Hz trains of biphasic, rectangular wave pulse pairs. The pulse pair consisted of a 50-μsec pulse, followed 100 μsec later by a 50-μsec pulse of opposite polarity. Intensities used in discrimination training ranged from 100 to 600 μA. Electrode impedance was continuously monitored during all stimulation periods. All electrodes were stereotoxically aimed at the mesencephalic central gray matter in the region of the dorsal raphe nucleus.

The first experiment (Hayes, 1975) employed the multiple-trial paradigm to study both the acquisition of discriminative responding in the procedure and the possible development of tolerance to the interoceptive cue value of electrical brain stimulation of the mesencephalic central gray matter of the rat. Previous experiments had shown that electrical stimulation of that region produces powerful analgesia (Liebeskind, Guilbaud, Besson, & Oliveras, 1973; Mayer & Liebeskind, 1974). Other work also showed that tolerance develops to the analgesic effect of brain stimulation, and this tolerance shows cross-tolerance with angesia produced by morphine (Hayes, 1975; Mayer & Hayes, 1975).

Initial training values (280–400 μA) of electrical stimulation were determined individually for each animal (N = 6) by slowly increasing stimulation intensities during early shaping sessions until behavioral signs of orienting and/or response disruption were evident to the experimenter. They were then reduced slowly until these responses disappeared, and this final current intensity was used during discrimination training.

As can be seen from Figure 15.2, the rapidly acquired stimulus control of brain stimulation was asymptotic after about 6 days of training and remained stable throughout the remaining 11 days.

Animals were then run at 50, 75, and 25% of their training intensities, in that order. Two days of retraining at original stimulation intensities intervened between 2 days of testing at reduced intensities. The incorporation of 2 days of testing allowed the study of the possibility of improvement over time of discrimination performance at reduced intensities (i.e., learning).

Figure 15.3 presents the results of this threshold assay averaged over the 2 days of training at each reduced intensity since performance on those days did not differ significantly. The values associated with the 100% intensity are the averages for the 2 days immediately preceding the threshold study.

After threshold assays, stimulation intensities for each animal were reduced to yield levels of discrimination performance that were just less than optimal in order to prevent ceiling effects from obviating the appearance of tolerance.

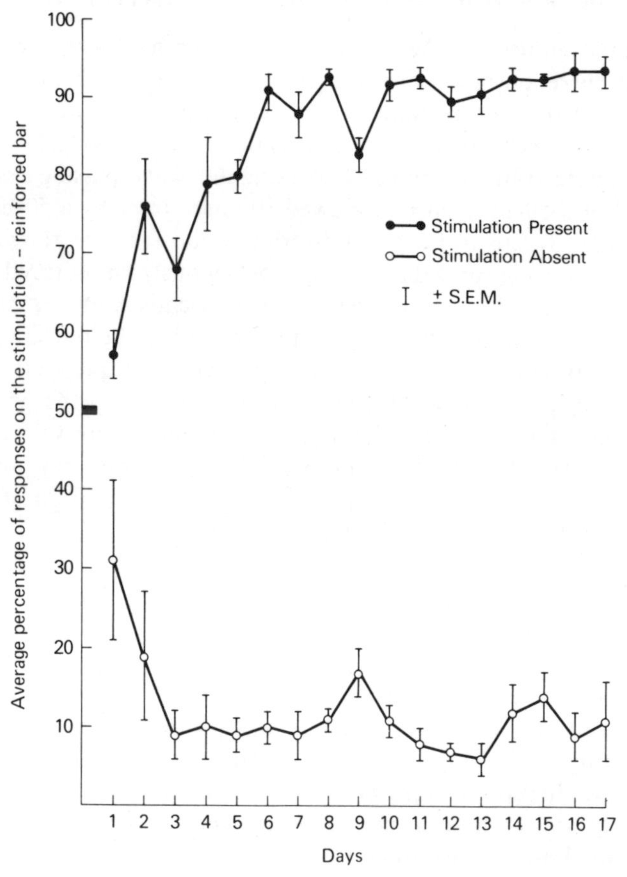

Figure 15.2 Acquisition of detection of presence or absence of brain stimulation.

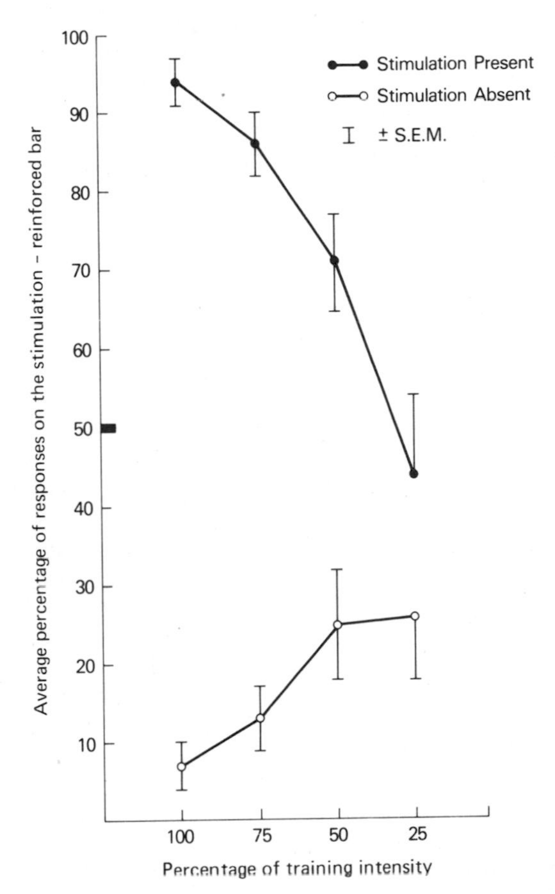

Figure 15.3 Assay of threshold for detection of brain stimulation.

Animals were then run for 5 days (C_1–C_5) to establish baseline values prior to initiation of tolerance-induction procedures.

On the fifth baseline day (C_5) and for the next 5 days (T_1–T_5), animals were stimulated at analgesic intensities (1–5 mA) after completion of their daily training sessions. These intensities had been determined for each animal individually prior to initiation of discrimination training. A total of 60 sec of stimulation was given on three 20-sec trials separated by 100 sec. After a final day of evaluation of discrimination performance during this tolerance induction procedure (T_6), animals were not run for 2 weeks and then retested (E_1). The deteriorating condition of the animals chronic electrodes prevented further study.

Figure 15.4 presents the results of this tolerance experiment. Tolerance to the interoceptive cue effect of brain stimulation was present by D_3 (C_5 vs. D_3; test of values of D'; $p < .01$; correlated τ test; $df = 5$), and no further decreases in discrimination performance were observed. After 2 weeks of no

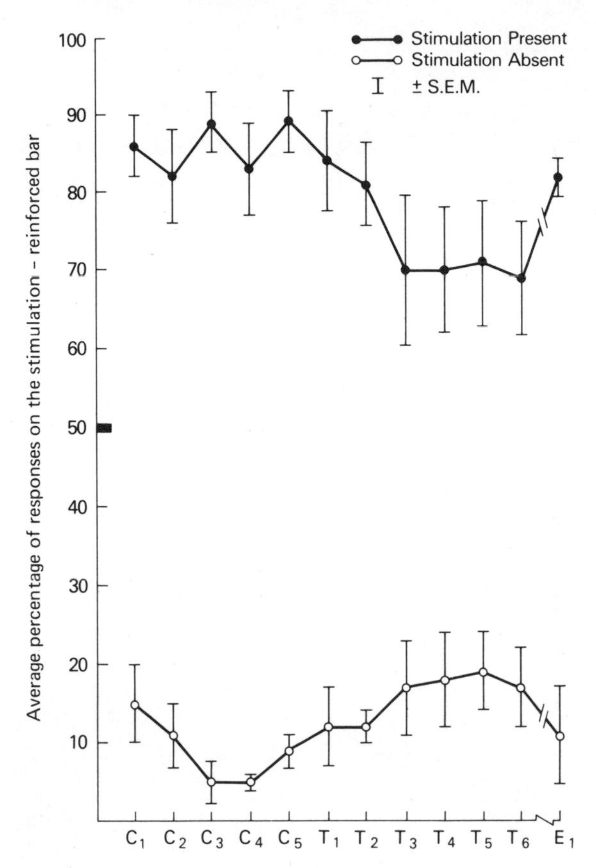

Figure 15.4 Tolerance to interoceptive cue effect of brain stimulation.

training, recovery was seen (D_6 vs. E_1; test of values of D′; $p > .05$; correlated τ test; $df = 5$) which approximated pretreatment levels of discrimination performance (C_5 vs. E_1; test of values of D′; $p > .20$; correlated τ test; $df = 5$).

The results of this experiment represent the first demonstration of tolerance to the discriminative cue effects of brain stimulation and further demonstrate the behavioral generality of tolerance to the effects of electrical stimulation in the region of the dorsal raphe nucleus. The time course for the development of tolerance to the cue effects of brain stimulation closely approximates that observed for tolerance to stimulation-produced analgesia (Hayes, 1975).

A possible alternative explanation for the decrement in discriminative responding is that analgesic intensities of brain stimulation used in the study produced tissue destruction at the site of stimulation. A previous publication (Mayer & Hayes, 1975) has dealt explicitly with this possibility, and several observations make this hypothesis untenable. It is sufficient here to note that electrolytic destruction of neural tissue can produce decrements in stimulation-produced responses, but no recovery is seen.

One might ask why no tolerance to cue effects of brain stimulation was seen during acquisition. As threshold assays demonstrated, however, training intensities of brain stimulation were clearly superthreshold, so some "tolerance" could have occurred without producing behavioral consequences. Moreover, to the extent that reduction in the discriminability of stimulation-produced cue occurred slowly over time, animals could have learned to compensate (e.g., attend more diligently) for this decrement. It is just this point—that decrements in response appropriateness were costly to the animal—that argues that the tolerance observed in this task was mediated by physiological and not behavioral phenomena.

It is also significant that on the average tolerance to the stimulus effects of brain stimulation was not complete and that some animals were always able to maintain a significant level of discrimination behavior. The average, of course, obscures the fact that while all 6 animals showed decreases in discriminatory responding during tolerance induction, values of D′ ranged from 19 to 85. It could be that this difference in degree of tolerance reflects differences in the subsets of cues produced by brain stimulation. That is, some animals' cues may have been mediated by neural phenomena to which tolerance occurs, while others may have attended to cues to which no tolerance develops under the treatment regiment imposed. Rigorous testing of this hypothesis awaits studies employing larger numbers of subjects necessary to expose viable relationships between electrode sites and behavioral data.

Finally, as we have pointed out previously (Mayer & Hayes, 1975), these results are consistent with the hypothesis that tolerance is not a CNS phenomenon restricted to pharmacological manipulations. Rather, tolerance may represent more general and endogenous neural processes that occur in response to activation by a wide range of events. The processes underlying tolerance could mediate responses to environmental conditions as well as organic pathological states and thereby potentially provide insights into the cyclicity of certain psychopathologies.

The second, previously reported experiment (Hirschhorn, Hayes, & Rosecrans, 1975) employed the single-trial paradigm to study both the acquisition of discriminative responding in the procedure and the possibility of generalization to lysergic acid diethylamide (LSD) and morphine. Numerous reports document the effects of LSD on the dorsal raphe nucleus, (e.g., Haigler & Aghajanian, 1974) and, as pointed out above, electrical stimulation of that region produces a morphine like analgesia.

Animals were able to learn the discrimination and reached asymptotic levels of performance after about 28 daily training sessions.

Subsequently, drug generalization tests were initiated. In these test procedures, the animals continued to receive brain stimulation and nonstimulation in the same double-alternation schedule. However, test sessions in which the drugs were administered were interposed among the regular training sessions. Test sessions were periods of 2 min during which no responses were reinforced.

Doses of 2.5, 5, and 10 mg/kg of morphine (i.p.) did not result in generalization. Rather, there was a dose-related decrease in D′, suggesting that the drug-produced cues differed from those associated with either the presence or the absence of brain stimulation.

Doses of 25, 50, and 100 μg/kg of LSD (i.p.) were also tested. In the presence of the 50 and 100 μg/kg doses, rats pressed the stimulation-appropriate bar a majority of the time even in the absence of brain stimulation, and the percentage of total responses on the stimulation-reinforced bar approached values observed in the presence of stimulation alone. These data were interpreted as indicating a substantial degree of generalization from the cues produced by brain stimulation in the region of the dorsal raphe nucleus to those produced by 50 and 100 μg/kg doses of LSD.

Unfortunately, subsequent efforts to replicate this experiment in our laboratory have proved disappointing. While data from individual animals show high degrees of generalization, the phenomenon has been statistically unreliable. The problem of cue multiplicity mentioned previously could contribute to this failure to achieve consistent generalization across animals. Perhaps the most direct and promising way of dealing with this problem would be to combine electrical stimulation with microinjection at the same locus in the same animal, an experiment that has yet to be done. This procedure employs existing technology and should make the population of neural tissue within each animal affected by both treatments as homogenous as possible, thereby maximizing the possibility of observing stimulus generalization.

REFERENCES

Deutch, J. A. Behavioral measurement of the neural refractory period and its application to intracranial self-stimulation. *J. Comp. Physiol. Psychol.*, 1964, *58*, 1.

Haigler, H. J., & Aghajanian, G. K. Lysergic acid diethylamide and serotonin: A comparison of effects on serotonergic neurons and neurons receiving serotongergic input. *J. Pharmacol. Exp. Therap.*, 1974, *188*, 688.

Hayes, R. L. Tolerance to analgesic and interoceptive cue effects of brain stimulation: Cross-tolerance between stimulation-produced and morphine analgesia. Unpublished doctoral dissertation. Virginia Commonwealth Univ., Richmond, 1975.

Hirschhorn, I. D., Hayes, R. L., & Rosecrans, J. A. Discriminative control of behavior by electrical stimulation of the dorsal raphe nucleus: Generalization to lysergic acid diethylamide (LSD). *Brain Res.*, 1975, *86*, 134.

Liebeskind, J. C., Guilbaud, G., Besson, J. M., & Oliveras, J. l. Analgesia produced by electrical stimulation of the periaqueductal gray matter in the cat: Behavioral observations and inhibitory effects on spinal interneurons. *Brain Res.*, 1973, *50*, 441.

Mayer, D. J., & Hayes, R. L. Stimulation produced analgesia: Development of tolerance and cross-tolerance to morphine. *Science,* 1975, *188,* 941.

Mayer, D. J., & Liebeskind, J. C. Analgesia produced by focal brain stimulation in the rat: An anatomical and behavioral analysis. *Brain Res.,* 1974, *68,* 73.

Swets, J. A., Tanner, W. P., & Birdsall, T. G. Decision processes in perception. *Psychol. Rev.,* 1961, *68,* 301.

16

Intragastric Self-Administration of Drugs by the Primate

Harold L. Altshuler

Texas Research Institute of Mental Sciences
and
Baylor College of Medicine
Texas Medical Center

Paul E. Phillips

Texas Research Institute of Mental Sciences

Progress in biomedical research has depended on the development of appropriate animal models, and drug abuse research is no exception. One of the conditions necessary for animal models of drug abuse is the selection of species whose central nervous system (CNS) sufficiently resembles that of man to allow generalizations. An animal model of drug abuse requires an animal that consumes its own drug and selects its own dosage.

The development of the subhuman primate model of drug self-administration was a significant advance in drug-abuse research. Before the technique was developed, drugs of abuse were administered to experimental animals either as injections, oral gavage, or in the animal's food or water. Dogs and most rodent species will readily consume drug-altered food and water and will choose food or water supplies that contain drugs with high abuse liability (i.e., morphine) in preference to undrugged food or water. Although this model is one of the most widely used in drug-abuse research, it has several limitations. Several drugs are inactive when consumed orally, for example, and many taste so aversive that the animals will not consume them voluntarily. The subhuman primate is highly suspicious of any adulteration of its food or water and usually will not drink water or eat food that tastes or smells strange.

A number of technical approaches have been used to provide a laboratory model of drug self-administration. Beach (1957), Myers, Stoltman, and Martin (1972), Nichols, Headlee, and Coppock (1956), Wikler, Martin, Pescor, and Eades (1963), and many other investigators have provided animals with food or drinking water containing the drugs of research interest. In several species, primarily rodents and more recently monkeys, the method has been moderately successful and has demonstrated that animals will voluntarily choose food or water containing dependence-producing drugs (Beach, 1957; Wikler *et al.*, 1963). Weeks (1962) expanded the armamentarium of laboratory animal models of drug abuse by reporting an intravenous self administration technique applicable to rats. His data demonstrated that rats equipped for intravenous self-administration of drugs will increase their drug intake, become physically dependent on the drug, and exhibit signs of opiate abstinence when the drug ceases to be available. Two years later, Thompson and Schuster (1964) reported an adaptation of Weeks' technique for use in rhesus monkeys. Their studies and the studies of Deneau, Yanagita, and Seevers (1969) and others (Balster & Schuster, 1973; Goldberg, Hoffmeister, Schlichting, & Wuttke, 1971; Hoffmeister & Schlichting, 1972; Schlichting, Goldberg, Wuttke, & Hoffmeister, 1970; Schuster & Thompson, 1969; Talley & Rosenblum, 1972; Wilson, Hitomi, & Schuster, 1971; Woods & Tessel, 1974; Yanagita, Deneau, & Seevers, 1965; Yanagita & Takahashi, 1973; Yanagita, 1973) used the technique extensively and demonstrated its value for studies related to drug abuse and particularly for the evaluation of the abuse liability of a variety of compounds. The data generated by these studies represent a classical body of information related to drug dependence and the pharmacology of drugs of abuse.

A great many studies have been conducted to delineate the psychological contingencies that modify self-administration behavior. A highly significant series of studies detailed the relationship between bar-pressing patterns and schedules of reinforcement (Goldberg *et al.*, 1971; Pickens & Thompson, 1968, 1972; Schuster & Thompson, 1969; Thompson & Pickens, 1970). Particularly relevant are the observations that monkeys will maintain fairly stable levels of day-to-day drug intake when the reinforcing drug is a CNS stimulant, such as cocaine, regardless of the complexity of the operant schedule or a wide range of concentrations of the reinforcing solution (Goldberg *et al.*, 1971; Pickens & Thompson, 1968; Thompson & Pickens, 1970; Woods & Tessel, 1974). Self-administration patterns of such other compounds as morphine and *d*-amphetamine are apparently less stable than those of cocaine, although monkeys have been shown to respond to many variations of schedule to maintain intake levels of these compounds. The barbiturates are a notable exception. Goldberg's group (1971) has shown that monkeys will not maintain stable pentobarbital intake levels when required to respond to an increasing fixed-ratio schedule, although they will maintain intake stability when drug-solution concentration is varied.

As early as 1966, Yanagita suggested the need for a technique that would allow monkeys to self-administer drugs into the gastrointestinal tract (Yanagita, 1968) to provide a better animal model for use in studies of drugs consumed orally by humans. Previous attempts to establish an oral self-administration model by providing primates with drinking water containing drugs had been unsuccessful because of the monkeys' suspiciousness and unwillingness to ingest such food or water (Myers *et al.*, 1972; Yanagita, 1968).

There are a number of other problems associated with experiments that involve drug administration to animals in food or water. For example, the behavioral changes that result from drug self-administration by this method are often intermixed with behavioral changes that may be caused by alterations in the animals' nutritional, fluid, or electrolyte balance. In addition, it is difficult to establish rigid behavioral criteria for self-administration via food or water and to manipulate the behavior in meaningful ways.

Since most attempts to induce monkeys to consume drug-laced fluids have met with failure, several investigators have devised systems to circumvent the problem (Myers *et al.*, 1972; Meisch, Henningfield, & Thompson, 1975). These manipulations, like those described for the rats, have many inherent problems related to alterations in the animals' fluid balance or nutritional state. Yanagita (1968) developed a system of intragastric self-administration by the subhuman primate using a permanently implanted nasogastric cannula. His technique was relatively successful, and he demonstrated that under special circumstances the animals would acquire and maintain self-administration behavior of the barbiturates, alcohol, and opiates. However, he reported some difficulties maintaining the nasogastric cannula for long periods of time. These difficulties were largely related to problems of infection at the implantation site.

At about the same time, Gotestam (1973), Trojniar, Cytawa, Frydrychavski, & Luszawska (1974), and Smith, Werner, & Davis (1974) developed procedures for intragastric self-administration by rats. Our laboratory developed an intragastric self-administration technique specifically for studies with subhuman primates (Altshuler, Deneau, Weaver, & Roach, 1974; Altshuler, Weaver, & Phillips, 1975a,b; Altshuler, Weaver, Phillips, & Burch, 1975). This report will describe the development and evaluation of the intragastric self-administration technique used in our laboratory.

METHODS

Intragastric Cannula

Figure 16.1 is a schematic diagram of the intragastric cannula used in our laboratory. The cannula is formed from silicone-rubber tubing, 50–55 cm in length (Silastic Dow Corning No. 602-205). Silicone-rubber adhesive (Silastic

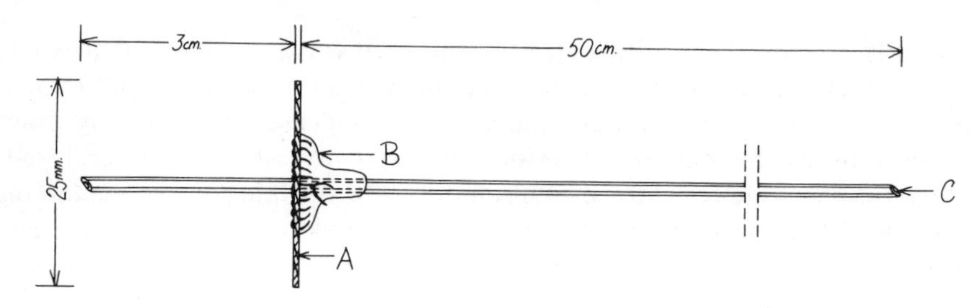

Figure 16.1. Schematic diagram of intragastric cannula used in simian intragastric self-administration preparation. *A*, 15 mm diameter Teflon patch; *B*, molded and trimmed silastic adhesive disc to which Teflon patch is sewn; *C*, silicone rubber tubing, .04 I.D. × .085 O.D.

adhesive Dow Corning No. 891) is used to mold a silicone-rubber disc about 4 cm from the end of the cannula. The disc is fabricated by attaching a small mass of adhesive to the cannula and allowing it to cure completely. Then the adhesive mass is trimmed to a disclike shape about 15 mm in diameter and 2 mm thick. A sheet of Teflon intracardiac patch material (C. R. Bard No. 3100) forms the interface between the silicone-rubber disc and the serosal surface of the stomach. The patch is sewn to the distal surface of the silicone-rubber disc with 4-0 surgical silk and trimmed into an oval shape, about 25 mm at its widest diameter. The entire cannula is steam-sterilized before implantation.

Surgical Procedures

The animals selected for intragastric self-administration were male and female rhesus monkeys weighing between 3.5 and 5 kg. The animals were screened for abnormalities, and control hematological and clinical biochemical values obtained. Food was withheld from the animals for about 18 hr and water for 3 hours before surgical intervention. The animals were anesthetized with pentobarbital sodium, 30 mg/kg intravenously (i.v.) and treated with atropine sulfate, 1 mg/kg subcutaneously (s.c.). Following the usual surgical preparations, an upper-left quadrant incision was made in the animal's abdomen. The viscera were packed away from the surgical field with warm saline-soaked sponges and the stomach exteriorized. A small incision was made in the upper portion of the greater curvature of the stomach and the end of the cannula inserted through the incision. The incision was then closed with a purse-string stitch of 4-0 chromic gut and the Teflon patch sewn to the stomach wall with 4-0 surgical silk using a continuous stitch (Figures 16.2, and 16.3). A small amount of blood was dripped over the Teflon patch to provide a substrate for healing.

The patency of the cannula and surgical area were checked by infusion of 10 cc of sterile saline through the cannula. The cannula was brought out of the peritoneal cavity via a small stab wound in the abdominal musculature slightly

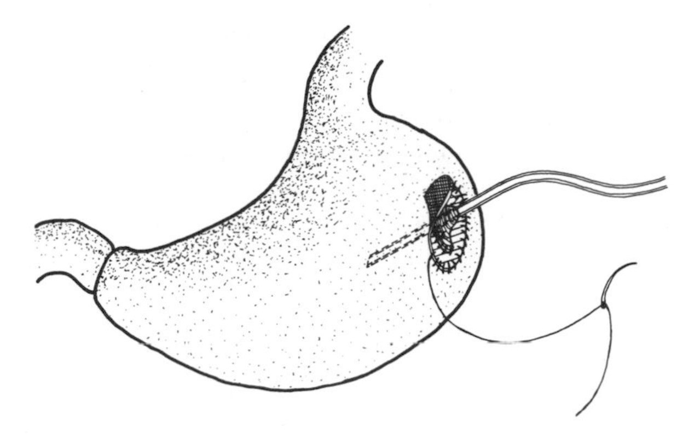

Figure 16.2. Sketch of gastric cannula implantation procedure. Illustrates Teflon patch partially sutured into stomach after end of cannula has been inserted through small incision in stomach wall.

lateral to the incision. The cannula was tunneled subcutaneously to a point between the shoulder blades, where it was brought out of the animal's body via a small incision in the skin. The abdominal incision was closed in 3 layers and the animal placed in a self-administration jacket or monkey harness, which in turn was connected to the cage by a flexible hose.

The animals were maintained on a regimen of antibiotics and atropine for about 1 week following the operation. The antibiotic regimen consisted of penicillin G, 300,000 units per day intramuscularly (i.m.); streptomycin sulfate, 75 mg/kg twice a day, i.m., and atropine sulfate, 0.1 mg/kg twice a day, s.c. The animals were fed a soft diet consisting of ground monkey chow, strained bananas, apples, and so forth for about one week. The cannula was flushed 4 times daily with 1 ml of saline during the recovery period.

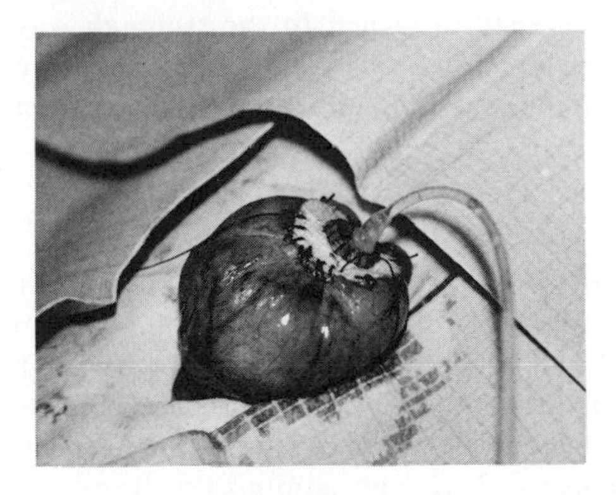

Figure 16.3. Photograph of intragastric cannula during implantation procedure when suturing of Teflon patch to stomach is almost completed.

Self-Administration Apparatus and Schedules

The self-administration apparatus used in these studies were similar in all respects to the systems used in a number of other laboratories that conduct research on intravenous self-administration (Deneau *et al.*, 1969; Schuster & Thompson, 1969). Although our experiments have included a number of different operant schedules, the data reported here will be only the data resulting from experiments with a fixed ratio 1 schedule (CRF).

RESULTS AND DISCUSSION

Animal Preparation

Twenty-one rhesus monkeys have been successfully prepared surgically with intragastric cannulae between January 1972 and January 1974. A small percentage of the animals developed postoperative infections that were easily treated. In only two animals in the series were there cannula failures at the implantation site. In both cases, it was possible to remove the remnants of the cannula, perform a partial gastrectomy, and restore the animal to normal health so that it could be used in our breeding program. Seven animals have remained functional in the experimental system for over a year; they have been exposed to many drugs and have served as excellent subjects for comparisons of the reinforcement properties of a variety of compounds.

Of the remaining 12, animals were removed from the system because of mechanical breakage of the cannula. As in other self-administration techniques involving subhuman primates, there were recurrent episodes of monkeys breaking their cannulae. The cannula breaks occurred in 2 general locations: along the subcutaneous path of the cannula or behind the silastic disc at the implantation site. The subcutaneous breaks were repaired by splicing new cannula material to the old. Breaks at the silastic disc were irreparable, and the portion of the cannula still attached to the stomach was removed surgically and the stomach repaired. In most cases, recovery from such surgery was uneventful, and the monkeys were added to our breeding program.

Self-Administration Behavior

Self-administration was acquired in 66% of the trials in which drugs were made available. Eleven drugs were evaluated in the system, and all but 3 were self-administered by at least one animal. The proclivity of the animals to self-administer drugs of various classes is summarized in Table 16.1.

One of the most interesting characteristics of intragastric self-administration is that the animals tend to self-administer drugs in a cyclical pattern characterized by several days of relatively low drug intake alternated with several days of relatively high drug intake. Despite recurrent "binges," there is

Table 16.1
Summary of Acquisition of Intragastric Self-Administration of Psychoactive Drugs

	Self-administration		
Drug	Trials	Acquisition	Percentage of acquisition
Ethanol	11	7	64
Pentobarbitol	5	4	80
Methaqualone	5	5	100
Methadone	4	4	100
d-Amphetamine	3	2	66
Chlordiazepoxide	2	1	50
Diazepam	2	0	0
Meprobamate	2	0	0
Chlorpromazine	2	0	0
Morphine	1	1	100
Cocaine	1	1	100
Total	38	25	66

an overriding pattern in which the animals increase the amount of drug taken on each successive binge so that their total drug intake tends to increase over time. In the cases of drugs that are known to produce tolerance and dependence, abstinence syndromes were observed when the animals were withdrawn from the drug.

Sedative–Hypnotic Drugs

Two drugs commonly prescribed for their sedative–hypnotic properties, pentobarbital and methaqualone, were evaluated by the intragastric self-administration model to determine whether or not these drugs would be self-administered and what the patterns of self-administration behavior would be. All of the exposures to pentobarbital sodium and methaqualone HCl resulted in acquisition of self-administration. Figure 16.4 illustrates a representative pentobarbital self-administration pattern for a single monkey. Figure 16.5 illustrates similar data for a different monkey exposed to methaqualone HCl. The cyclical self-administration pattern is obvious in both sets of data, although differences between the drugs are apparent. The recurrent binges occurred with greater frequency in the animals administering methaqualone than they did in the animals exposed to pentobarbital. We speculate that the differences are related to differences in rate of metabolism of the two drugs and that methaqualone is metabolized more rapidly by monkeys than is pentobarbital. This could account for the frequent binges of short duration observed with methaqualone. Biochemical studies are required to confirm this hypothesis.

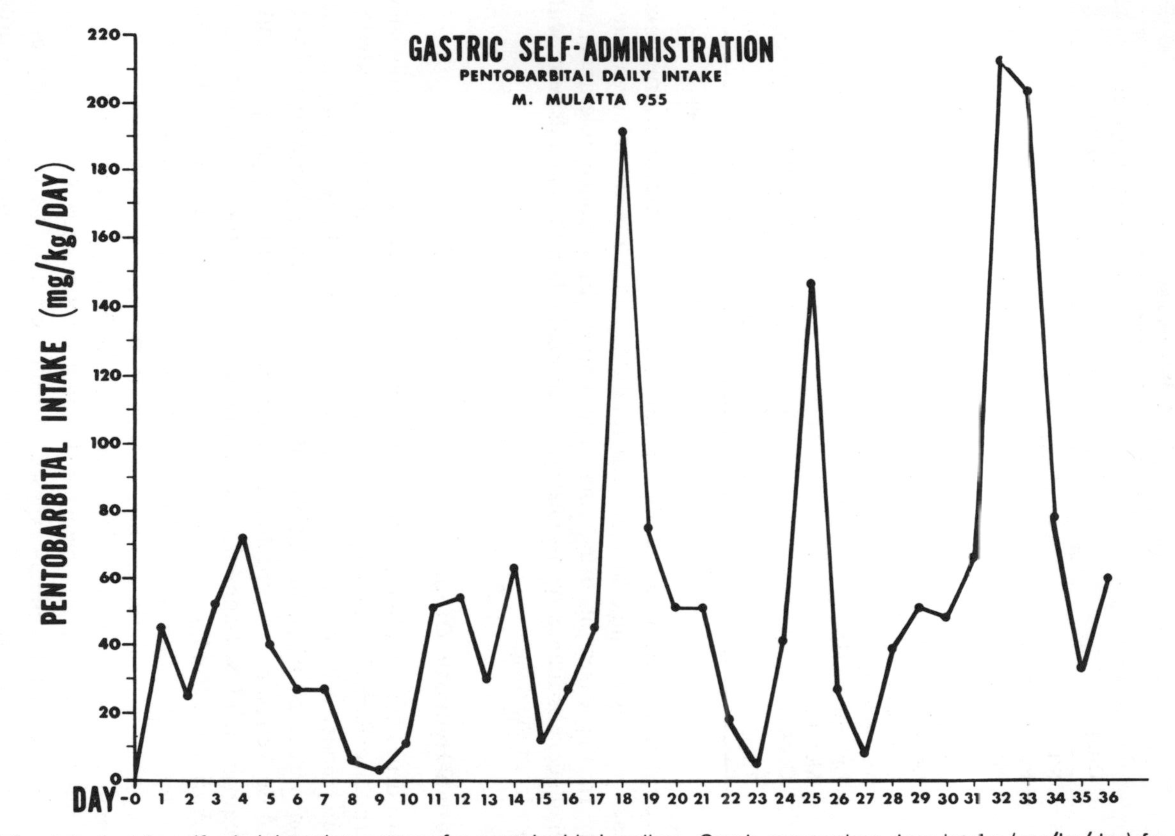

Figure 16.4. Intragastric self-administration pattern for pentobarbital sodium. Graph summarizes drug intake (mg/kg/day) for 36 days of drug exposure for representative monkey. Abcissa: day of experiment. Ordinate: daily drug dose.

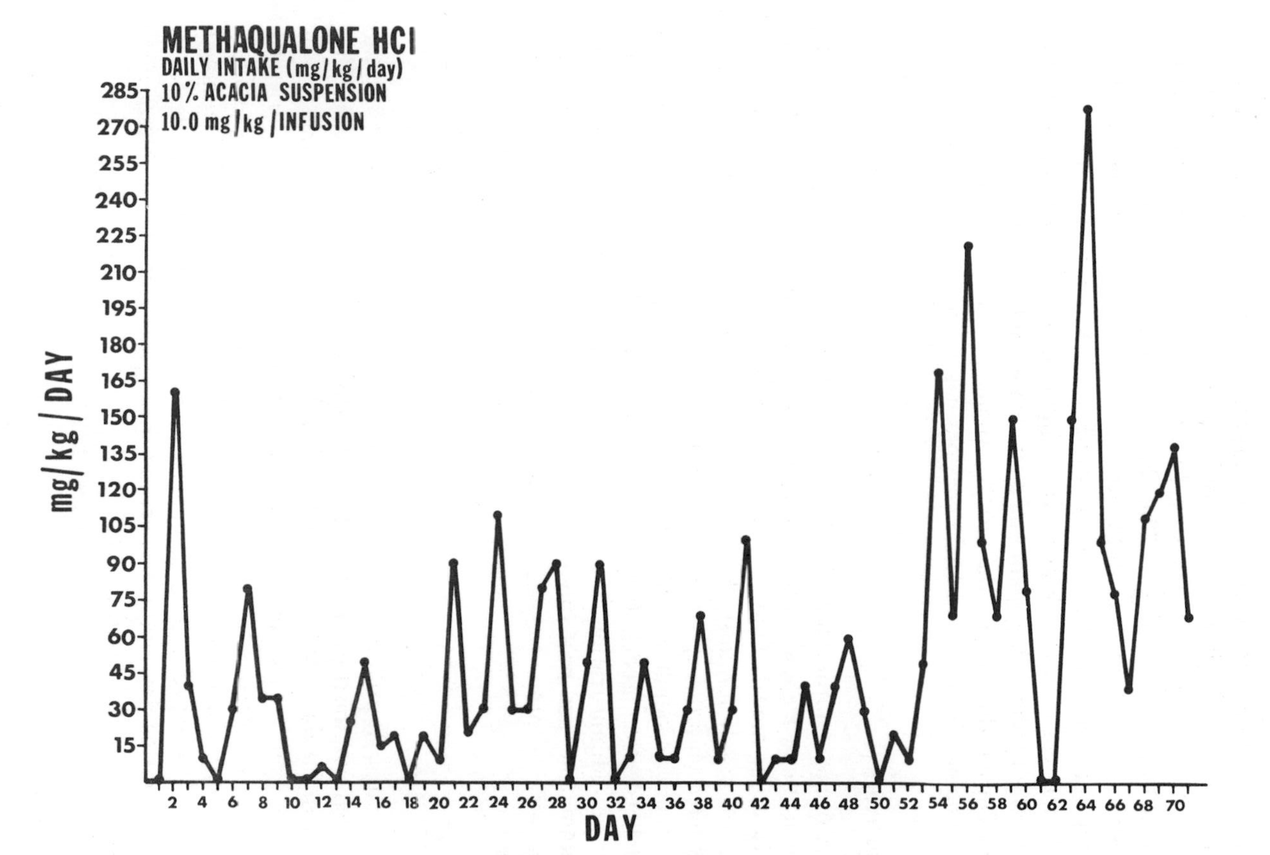

Figure 16.5. Intragastric self-administration pattern for methaqualone hydrochloride. Graph summarizes methaqualone hydrochloride drug intake during intragastric self-administration by representative monkey. Drug was suspended in 10% acacia and delivered as 10 mg/kg/dose infusion. Ordinate: drug intake (mg/kg/day). Abscissa: day of experiment.

Ethanol

We have attempted to establish intragastric self-administration of ethanol in 11 trails, 64% of which were successful. Animals that initiate and maintain ethanol self-administration behavior do so in the cyclical, bingelike pattern similar to the other drugs. The animals that self-administer ethanol intragastrically seem to have difficulty maintaining control of the amount of drug they consume, and they tend to overdose to the point of unconsciousness and, in one case, death. Figure 16.6. summarizes the self-administration behavior of an animal that maintained ethanol self-administration for almost 6 months. During this period, the monkey frequently consumed the drug to the point of unconsciousness and on 2 occasions consumed almost fatal overdoses.

Tranquilizers

Three drugs, chlordiazepoxide, diazepam, and meprobamate, generally categorized as minor tranquilizers, were evaluated by the monkey intragastric self-administration model. Only chlordiazepoxide was reinforcing enough to be maintained in self-administration for 30 days or longer. One monkey has self-administered chlordiazepoxide for longer than 6 months.

One major tranquilizer, chlorpromazine, was evaluated in the intragastric self-administration model. The two animals exposed to chlorpromazine did not initiate self-administration. We concluded on the basis of those studies that chlorpromazine, and probably drugs with a similar mechanism of action, have low reinforcing properties in the intragastric self-administration model in monkeys.

Opiates

Two opiates, morphine sulfate and methadone HCl, were evaluated with this model. Self-administration of both drugs was acquired and maintained for longer than 30 days. When the animals were withdrawn from either of the compounds, most signs of the opiate abstinence syndrome were observed. Figure 16.7 summarizes the self-administration pattern for one of the animals that self-administered methadone HCl. It is interesting to note that although this animal developed a cyclical pattern of self-administration, the degree of stability of day-to-day intake was greater with methadone than with any of the other compounds evaluated to date.

Stimulants

Two CNS stimulants, *d*-amphetamine sulfate and cocaine HCl, were evaluated by the intragastric self-administration model. Self-administration of both compounds was initiated within 10 days of initial exposure. Two-thirds of

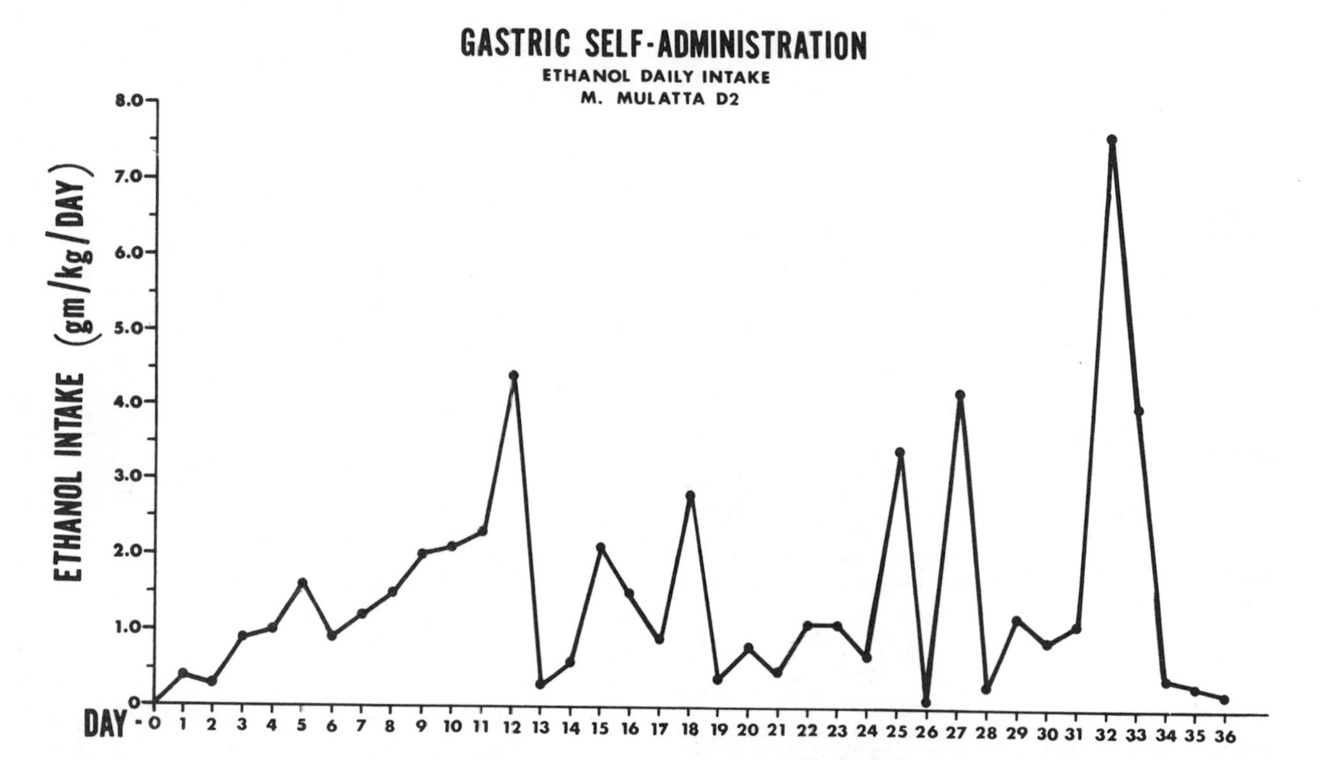

Figure 16.6. Intragastric self-administration pattern of ethanol. Graph summarizes 36 days of intragastric self-administration of ethanol by representative monkey. Ordinate: daily drug dose shown as gm/kg/day. Abscissa: day of experiment.

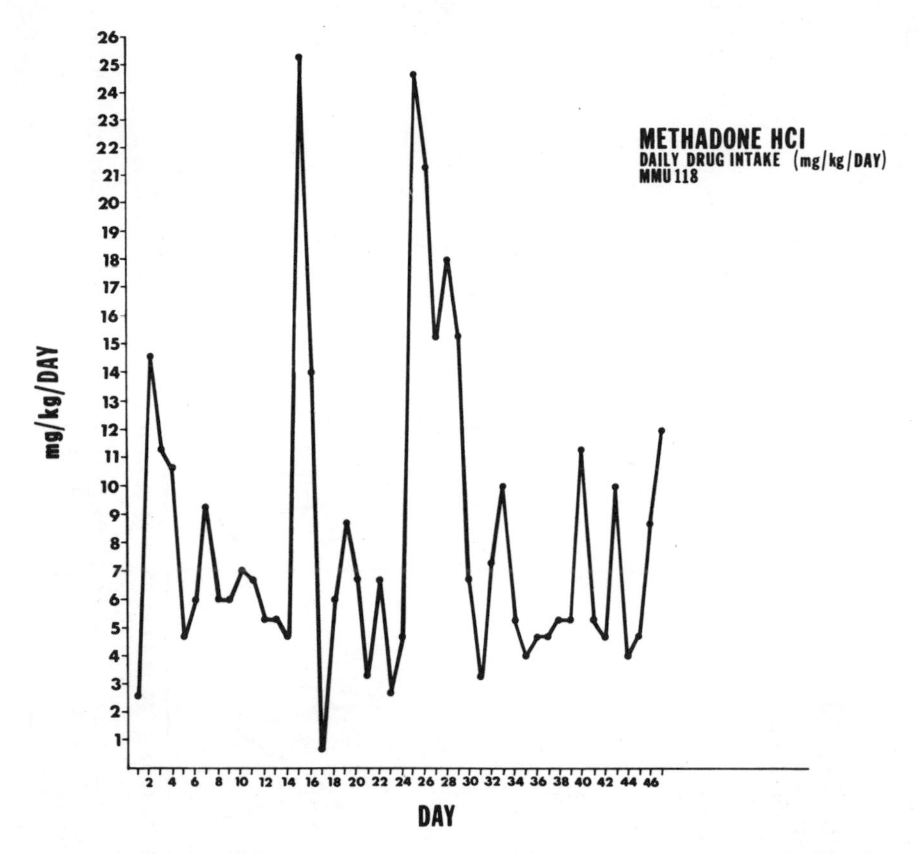

Figure 16.7. Intragastric self-administration pattern of methadone hydrochloride. Graph summarizes methadone hydrochloride daily drug intake (mg/kg/day) by representative monkey. Ordinate: drug dose in mg/kg/day. Abscissa: day of experiment. Note that except for 2 days of extremely high intake, daily methadone dose is relatively stable.

the animals that had *d*-amphetamine sulfate available initiated and maintained self-administration. Figure 16.8 summarizes the *d*-amphetamine self-administration behavior of one of the monkeys.

It was believed for some years that cocaine HCl was metabolized completely in the stomach following oral or intragastric administration. The theory was, therefore, that the drug, when consumed orally, is not positively reinforcing to the user. One monkey exposed to cocaine (Figure 16.9) demonstrated, however, that when sufficiently large doses of cocaine are consumed intragastrically, the drug has enough reinforcement properties to allow the initiation and maintenance of self-administration behavior. We postulate that when sufficiently large doses are taken into the alimentary canal, a relatively small percentage is absorbed from the gastrointestinal tract as either parent compound or an active metabolite, that the absorbed compound retains reinforcement properties that promote self-administration.

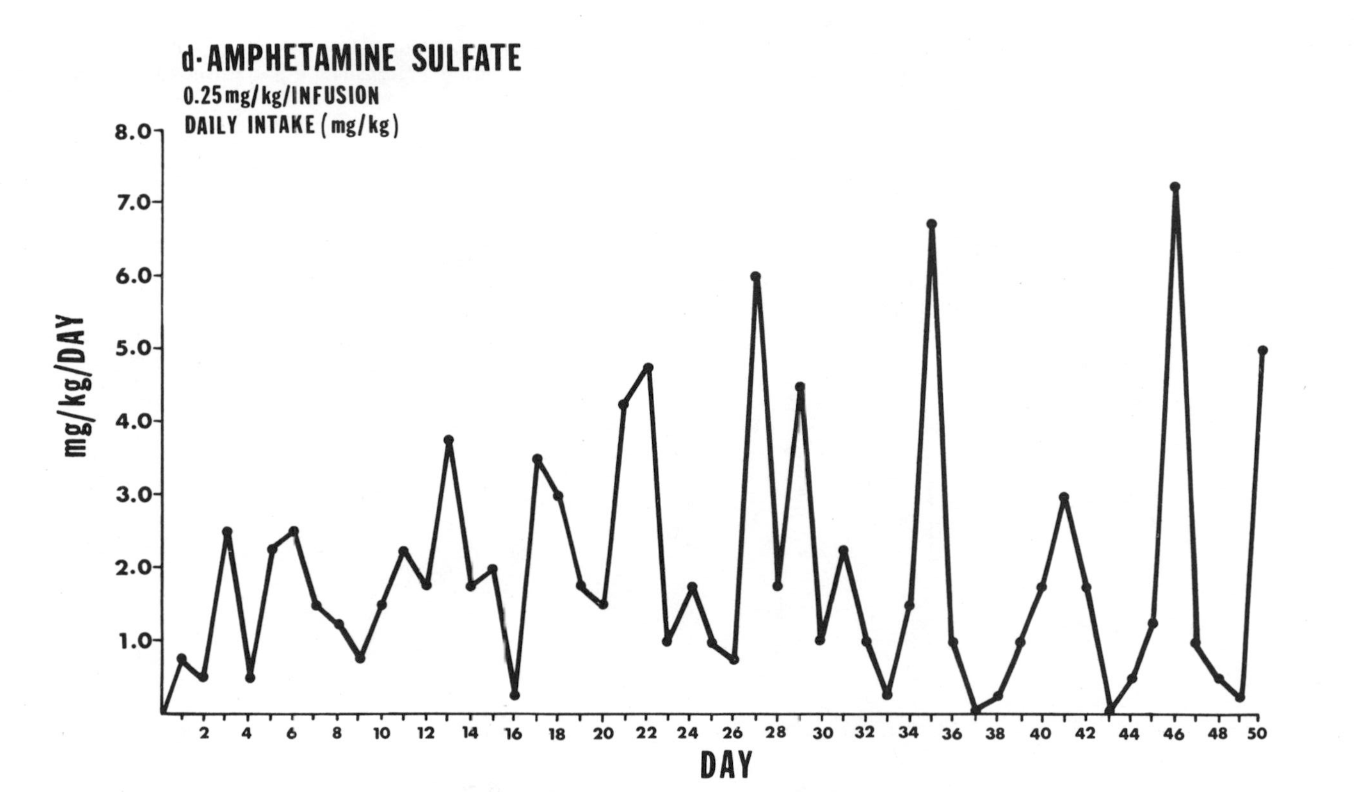

Figure 16.8. Intragastric self-administration pattern of *d*-amphetamine sulfate. Graph summarizes daily drug intake (mg/kg/day) of *d*-amphetamine sulfate by single monkey. Ordinate: daily dose as mg/kg/day. Abscissa: day of experiment.

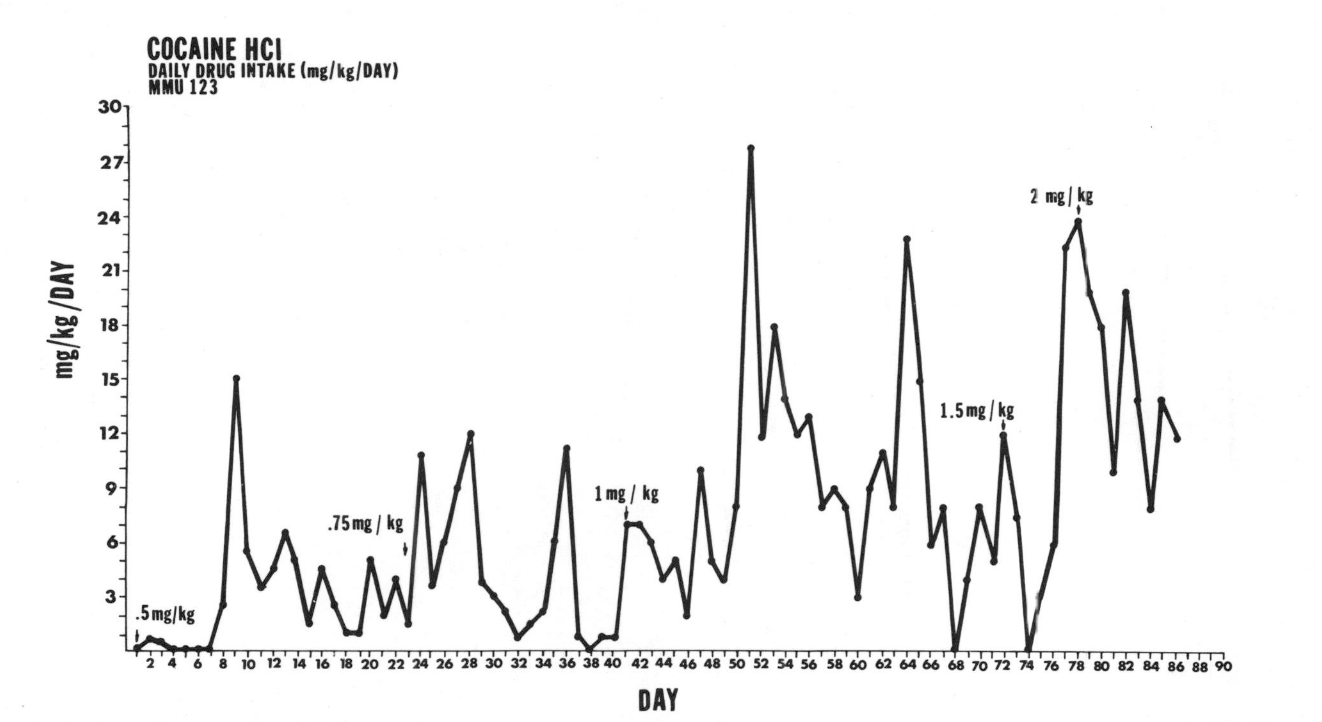

Figure 16.9. Intragastric self-administration pattern of cocaine hydrochloride. Graph summarizes 90 days of intragastric self-administration of cocaine hydrochloride. Drug was delivered intragastrically as saline solution of varying concentrations as indicated. Ordinate: drug dose (mg/kg/day). Abscissa: day of experiment. Note that although there is a general trend toward an increase in amount of drug consumed over the 3-month experiment, increase is largely related to higher concentration of drug dose per infusion.

SUMMARY AND CONCLUSIONS

The studies outlined in this report were begun in an attempt to develop an animal model of drug abuse that closely resembles the mode of drug abuse most widespread in humans, the oral consumption of centrally active drugs. Our data have demonstrated that intragastric self-administration by monkeys is a useful model of that condition. The paradigm provides one of the few laboratory systems in which animals exhibit drug-seeking behavior for ethanol, a model that has been sought by alcoholism research workers for a number of years. The system has the advantage of providing an animal that exhibits such behavior without much experimental manipulation that might alter the animal's physiological or biochemical state.

Our first 2 years of research with intragastric self-administration demonstrated that the model is durable, maintaining animal subjects in good health for periods of a year or longer. Such animals are invaluable in experimental designs in which it is desirable that a single animal serve as its own control or in which the long-term consequences of drug self-administration are being studied.

Most of the psychological principles that underlie responding in the intragastric self-administration model are similar to the principles that underlie responding for intravenous self-administration. The major difference between intragastric self-administration and intravenous self-administration is that the perception of the reinforcement resulting from intragastric self-administration of a drug is delayed; the intragastric self-administration model is therefore essentially a delayed reinforcement paradigm. It was thought that acquisition of self-administration in such a paradigm would be extremely difficult, but these studies demonstrate that rhesus monkeys quite successfully learn the behavior despite the delay in reinforcement.

Intravenous self-administration provides a much more rapid and sensitive test for abuse liability of most compounds than does the model presented here, but it does not provide an animal model that mimics the most common form of drug abuse by the human, the oral consumption of centrally active drugs for nonmedical reasons. The intragastric self-administration model, while lacking some of the sensitivity and speed of the intravenous model, fills a need for an animal model of the oral self-administration of drugs and adds a dimension to self-administration studies that is not possible with the intravenous models. Although there is a great deal of similarity in the drug preferences observed with each model, it is interesting that monkeys will self-administer alcohol intragastrically more predictably than they will intravenously. Any speculation about the mechanisms that might underlie that difference would be premature.

The only other situation in which we have demonstrated intragastric self-administration to be substantially different from intravenous self-administration is in the case of insoluble compounds such as methaqualone, in which the intragastric model allows study of such compounds that were impossible with the intravenous model.

It would be interesting to evaluate comparatively how accurately monkeys can form drug discriminations when the drugs are self-administered intravenously or intragastrically. Such studies would be quite difficult since even with the intravenous self-administration model there is a marked tendency for monkeys to perseverate on a single bar or schedule. We conducted preliminary experiments in an attempt to establish a double-bar drug discrimination of methadone and alcohol with the intragastric self-administration paradigm. Although we were successful in producing excellent stimulus control and self-administration of both ethanol or methadone, the animals were not able to perform appropriate discriminations. A great deal of additional research is required to evalute the degree to which state-dependent learning controls self-administration behavior in general and more specifically drug discriminations. An understanding of such mechanisms will provide a significant increase in our understanding of parameters that underlie drug preferences by drug users, both human and subhumans.

In summary, our studies have demonstrated that intragastric self-administration is a useful animal model of drug abuse. We do not expect it to supplant intravenous self-administration models in general application or for the primary evaluation of the abuse liability of new compounds. We are convinced, however, that the model is a useful adjunct to existing methods and it provides important added information on the abuse liability and psychological and pharmacological contingencies associated with self-administration of the drugs that the human abuses by the oral route.

ACKNOWLEDGMENTS

For capable technical assistance we are indebted to S. S. Weaver, D. S. Sanders, Mary N. Hubler, Kenneth Davis, Linda Talley, and to Eileen Ponton for typing.

REFERENCES

Altshuler, H. L., Deneau, G. A., Weaver, S. S., & Roach, M. K. Intragastric self-administration of psychoactive drugs in the rhesus monkey. *Pharmacologist*, 1974, *16*, 238.

Altshuler, H. L., Weaver, S. S., & Phillips, P. E. Intragastric self-administration of drugs in the subhuman primate. *Federation Proc.*, 1975, *34*, 3211. (a)

Altshuler, H. L., Weaver, S. S., & Phillips, P. E. Gastric self-administration of psychoactive drugs in monkeys. *Life Sci.*, 1975, *17*(6), 883–889. (b)

Altshuler, H. L., Weaver, S. S., Phillips, P. E., & Burch, N. R. Gastric self-administration in monkeys: Neurophysiological correlates and recent developments. *Proc. West Pharmacol. Soc.*, 1975, *18*, 58.

Balster, R. L., & Schuster, C. R. A comparison of *d*-amphetamine, *l*-amphetamine, and methamphetamine self-administration in rhesus monkeys. *Pharmacol. Biochem. Behav.*, 1973, *1*, 67.

Beach, H. D. Morphine addiction in rats. *Can. J. Psychol.*, 1957, *11*, 104.

Deneau, G., Yanagita, T., & Seevers, M. H. Self-administration of psychoactive substances by the monkey: A measure of psychological dependence. *Psychopharmacologia*, 1969, *16*, 30.

Goldberg, S. R., Hoffmeister, F., Schlichting, U. U., & Wuttke, W. A comparison of pentobarbital and cocaine self-administration in rhesus monkeys: Effects of dose and fixed-ratio parameter. *J. Pharmacol. Exp. Therap.*, 1971, *179*, 277.

Gotestam, K. G. Intragastric self-administration of medazepam in rats. *Psychopharmacologia*, 1973, *28*, 87.

Hoffmeister, F., & Schlichting, U. U. Reinforcing properties of some opiates and opioids in rhesus monkeys with histories of cocaine and codeine self-administration. *Psychopharmacologia*, 1972, *23*, 55.

Meisch, R. A., Henningfield, J. E., & Thompson, T. Establishment of ethanol as a reinforcer for rhesus monkeys via the oral route: Initial results. In M. M. Gross (Ed.), *Alcohol intoxication and withdrawal*. New York: Plenum Press, 1975.

Myers, R. D., Stoltman, W. P., & Martin, G. E. Effects of ethanol dependence induced artificially in the rhesus monkey on the subsequent preference for ethyl alcohol. *Physiol. Behav.*, 1972, *9*, 43–48.

Nichols, J. R., Headlee, C. P., & Coppock, W. H. Drug addiction. I. Addiction by escape training. *J. Am. Pharm. Assoc.*, 1956, *45*, 788.

Pickens, R., & Thompson, T. Cocaine-reinforced behavior in rats: Effects of reinforcement magnitude and fixed-ratio size. *J. Pharmacol. Exp. Therap.*, 1968, *161*, 122.

Pickens, R., & Thompson, T. Simple schedules of drug self-administration in animals. In *Drug addiction: Experimental pharmacology*. Vol. 1. Mount Kisco, New York: Futura Publishing, 1972.

Schlichting, U., Goldberg, S., Wuttke, W., & Hoffmeister, F. Self-administration of different classes of psychotropic drugs by rhesus monkeys as compared to self-administration of cocaine. Presented to the International Institute on the Prevention and Treatment of Drug Dependence, Lausanne, Switzerland, June, 1970.

Schuster, C. R., & Thompson, T. Self-administration of and behavioral dependence on drugs. *Ann. Rev. Pharmacol.*, 1969, *9*, 483.

Smith, S. G., Werner, T. E., & Davis, W. M. Morphine and ethanol: Intragastric and intravenous self-administration. *Proc. Soc. Neurosci.*, 1974, *4*, 645.

Talley, W. H., & Rosenblum, I. Self-administration of dextropropoxyphene by rhesus monkeys to the point of toxicity. *Psychopharmacologia*, 1972, *27*, 179.

Thompson, T., & Pickens, R. Behavioral variables influencing drug self-administration. In R. T. Harris, W. M. McIsaac, & C. R. Schuster, Jr., (Eds.), *Drug dependence*. Austin: Univ. of Texas Press, 1970.

Thompson, T., & Schuster, C. R. Morphine self-administration, food-reinforced, and avoidance behaviors in rhesus monkeys. *Psychopharmacologia*, 1964, *5*, 87.

Trojniar, W., Cytawa, J., Frydrychowski, A., & Luszawska, D. Intragastric self-administration of morphine as a measure of addiction. *Psychopharmacologia*, 1974, *37*, 359.

Weeks, J. R. Experimental morphine addiction: Method for automatic intravenous injections in unrestrained rats. *Science*, 1962, *138*, 143.

Wikler, A., Martin, W. R., Pescor, F. T., & Eades, C. G. Factors regulating oral consumption of an opioid (Etonitazine) by morphine-addicted rats. *Psychopharmacologia*, 1963, *5*, 55.

Wilson, M. C., Hitomi, M., & Schuster, C. R. Psychomotor stimulant self-administration as a function of dosage per injection in the rhesus monkey. *Psychopharmacologia*, 1971, *22*, 271.

Woods, J. H., & Tessel, R. E. Fenfluramine: Amphetamine congener that fails to maintain drug-taking behavior in the rhesus monkey. *Science*, 1974, *185*, 1067.

Yanagita, T. A technique for self-administration of water-insoluble drugs to monkeys by means of chronically implanted stomach catheters. *Bulletin of the Committee on the Problems of Drug Dependence*, 1968, 5631–5640.

Yanagita, T. An experimental framework for evaluation of dependence liability of various types of drug in monkeys. *Bull. Narcotics*, 1973, *25*, 7.

Yanagita, T., Deneau, G. A., & Seevers, M. H. Evaluation of pharmacologic agents in the monkey by long term intravenous self or programmed administration. *Proceedings of the 23rd International Congress of the Physiological Sciences,* 1965, 453.

Yanagita, T., & Takahashi, S. Dependence liability of several sedative–hypnotic agents evaluated in monkeys. *J. Pharmacol. Exp. Ther.,* 1973, *185,* 307.

State Dependent Phenomena

17

Major Theories of State Dependent Learning[1]

Donald A. Overton
Temple University School of Medicine and
Eastern Pennsylvania Psychiatric Institute

State dependent learning (SDL) produced by drugs[2] has excited considerable interest during the last 40 years. Various theories and hypotheses have been proposed regarding the causes of SDL and of drug discriminations.[3] This chapter will review these theories. In general, the review will be comprehensive rather than critical, as the available data do not allow an evaluation of most theories. Nonetheless, the relationship of data to theory will be discussed in cases in which relevant data exist.

The chapter contains three major divisions. The first seven sections describe all the major theories or mechanisms that have been postulated to produce

[1] This work was supported by NIMH Grant MH25136, by Temple University Health Sciences Center and by Eastern Pennsylvania Psychiatric Institute, Office of Mental Health, Department of Public Welfare, Commonwealth of Pennsylvania.

[2] When SDL occurs, a behavioral response learned while an animal is drugged will thereafter be performed well whenever the animal is drugged again, but will be performed poorly or not at all during tests without drug. Conversely, a response learned while undrugged will be adequately performed only without drug and will be performed poorly when drug is present. Engrams acquired in one drug state are said to be dissociated, and hence irretrievable in another drug state. Descriptively, occurrence of the learned response is state dependent or contingent upon the presence of the drug conditions that existed at the time of initial acquisition.

[3] In a task allowing two alternative responses, rats can be required to perform one response when drugged and the second response when undrugged. For example, in a T-maze they may have to turn right when drugged and left when undrugged. A drug discrimination is said to have been learned when the animals reliably perform the correct response in each drug state. In general, such discriminations are easily learned.

the phenomenon of SDL. The following four sections discuss various demonstrated or hypothesized types of SDL that appear to have important theoretical ramifications. Finally, the last two sections describe some of the more interesting postulated consequences of SDL in humans, although for the most part these have not yet been investigated empirically.

Since this chapter is about theories, I will not review the central and well-demonstrated factual properties of drug-induced SDL and drug discriminations. The reader must be referred elsewhere for such summaries. Unfortunately, the only comprehensive review of these areas was published some time ago (Overton, 1968a), and many reports have appeared subsequently. There have been several selective reviews of drug-discrimination work that supplement those appearing in this book (Barry, 1974; Lal, 1977; Overton, 1973, 1974; Schuster & Balster, 1977; Thompson & Pickens, 1971). Also, the reader may be interested in a previous theoretical review by Bliss (1974).

MAJOR HISTORICAL THEMES

Clinical Dissociative Phenomena

Early antecedents to our current interest in dissociation are found in clinical reports of dissociative phenomena. The clinical use of the term "dissociation" is relatively unselective, and such diverse phenomena as thought disorder, delirium, hallucinations, fugue states, and multiple personality have all been referred to as dissociation. Clinical manifestations that appear analogous to SDL are those involving parallel dissociation. In parallel dissociation, there is indication that two or more memories coexist simultaneously, although only one may be represented in conscious experience. Clinical examples include cases of multiple personality, fugue states, and multiple streams of thought. These parallel dissociations can be contrasted to temporal dissociations such as drug-induced delirium, in which temporally sequential events become nonassociated probably because of the severe memory disorder induced by the drug.

Janet's work (1965), originally published in 1907, presents case histories of multiple personality and fugue along with detailed descriptions of the disorders of memory he observed. Similar cases have been reported by other clinicians (Abse, 1950, 1966; Hart, 1926; Thigpen & Cleckley, 1957), and recently operant techniques have been employed to elucidate the memory disorder present in such patients (Dougherty, 1977). Episodes of dissociation do not seem to be mediated by major changes in chemical state because no major electrophysiological changes were observed concurrent with shifts between different "personalities" in a patient suffering from multiple personality disorder (Ludwig, Brandsma, Wilbur, Bendfeldt & Jameson, 1972). Nonetheless, the mechanisms underlying these clinical dissociations may at least overlap with those responsible for drug dissociations.

Multiple streams of thought provide the most mundane example of parallel dissociation. Here, the intrusion of an affect, idea, or behavior unrelated to the conscious stream of thought may alert the observer to other trains of thought occurring simultaneously. It should be noted that in these cases the dissociated experience may not be pathological.

Parallel dissociation has been used to explain several psychological disorders. For example, Freud's model of neurosis implicates the repression (i.e., dissociation) of libidinal impulses from consciousness. The hysterical personality has been linked with a reduced field of conscious attention, perhaps caused by shifting massive repression (Abse, 1966; Chodoff & Lyons, 1957; Janet, 1965). Even some alcohol-induced "blackouts" are attributed to SDL, as memories for the blacked-out period can be retrieved under the proper interview conditions (Goodwin, 1971; Goodwin & Hill, 1973). However most blackouts reflect a simple disruption of memory consolidation resulting in a permanent amnesia.

A striking example of parallel dissociation is provided by the human split-brain preparation studied by Sperry and his colleagues (Gazzaniga & Hillyard, 1971). These patients convincingly illustrate that the brain need not necessarily function as a single harmoniously organized unit. Instead, anatomically separated portions of the brain independently process incoming events, often arriving at different conclusions and attempting to initiate different behaviors. It remains to be shown whether drugs achieve dissociation via analogous mechanisms, as was suggested by Girden and Culler (1937).

Dissociation of Memory by Drugs

The history of experimental work on drug-induced dissociation of memory begins with a report of symmetrical dissociation observed in dogs paralyzed with erythroidine or raw curare extract (Girden & Culler, 1937). Girden followed up this discovery with a series of subsequent investigations (Girden, 1940, 1942a,b,c, 1947). Although his work was described in contemporary textbooks (Morgan & Stellar, 1950) and had striking implications, no serious replication attempt was made until 20 years later (Case & Funderbunk, 1947; Gardner, 1961; Gardner & McCullough, 1962). From our current vantage point, we may question the creditability of Girden's reports on 2 grounds. First, his reports, together with the replication studies of Gardner, contain a total of only 4 experimental subjects that showed SDL. Second, Girden's choice of curare was unfortunate; the paralyzing action of curare interfered with the behavioral response that he used to measure dissociation.

In the early 1960s, several new reports of dissociation appeared (Holmgren, 1964; Otis, 1964; Overton, 1961, 1964; Sachs, 1961, 1962). Most of these investigators invoked various neurological (nonsensory) explanations for their results. The report by Overton (1964) appears to have been most influential, probably because the results of that study showed the most complete amnesias.

Between 1966 and 1974, many SDL studies appeared in literature using the 2 × 2 factorial design suggested by Grossman and Miller (1961). The apparent goal of most of these studies was a simple demonstration of the dissociation effect. Unfortunately, the information yield from these experiments was small because the 2 × 2 design allowed various drug effects to be mistaken for one another. Finally, it was recognized that these design limitations are intrinsic in 2 × 2 experiments and essentially unavoidable (Overton, 1974). Subsequently, there has been a decrease in the number of studies using the 2 × 2 design. At present, the 2 × 2 is used primarily for studies in human subjects in which it is impractical to administer drugs repeatedly (as required by most alternative designs) and in those animal studies in which the SDL-inducing manipulation cannot be repeated.

Stimulus Effects of Drugs

The ability of drugs to act as discriminative conditions appears to have been first appreciated as the result of a study by Conger (1951). Although primarily interested in the fear-reducing effects of alcohol, he decided to use a drug discrimination training paradigm and correctly interpreted the resulting discriminative control. Theorists immediately recognized that the stimulus effects of drugs might produce generalization decrements as well as discriminative control. Several investigators obtained results suggesting such generalization decrements, and others discussed the theoretical and methodological consequences of such effects (Barry, Miller, & Tidd, 1962a; Barry, Wagner, & Miller, 1962b; Belleville, 1964; Grossman & Miller, 1961; Heistad, 1957; Heistad & Torres, 1959; Hunt, 1956; Miller & Barry, 1960).

Overton (1964) concluded that dissociation and drug discrimination were two manifestations of the same process. With high drug doses, response control by drug state was obtained without discriminative training. Lower doses could acquire control of responding only after prolonged discrimination training. A linear dose–effect curve was obtained connecting obvious SDL at high drug doses with slowly acquired discriminative control at low doses. Overton argued that a single drug effect of varying magnitude produced the entire dose–effect curve (Overton, 1964, 1974). Although different drug effects could conceivably operate at different dose levels, the most parsimonious explanation was that of a single effect. This conclusion is disputed by some investigators (e.g., Colpaert, Niemegeers, Kuyps & Janssen, 1975; Colpaert, Niemegeers, & Janssen, 1976). However, the present review will assume that drug discriminations and SDL result from the same drug effects.

Drug-discrimination studies were less frequent than 2 × 2 SDL studies until 1972. By this time, it became apparent that drug-discrimination procedures might be used as tools to answer various pharmacological questions, and the frequency of such studies has increased dramatically. Although the mechanism for drug discriminations is still uncertain, it is now known that virtually all centrally acting drugs are discriminable (Overton & Batta, 1977) and that more than 20 discriminably different categories of drugs exist (Overton, 1976). Thus, drug-discrimination procedures can be used for drug categorization and for the investigation of other pharmacological issues (Cameron & Appel, 1973; Huang & Ho, 1974; Shannon & Holtzman, 1975; Winter, 1974).

STIMULUS THEORIES OF SDL

Stimulus interpretations of SDL are widely accepted, and drug-discrimination data fit easily into the explanatory framework developed in connection with drive-discrimination studies and other investigations of interoceptive stimuli (Bailey, 1955; Bykov, 1957; Webb, 1955). Three variants of these stimulus theories can be distinguished:

1. Drugs Produce Interoceptive Stimuli That Cause SDL

The notion is that SDL and drug discriminations must reflect some variety of stimulus control and that the relevant sensory stimuli enter the brain via the classical sensory pathways (Heistad, 1957; Heistad & Torres, 1959). Postulated internal stimuli include dry mouth, altered gastric or intestinal events, cardiovascular events, and proprioceptive cues induced by ataxia. Because SDL is produced by centrally acting drugs, it is postulated that these cause peripheral changes in muscle tone, for example, which in turn produce altered sensory feedback.

2. Drugs Modify the Perception of Internal or External Stimuli

According to this theory, drugs do not produce stimuli but instead alter the perceptual processing of stimuli, for example, by blurring vision or by making the rat analgesic (Feldman, 1968; Overton, 1968b).

3. SDL Is Mediated by Changes in Naturally Occurring Internal States

This theory proposes that drug effects mimic naturally occurring internal states (e.g., arousal, fatigue, hunger, thirst, sexual arousal), which in turn acquire response control. Specialized neural mechanisms probably exist, allowing such states to control behavior efficiently, and SDL might reasonably result if drugs activated the same mechanisms. The neural pathways through which internal states acquire control are unspecified.

Empirical Tests of Stimulus Theories

The question of whether SDL and drug discriminations are mediated by sensory stimuli was first investigated by quantitatively comparing the discriminability of drugs to that of sensory stimuli. The rationale was that if drug discriminations were mediated by induced sensory stimuli, then directly imposed sensory stimuli should acquire control of responding at least as rapidly as drugs. In a T-maze task, Overton (1964) reported that pentobarbital was more discriminable than visual stimuli, auditory stimuli, or the drive states induced by hunger or thirst, and from this result he inferred that the drug discriminations were not mediated by sensory stimuli. However, using a less discriminable drug in an operant task, Balster (1970) found drug stimuli less discriminable than sensory stimuli, and he suggested that stimuli did mediate drug discriminations. Finally, Kilbey found sensory and drug stimuli that were equally discriminable (Kilbey, Harris, & Aigner, 1971).

In human subjects, changes in environmental stimuli have been shown to impair recall of learned materials, and these effects are sometimes strong if major stimulus changes occur (Bilodeau & Schlosberg, 1951; Dallett & Wilcox, 1968; Godden & Baddeley, 1975; Greenspoon & Ranyard, 1957; Nagge, 1935). Analogously, Moffett and Ettlinger (1967) showed that after monkeys learned a response in total darkness, they would remember it poorly when the lights were turned on. Equally strong effects of changes in contextual cues have been observed in rats (Overton, 1971; Zentall, 1970). Hence, extreme sensory changes do have the capacity to produce SDL-like effects.

Overall, this approach has not yielded definitive results. Most stimulus context changes do not produce generalization decrements large enough to result in a loss of learned responses; similarly, most drug state changes do not result in total dissociation. In both cases, discriminative training can make a response strongly conditional on a particular drug or stimulus condition. The data do not demonstrate either that sensory control is more readily established than drug control nor that the opposite is true. Hence, they provide no compelling evidence either supporting or opposing the idea that SDL is mediated by drug-induced stimuli.

A second approach to testing the role of sensory stimuli as mediators for drug discriminations has been to ablate sensory modalities postulated to provide the mediating stimuli. For example, if drugs were discriminable because they blurred vision, then blinded rats should be unable to learn drug discriminations. However, blind rats have been shown to learn drug discriminations as rapidly as sighted rats, and if sighted rats learn a drug discrimination and are then blinded, only a transient disruption of discrimination results (Overton, 1968b, 1971). These data suggest that drug-induced alterations in visual stimuli are not responsible for drug discriminations even if such alterations do occur. Similarly, olfactory bulb ablations do not interfere with the acquisition of drug discriminations (Overton, unpublished observations). Feldman (1968) suggested that drugs were discriminable because they attenuated the effects of electric shock. However, experiments have shown that after rats learn a drug discrimination, they make drug-appropriate choices despite wide variations in shock level (Overton, 1968b, 1971; Rosecrans,

Goodloe, Bennett, & Hirschhorn, 1973). This finding suggests that analgesia is not responsible for drug discriminability. In summary, sensory ablation studies have not identified any sensory modality that is indispensable to the formation of drug discriminations.

A third approach to determining whether sensory events mediate drug discriminations has been to directly induce bodily changes similar to those induced by discriminable drugs. It is well established that some bodily events are readily discriminated (Bykov, 1957). These experiments tested whether the interoceptive changes induced by drugs are highly discriminable. For example, gallamine and chlorpromazine produce ataxia and muscle flaccidity similar to that produced by pentobarbital. However, neither of these drugs had readily discriminable effects, suggesting that muscular relaxation does not provide mediating cues for the discriminability of pentobarbital (Overton, 1964, 1966). More precise comparisons are possible when peripherally and centrally acting forms of a drug are available. Thus, the peripherally acting forms of atropine and scopolamine lack the discriminable effects produced by the tertiary centrally acting forms. Tetraethylammonium chloride is nondiscriminable, whereas mecamylamine has discriminable effects. Centrally acting nicotine is discriminable, whereas nicotine isomethonium iodide is not (Jones, 1972; Overton, 1966, 1969, 1971, 1974; Schechter & Rosecrans, 1971, 1972). These studies show that it is the central rather than the peripheral effects of the tested drugs that are discriminable. They also suggest that mediating stimuli, if any, probably do not originate in the autonomic nervous system inasmuch as major pharmacologically induced alterations in autonomic activity are not readily discriminated.

Although this research has failed to identify a mediating stimulus for drug SDL, it has not been wasted. Several proposed mediators, which appeared a priori to be plausible, have been discarded. Hence, the work has reduced the number of possible explanations for drug discriminability.

AFFECT-SPECIFIC RECALL AS A MECHANISM FOR SDL

Affects generally are not considered to be important retrieval cues. However, this theory postulates that affects are powerful memory cues and that drugs produce SDL by inducing affects. Drugs are assumed to alter the subject's position on continua such as the euphoria–dysphoria dimension. The neural structures responsible for such affects are unspecified, although commonly assumed to be limbic. Conscious emotional experience may or may not be regarded as important.

A key role for drug-induced affects as mediators of SDL was first suggested by Overton (1973) because of the apparent correlation between discriminability and abuse liability and because drug abuse was assumed to be caused by the positive affective consequences of self-administration. Although the proposed correlation between abuse liability and discriminability has not received empirical support (Overton & Batta, 1977), the resulting theory can now be supported on other grounds.

Naturally Occurring Affects

Using human subjects, Macht (1976) reported SDL effects when affect was altered by "fear-of-electric-shock"; these SDL effects were small. In manic–depressive patients, SDL retention deficits were observed when patients entered or exited from manic episodes (Weingartner, Miller, & Murphy, 1977). If manic–depressive patients show more SDL produced by a given change in affect than do normal subjects, this finding might reflect a significant factor in the etiology of manic-depressive illness. Obviously, mood swings will be more extreme if the patient, when depressed, primarily remembers events from previous depressive episodes and is relatively amnesic for prior normal and manic epochs. Recently, Weingartner has collected subjective reports of "state" from subjects in a drug SDL experiment. The degree of change in reported subjective state accurately predicted the amount of memory loss due to SDL (Weingartner & Murphy, 1977).

Although preliminary, these reports certainly encourage further investigation of the cueing effects of affects and of the possible role of such effects in mediating drug SDL.

Electrical Brain Stimulation

Studies on the discriminability of electrical brain stimulation provide indirect evidence for the cueing effects of induced affects, as stimulation of limbic sites is highly discriminable. Ellen and Powell (1966) showed that septal stimulation could act as a discriminative stimulus in a hunger-motivated bar-pressing task. More recently, both Mayer and Silverberg (personal communications) have established discriminations based on stimulation of the raphe nucleus and the septal area, respectively. It is too early to draw conclusions from these studies, as comparative data from nonlimbic structures is lacking. However, it may be that limbic stimulation produces stronger SDL effects than does stimulation of other brain sites, indicating that affects have significant cue properties.

Stutz and his collaborators (Butcher & Stutz, 1969; Stutz, 1968; Stutz, Butcher, & Rossi, 1969) related the discriminability of brain stimulation to its positively reinforcing effects: They showed that after a fear response had been conditioned to either hippocampal or septal stimulation, fear would generalize to stimulation of the other structure only if stimulation of both electrodes was positively reinforcing. Also, in a two-bar operant discrimination task, they found that rats could very rapidly discriminate septal from hippocampal stimulation when one electrode was reinforcing and the other was not; however, discriminations were formed slowly if both electrodes were reinforcing or if both were nonreinforcing. From these findings, they concluded that positively reinforcing sites must share a similar cueing effect. Nonetheless, after sufficient training, rats discriminated between positively reinforcing septal and hippocampal stimulation with stimulation intensities that produced matched self-stimulation rates. This showed that the cue effects of the positively reinforcing electrodes were not identical, even though they apparently shared many properties.

ANATOMICAL ABLATION THEORIES

The anatomical ablation theories rest on the assumption that some brain structures can be made nonfunctional by drug. If so, learning under drug must be performed by the remaining functional brain structures. This learning might be dissociated in the no-drug state when the brain returned to its normal mode of functioning.

The original theory of this type was proposed by Girden and Culler (1937) and allowed only one dissociated state. Drug was assumed to functionally decorticate the animal. Thus, conditioned responses (CRs) acquired while drugged were learned by functioning subcortical structures and could not be performed without drug because the cortex would then inhibit the trained subcortical structures. CRs acquired without drug were learned only by the cortex and were unavailable when the cortex was depressed by drug.

Revisions of this theory have been proposed to accommodate the existence of multiple dissociated states. One revision states that each drug produces its own pattern of partial ablations, with the residual brain mediating learning under each drug. Changes in the array of functioning brain areas resulting from changes in drug state are assumed to impair performance. Hence, drug states are dissociated from one another to a variable degree. Another essentially equivalent revision postulates that impulses mediating learned responses have to follow different routes through the brain under each drug to avoid ablated areas (Nielson, Justesen, & Porter, 1968). These revised theories have not been tested.

The Girden hypothesis predicts symmetrical SDL because it assumes inhibition of the subcortex by the cortex. If this assumption is not made, the prediction is less clear. Training without drug would educate both cortex and subcortex and would be recalled under drug and no-drug conditions. Training with drug would only educate the subcortex, which might or might not control behavior when the animal was undrugged.

Berger's model (Berger & Stein, 1969) for asymmetrical SDL[4] was developed by removing the inhibition assumption from the Girden theory. Berger's model also incorporated another important change, as in it the

[4] Asymmetrical dissociation is said to occur if responding generalizes from the no drug to the drug state but fails to generalize from the drug to the no drug state.

ablated regions were specified in terms of their specific sensitivity to drug and were not assigned a neuroanatomical locus. The Berger theory can be outlined as follows: Drug disables a majority of the synapses normally involved in learning fear. The remaining drug-resistant synapses learn the response. Without drug, the uneducated synapses are in the majority. Hence, the learned response is not observed. Due to the "majority rule" assumption, the model predicts asymmetrical dissociation. Pert and Avis (1974) have applied the same theory to the dissociation between mecamylamine, scopolamine and saline in a more explicit analysis of the two-drug case. However they were forced to abandon the strict "majority rule" assumption in order to make the theory fit their data. By extrapolation, it is obvious that this type of "majority rule" theory will not easily predict results when more than two drugs are involved.

Recently, Duncan and Copeland (1975) performed the first actual test of an ablation theory of SDL since Girden's (1940) study. They showed that procaine injected into the septal area caused SDL effects. It appears that such experiments might aid in understanding the mechanism responsible for SDL, and perhaps another 35 years will not elapse before the next experiment of this type.

NEUROLOGICAL THEORIES NOT INVOLVING ABLATION

Most central nervous system (CNS) theories for SDL attribute dissociation to a drug-induced disruption of normal brain function. The following theories do not:

1. Cell Assemblies Are Intrinsically State Specific

According to Hebb (1949), cell assemblies are virtually self-exciting patterns of nerve discharge. The initial creation of a cell assembly requires modification of the excitability of various synapses within the cell assembly until it regularly discharges when appropriate stimulus conditions are met. I conducted my first SDL experiment because it appeared to me that cell assemblies would be state dependent; that is, that the synaptic changes that were appropriate to establish cell assembles in one drug condition would not be appropriate to allow that cell assembly to operate efficiently under markedly different conditions of neural excitability such as those induced by drugs. Further, it appeared that new cell assemblies could be organized under these abnormal conditions and that these drug-state cell assemblies would function optimally only when drug was present (Overton, 1961, 1964). The same idea had previously occurred to Hebb (1949, p. 201), who used it to explain the reappearance of learned food-seeking behaviors whenever blood sugar dropped to a low level. A disadvantage of the theory is that it does not specify the locations in brain in which cell assemblies occur; thus, the theory is difficult to test. Nonetheless, several investigators have been influenced by this model (John, 1967; Nielson, 1968; Pusakulich & Nielson, 1972).

There is some similarity between the cell assembly theory of SDL and the "metastable states" of brain function discussed by Ashby (1954, p. 74). Ashby emphasized the ability of the brain to remain in one or another operative mode for a period of time, while excluding other activities. Metastable states were viewed as relatively discrete and discontinuous events. Similarly, the brain is presumed to switch from one cell assembly to another. Some SDL researchers have suggested that the transition between the dissociated state and the no-drug state is abrupt, indicating discontinuity between the states. This inference seems to be based on the observation that animals in a go/no-go task, while gradually shifting from one drug state to another, suddenly begin to respond as if they had just "switched" into a new condition (Girden & Culler, 1937; Pusakulich & Nielson, 1972). However, a much larger array of data supports the inference that the transition from one dissociated drug state into another is a gradual transition along a continuum with cell assemblies gradually becoming more and more disorganized as the internal milieu shifts farther and farther away from the training state (Overton, 1964). These data suggest that drug-induced states and metastable states may have little in common.

2. *Drugs Induce Abnormal Output from "Drug Receptor" Sites in Brain*

This theory suggests that some receptor sites have especially effective interconnections with the rest of the brain so that abnormal efferent activity departing from these sites can propagate to the regions of the brain where learning occurs. The abnormal efferent output from the receptor site is considered to act more or less like an afferent stimulus upon its arrival at a distal part of the brain. Responses learned under drug become contingent upon this abnormal activity for their occurrence. This is an unusual variant of stimulus theory; the stimulus is assumed to arise entirely within the brain, it is not necessarily propagated via a sensory pathway, and it may not be perceived as a sensation. Overton (1966) introduced the model and suggested that the ascending reticular formation might have the requisite connections to other regions of the brain where learning occurs. The theory appears to be readily testable if electrical stimulation can produce the required local changes in state.

3. *Drug-Induced Alteration of Neuronal Excitability Causes SDL*

SDL was hypothesized to result from changes in neuronal excitability by Pusakulich and Nielson (1972), who tested this hypothesis using electrical brain stimulation as a conditioned stimulus (CS). They were unable to demonstrate altered excitability of the neurons adjacent to their stimulating electrodes during drug-induced dissociation. This is an incomplete theory, as no mechanism was ever specified by which altered excitability might produce SDL.

FUNCTIONAL ABLATION THEORIES

Three theories of SDL focus on the behavioral and cognitive consequences of drug action instead of emphasizing the anatomical locus of drug effects. Two of these theories are specific, and the third is extremely general.

1. SDL Occurs When Habituation to Novelty Is Blocked by Drug

To explain SDL-like effects in the shuttlebox avoidance task, Sachs (1967) postulated that animals under drug learn rapidly without exploring the task. When drug is removed, exploration occurs, and it interferes with the learned avoidance response. An apparent SDL decrement results. In the reverse direction, rats trained without drug show only a small response decrement when drug is introduced, based on the depressant effects of the drug (Sachs, Weingarten, & Klein, 1966). This is a reductionist theory. Essentially, it postulates that SDL does not exist as a bona fide phenomenon and that the apparent demonstrations of SDL reflect poorly understood combinations of other drug effects. The basic postulate that drug blocks habituation was not supported by Iwahara and Sakama (1972); they observed normal habituation under chlordiazepoxide.

2. Position Learning Is Blocked by Drug so Response Learning Occurs

Undrugged rats learn position habits and apparently form a cognitive map of the maze. Drugged rats are too obtunded to learn position habits and solve the task by learning a specific response pattern. Neither solution is attempted in the other state, causing SDL. This theory was used by Pusakulich (1974) to explain data collected with pentobarbital and in its present form allows only two dissociated states.

3. SDL Is Induced by Addition or Deletion of Brain Functions

Overton (1973) suggested that even though the same overt response is learned with and without drug, the drugged animal may use a different strategy or encoding process to learn the task than it would use when undrugged. The idea is that by ablating certain brain functions, the drug essentially changes the task. A response that was learned without drug cannot be performed when drugged because part of the requisite brain functions that were utilized in learning the response are now missing. Similarly, a response learned while drugged may not be performed when undrugged because the animal's more complete perception of the task fails to provide the specific stimulus array required to elicit the response. When dissociations among several drugs are involved, each drug is assumed to delete a specific set of brain functions, and the task is mastered in each state using the remaining brain functions. Addition of functions causes altered perception of the task, and a response decrement often results. Training transfers between two drugs only if they delete a similar set of brain functions.

Obviously, this theory is more general than the specific proposals by Pusakulich and Sachs. It no longer specifies the particular brain functions involved in producing SDL, and it is probably untestable until they are specified. However, it can accommodate numerous dissimilar drugs and may predict either symmetrical or asymmetrical SDL, depending on the precise relationship between the affected brain functions and the requirements of the task. After one specifies the brain functions impaired by drugs A and B along with the cognitive functions required in Tasks X, Y, and Z, it is possible to predict whether SDL will be produced by Drug A or B in Task X, Y, or Z. For example, suppose that Drug A impairs habituation and Drug B degrades visual perception. Also, assume that Task X normally involves habituation prior to task learning, Task Y usually involves visual perception, and Task Z involves both habituation and visual perception. Finally, and most important, assume that X and Z can be learned in spite of the fact that habituation has been blocked by Drug A. Similarly, assume that Tasks Y and Z can be mastered even with impaired visual processing. State changes from A to N (N = no drug) or from N to A should produce SDL in X and in Z. Similarly, transitions from B to N or from N to B should produce SDL in Y and Z. Task Y should be unaffected by drug A and task X unaffected by B. Transitions from A to B or B to A should produce SDL in X, Y, and Z. No prediction is made as to the effect of each drug on initial task acquisition. This will depend on how easily alternate strategies can be substituted for the "usual" solution when it is rendered unavailable by drug action.

Strategies for testing this theory involve manipulations of psychological variables rather than of neuroanatomical or neurochemical systems. In order to predict whether SDL will occur in a task, it is necessary to identify the cognitive functions involved in learning a task and the effect of drug X on these functions. However, no knowledge of the location or chemical nature of drug action within the brain is required. All of the functional ablation theories are similar in that the process of testing and improving them can be accomplished primarily by behavioral research. However, such research requires a rather sophisticated knowledge of the cognitive functions utilized during task learning.

THEORIES DERIVED FROM HUMAN RESEARCH

For several years, SDL studies in human subjects were exclusively concerned with alcohol-induced SDL. These studies (reviewed by Overton,

1972a) were rather empirical in nature, designed to test whether alcohol would produce SDL and to investigate whether a relationship might exist between dissociation and the addictive process. More recently, similar studies have tested marihuana and other abused drugs.

The possibility that SDL may be induced by clinically used drugs has received relatively little attention in spite of the early suggestion by Heistad (1957) that such effects might be significant. Due to this lack of research, we are only able to estimate the SDL efficacy of most drugs by inference from animal experiments (Overton, 1976). An exception exists in the case of stimulants as used for the treatment of hyperactivity in children; this single clinical instance has received adequate investigation by Sprague and others. (Aman & Sprague, 1974; Goodwin, 1974; Goodwin, Powell, Hill, Lieberman, & Viamonte, 1974; Goodwin, Powell, Bremer, Hoine, & Stern, 1969; Ley *et al.*, 1972; Powell, Goodwin, Janes, & Hoine, 1971; Rickles, Cohen, Whitaker, & McIntyre, 1973; Sprague, 1972; Stillman, Weingartner, Wyatt, Gillin, & Eich, 1974; Storm & Caird, 1967; Storm & Smart, 1965; Swanson & Kinsbourne, 1976). Throughout the literature on SDL in human subjects, there is a relative absence of theorization regarding the causes of SDL; this is understandable in view of the difficulty that investigators have experienced in even obtaining adequate demonstrations of the existence of dissociation.

Verbal Learning Concepts

Investigators of verbal learning have recently become interested in SDL. They have an already established nomenclature for describing the effects on memory and retrieval of events at the time of encoding and at the time of recall (Tulving & Madigan, 1970). That drugs may constitute effective retrieval cues has apparently fit easily into their system of thinking, and several recent SDL studies have been heavily influenced by these verbal learning concepts. In these studies, the relative effectiveness of free recall, category cued recall, and recognition has been compared in 2×2 studies that measured the effect of one or more of these factors on recall and on drug-induced SDL. The results apparently show that SDL can be overridden if sufficiently effective retrieval cues are present, and this appears not to be a ceiling effect (artifact) in at least some studies (Eich, Weingartner, Stillman, & Gillin, 1975; Keane & Lisman, 1976; Petersen, 1974). Similar studies have also been conducted in animals. For example, SDL can be eliminated by overtraining (Bliss, 1972; Iwahara & Noguchi, 1972, 1974), if retraining instead of extinction testing is employed to test for recall, or if fear or anxiety is induced by a CS previously paired with electric shock (Connelly, Connelly, & Epps, 1973; Connelly, Connelly, & Phifer, 1975).

The verbal learning studies do not appear to involve any new theory regarding the mechanism of SDL. Drugs are called retrieval cues instead of stimuli, which is a change in nomenclature but apparently not in meaning. It is interesting that the properties of contextual cues have sometimes been interpreted in such a way as to predict asymmetrical dissociation rather than symmetrical SDL. Asymmetrical dissociation has posed a problem for SDL theorists for several years (Overton, 1968a), as stimulus generalization decrements are usually postulated to be symmetrical. Two rather concrete theories regarding possible mechanisms for SDL have derived from the work of Weingartner as follows:

1. Drug-Induced Alteration of Associative Mediators Causes SDL

Weingartner and Faillace (1971) showed that alcohol modifies the spontaneous word associations of human subjects. The theory suggests that such spontaneous associates are used to form mnemonics, which subsequently assist in retrieval. However, since the unlearned associates are state dependent, the learned mnemonics will also be dissociated by changes in state.

2. Drug-Induced Changes in Scan Patterns Cause SDL

It is observed that drugs alter the visual scanning movements with which subjects look at pictures (Weingartner, this volume). The theory states that these scanning patterns become encoded as part of the engram and that recognition or retrieval requires a reinstatement of the abnormal scanning patterns induced by drug at the time of learning.

It is unlikely that either of these proposed mechanisms is responsible for all of the observed instances of SDL in humans. However, a serious alternative is the possibility that SDL is produced by a multitude of specific changes in encoding and retrieval processes, with no specific mechanism being overwhelmingly important. If so, the mechanism responsible for SDL may be essentially undiscoverable. Recall that a similar impasse was reached several years ago with regard to the stimulus theories for SDL in animal subjects. In that instance, it was concluded that even if a variety of sensory changes did simultaneously mediate SDL, it might be impossible to convincingly demonstrate the contributions of any individual modality to the production of SDL unless one particular modality or change was overwhelmingly important (Overton, 1971). For similar reasons, it may be difficult or impossible to determine the cause of SDL in human subjects.

THEORETICAL CONSEQUENCES OF NONPHARMACOLOGICAL SDL

Several nonpharmacological manipulations have been shown to produce SDL-like effects. We have already discussed the possibility of affect-specific recall. The following paragraphs will describe other instances of nonpharmacological SDL and discuss their theoretical significance.

Electroconvulsive Shock (ECS)

Learning that occurs during the postictal period after a convulsion is state-dependent. The convulsion can be elicited by shock applied to the whole brain (Overton, Ercole, & Dutta, 1976; Thompson & Neely, 1970). Alternatively, kindled convulsions elicited by amygdala stimulation may be employed (McIntyre & Reichert, 1971; Wann, 1971). There has been considerable concern that the SDL effects of ECS might be confounded with its effects on memory consolidation, and investigators have also studied the relationship of ECS–SDL to drug SDL (Gardner, Glick, & Jarvik, 1972; Mayse & DeVietti, 1971; Miller, Malinowski, Puk, & Springer, 1972; Overton *et al.*, 1976; Thompson & Grossman, 1972). The occurrence of SDL after ECS is theoretically important because it provides the best available evidence that SDL is basically a property of brain rather than of drugs. Also, the occurrence of ECS-SDL suggests that a variety of other toxic brain states might similarly produce SDL.

Rapid Eye Movement (REM) Sleep

Learning during REM sleep is dissociated. Evans reported that sleeping subjects were able to hear verbally administered suggestions during REM sleep and could subsequently respond as suggested, indicating that learning had occurred. Responding in accordance with the suggestion was only observed during REM sleep, and subjects were amnesic for the whole procedure when awake. Although the verbal suggestion training procedure employed in this experiment did not allow Evans to monitor the learning process very adequately, it is apparent that some learning did occur and that retrieval was state-dependent (Evans, 1972; Evans, Gustafson, O'Connell, Orne, & Shor, 1966, 1969, 1970). Obviously, REM-SDL provides an attractive alternative to repression as an explanation for the rather universally observed difficulty in recalling dreams.

REM–SDL is important because it demonstrates SDL produced by a normally occurring altered state accompanied by essentially normal physiological conditions. The REM–SDL was probably caused by the altered level of arousal and cognitive functioning that accompanies REM sleep. Although circadian chemical variations do exist, the REM sleep SDL effect was presumably not mediated by these rhythms, or else it would not occur if subjects slept during the day. It is not known whether SDL might occur in stages of sleep other than REM, and this should be studied.

Lesions

It has been suggested that brain lesions might produce SDL-like effects and that the gradual recovery of function often observed after lesions might involve changes comparable to task relearning after a dissociative state change. Reversible brain lesions produced by local cooling or drug application appear to provide a convenient method for studying this possibility, and the literature contains one such study in which Duncan and Copeland (1975) observed SDL effects after procaine infusion into the septal region of rats. Except for this study, the question is uninvestigated.

Spreading Depression

Cortical spreading depression produced by potassium chloride applied to the cortex is considered to functionally ablate the cortex. Schneider reported that spreading depression produced SDL in operant bar-pressing tasks and in active- and passive-avoidance tasks, although the effects that he observed were rather weak (Schneider, 1966, 1967, 1968, Schneider & Ebbsen, 1967; Schneider & Kay, 1968; Schneider & Hamburg, 1966). At face value, this preparation appears to provide verification of Girden's hypothesis that functional decortication will produce dissociation. However, the preparation is less than optimal because the degree of depression constantly waxes and wanes, with successive waves of depression sweeping over the cortex interspersed with periods of relative normalcy (Freedman, 1969; Freedman, Pote, Butcher, & Suboski, 1968). Additionally, it has been reported that rats can differentiate between various concentrations of cortically applied potassium, and this suggests that the SDL effect may be produced directly by the cannulated potassium rather than by the resulting cortical spreading depression (Langford, Freedman, & Whitman, 1971; Reed & Trowill, 1969). In summary, the spreading depression studies do not contribute very much to our understanding of the possible SDL consequences of brain lesions.

Circadian SDL

Circadian SDL now appears to be established as a phenomenon. At least in some experimental tasks, responses learned at a particular time of day are most efficiently performed at that time on subsequent days and less efficiently performed at other times of day. Surprisingly, periods of optimal recall reoccur every 12 hr rather than every 24 hr (Holloway, this volume; Holloway & Wansley, 1973a,b; Strobel, 1967; Wansley & Holloway, 1975).

Although the phenomenon of circadian SDL may be of considerable importance, its current contribution to theory is minimal because of our inability to specify its physiological causes. In most SDL situations, the nature of the altered state can be specified (usually chemically), and the theoretical problem is that of determining what characteristics of this altered state give rise to state-dependent retrieval. However, in the case of time-of-day SDL, even the very existence of an altered state with a 12-hr periodicity might be disputed. Possibly some specific neural mechanism rather than a diffusely altered state is responsible for this SDL-like effect.

Stress-Induced SDL

It has been proposed by several authors that stress or anxiety might produce SDL and that engrams formed during periods of stress would be poorly recalled thereafter except when the stress condition was reinstated (Cherkashin and Azarashvili, 1972; Klein, 1972; Klein & Spear, 1970; Overton, 1968a; Spear, 1970; Spear, Klein, & Riley, 1971). This is an intriguing hypothesis, but no very adequate experimental evidence has been forthcoming to support it.

Drive-State SDL

Discriminative control by drive states and generalization decrements produced by changes in drive state were among the major historical antecedents of current work on SDL. One theory proposes that drugs produce SDL by inducing changes in naturally occurring drive states that in turn directly exercise response control (Overton, 1971). Although it is possible that drives and drugs do acquire discriminative control via the same general type of mechanism, there are no reported instances in which a full dissociation of learning has been produced by changes in drive state (Otis, 1956; Peck & Ader, 1974; Webb, 1955). This suggests that the cueing properties of drive states are not sufficiently strong to mediate drug SDL.

REM-Deprivation SDL

Deprivation of REM sleep in rats produces sequelae such that responses learned immediately after a period of REM deprivation are state dependent. Whether this effect is caused specifically by REM deprivation or would also be induced by other fatigue-induced procedures is untested (Huang, 1973; Joy & Prinz, 1969; Kruglikov, Aleksandrovskaya, & Dish, 1975).

The SDL effects produced by normal physiological shifts (REM sleep, fatigue, thirst) provide an interesting contrast to drug SDL effects that only occur when relatively high doses produce decidedly nonphysiological conditions. It will be extremely interesting if the mechanisms for some of the naturally occurring SDLs can be worked out. It appears unlikely that very similar mechanisms are involved in all of the SDL-like effects described above. For that matter, even different drugs may achieve response control via significantly different mechanisms. However, until specific mechanisms are delineated, it is probably useful to categorize all of these SDL-like effects together, and some may indeed be interrelated (e.g., see Huang, 1973).

DESCRIPTIVE "THEORIES" ABOUT SDL

Some SDL "theories" are simply descriptive statements about properties of the phenomena. They incorporate no proposed mechanism that might explain the occurrence of SDL. For example:

1. Drug states are points on a continuum produced by various dosages of a drug. Transitions between drug states are gradual rather than discontinuous (Overton, 1964).
2. Movement initiation dissociates, whereas response choice generalizes across changes in drug state in a shock avoidance task (Bindra & Reichert, 1966, 1967; not found by Iwahara & Sugimura, 1970, or by Iwahara & Matsushita, 1971).
3. Overtraining causes learned responses to generalize across dissociative barriers even though minimally learned material may simultaneously fail to generalize into the test drug condition (Bliss, 1972; Iwahara & Noguchi, 1972, 1974).
4. In human subjects, category cueing or recognition testing will eliminate SDL effects that are obvious during free recall tests of verbal material (Eich *et al.*, 1975).

Obviously, all these statements represent generalizations describing experimental observations rather than proposed mechanisms explaining the results. Although sometimes referred to as theories, they are more properly regarded as facts.

DRUG AS CS OR US

A drug can act as a conditioned stimulus (CS) and elicit a classically conditioned response originally produced by some other event. For example, acetylcholine or epinephrine can elicit conditioned leg flexion responses in dogs (Cook, Davidson, Davis, & Kelleher, 1960). However, there are relatively few studies of this type (Franks, 1958), allowing one to question the generality of this phenomena.

A drug may also act as an unconditioned stimulus (US) and part or all of its actions may become conditioned responses (CRs); conditioned motionlessness, hyperactivity, stereotyped behavior, EEG reactions, cardiac reactions, hypoglycemia, immunological reactions, fear reduction, salivation, altered dopamine metabolism, suppression or augmentation of bar pressing, and suppression or production of opiate withdrawal reactions have all been reported (Alvarez-Buylla & Carrasco-Zanini, 1960; Cameron & Appel, 1972; Crisler, 1930; Crowder, Smith, Davis, Noel, & Coussens, 1972; Ellinwood, 1971; Hecht, Baumann, & Hecht, 1967; Kamano, 1973; Lal & Drawbaugh, 1973; Lal, Reddy, & Roffman, 1972; Lang, Brown, Gershon, & Korol, 1966; O'Brien, O'Brien, Mintz, & Brady, 1975/6; Perez-Cruet, 1971, 1974; Pickens & Dougherty, 1971; Razran, 1933; Russek & Pina, 1962; Takahashi, Nagayama, Kido, & Morita, 1974; Tilson & Rech, 1973). Occasionally, the CR appears to be opposite to the UR (perhaps a concurrent compensatory reaction), although this is less frequently observed (Korol, Sletten, & Brown, 1966; Siegel, 1972, 1975).

These phenomena fit easily into the Pavlovian rubric of conditioning with internal events as CS and/or US. Current investigators appear to be content with the Pavlovian nomenclature, and there is virtually no theoretical controversy connected with these experiments. However, the relationship of the drug actions that act as US to the usual pharmacological actions of the same drugs has not been systematically investigated, and further research will be required to determine whether a simple relationship exists. In relation to SDL, the primary theoretical question raised by these phenomena is whether the same drug effects are involved when a drug acts as CS or US and when it acts as a discriminative stimulus or produces SDL. To date, no one appears to have figured out how to test this possibility.

THEORETICAL IMPLICATIONS OF POST-TRIAL SDL

It has been claimed that the presence of drug during memory consolidation is a necessary *and* sufficient condition for SDL. If so, SDL might occur if drug were administered immediately after the training trial, as long as the brain was drugged by the time when memory consolidation took place. This idea was first investigated by John, Bartlett, and Sachs (John, 1967, p. 83) and has been repeatedly reinvestigated since then. It is not really a theory of SDL since no mechanism for SDL is described. It is, however, a significant assertion about the properties of SDL. Post-trial SDL, if it exists, is not a simple extension of our concepts about pretrial SDL. In order for post-trial SDL to occur, short-term memory must generalize across changes in drug state without SDL decrements. This is contrary to the predictions of many current SDL theories and sets post-trial SDL apart as a separate phenomen, requiring a significant revision in our concepts of SDL.

To date, most post-trial SDL studies suffer from an unnoticed methodological problem, as the drug state has been allowed to shift during the presumed consolidation period. For example, the consolidation period in a passive avoidance task is reported to be several hours long. However, the SDL effects of ECS and pentobarbital wear off gradually and are virtually gone after 120 min (Overton *et al.,* 1976). Thus, the predicted effect of a post-trial injection is that consolidation should be "smeared" across a range of drug dosages, including the no-drug state. Any particular drug state re-established during retrieval testing should be only partially effective in producing retrieval (e.g., Thompson & Neely, 1970), and very strong SDL effects are not predicted unless a uniform drug state has been maintained during the consolidation period. A regimen of repeated small supplemental injections could accomplish this goal, but this procedure has not been used in any post-trial SDL studies to date.

Because of the theoretical significance of post-trial SDL, we will review briefly the evidence regarding this phenomenon. In John's experiments, cats were trained for 20 trials in a shuttlebox avoidance task and then received intraventricular calcium or potassium. Sachs (1961, 1962) had previously noted weak SDL when such injections were administered before training in this task. Post-trial SDL was not observed by John, perhaps because the interval between onset of training and injection was 20 min.

Reichert (1968, 1971) reported that post-trial SDL did occur, and he developed a two-phase training procedure for demonstrating this effect in a T-maze task. In phase one, right-turn trials (R) and left-turn trials (L) alternate; R trials are followed by injection, and L trials are not. Hence, the "right turn" engram should be consolidated under D (drug) even though the animal has no opportunity to express this engram. The L engram should be consolidated in the no-drug (N) state. In phase two, pretrial injections are used in a regular drug discrimination paradigm, and the subjects are split into two groups. Group 1 learns N–L, D–R, which is facilitated by any previously established state-specific engrams. Group 2 learns N–R, D–L (opposite to the reactions previously established). Using intraperitoneal injections of pentobarbital or scopolamine, Reichert found that Group 1 learned slightly faster than Group 2, providing evidence for weak post-trial SDL effects. In the absence of post-trial SDL, both groups should have learned at an equal speed.

Recently, this investigator used the same procedure to test whether ECS administered 1 sec after T-maze training trials would produce SDL (Overton, unpublished observation); the data clearly showed an absence of appreciable post-trial SDL (details of procedure similar to those described in Overton *et al.*, 1976). The shock escape T-maze is probably a poor task in which to investigate such effects, as the consolidation period in this task is very short (Overton, unpublished observation). Hence post-trial manipulations have to be applied very promptly after the training trial, and negative results obtained in the T-maze do not necessarily refute an earlier report of weak SDL induced by post-trial ECS in a passive avoidance task (Thompson & Neely, 1970).

Wright and Chute have reported very robust post-trial SDL, thus reactivating interest in the phenomenon (Chute & Wright, 1973; Wright & Chute, 1973). Their data strikingly suggest that post-trial SDL may have occurred. However, the demonstration is less than convincing for the following reasons: (*1*) They use a 2 × 2 design, with response failure indicating amnesia; such procedures are very likely to produce artifactual results (Overton, 1974). (*2*) Ceiling and floor effects were apparent in some of their data (Wright, 1974), and some of the observed amnesias were more profound than predicted, in view of the shifting drug state presumably achieved during consolidation. (*3*) Reports of successful replication have not appeared. Indeed, Settle (1973), working in the same laboratory, was unable to replicate one of their studies even before it was published (Wright, Chute, & McCollum, 1974).

In view of the great theoretical importance of post-trial SDL, a very convincing and unequivocal demonstration of its existence would be desirable. Presently, available data appear to fall short of this goal.

THEORIES RELATING SDL TO DRUG ABUSE

Both drug abuse and SDL may be caused by "consciously perceived" drug effects, and several theories have been developed that specifically relate SDL to drug abuse. These theories do not describe a mechanism for SDL but instead describe hypothetical processes by which SDL and abuse may be related to one another.

1. SDL and Abuse Are Caused by the Same Drug Effects

For example, both SDL and abuse may result from drug-induced affects. It follows from this assumption that discriminability and abuse liability might be highly correlated and that measurements of discriminability might predict abuse liability (Overton, 1969, 1973). This has been recently disproved (Overton & Batta, 1977).

2. Drug Users Develop a Repertoire of Drug-Specific Behaviors

Such drug-linked behaviors might be more reinforcing than normal undrugged behaviors; hence, the subject increasingly might use drugs in order to gain access to the drug-specific response repertoire rather than because of any intrinsically reinforcing drug effects. This theory comes in two versions. The first postulates that a unique drug response repertoire is developed because environmental reinforcement contingencies are changed while the subject is drugged, for example, because other people treat an intoxicated person differently than they treat an undrugged person. The second version proposes that drug alters the user's sensitivity to reinforcement so that reinforcement contingencies are effectively changed even in the absence of any real change in the external environment. The theory provides a mechanism by which drugs initially possessing few reinforcing effects can become more reinforcing after prolonged use because of the desirability of behaviors linked to the drug condition (Harris & Balster, 1970; Overton, 1972a).

3. Drug Acts as a CS for Positively Reinforcing Acquired Drug Effects

Because positively reinforcing reactions become conditioned to the drug, it becomes more reinforcing and more abused. For example, if experiences with an initially neutral drug are paired with positive social reinforcement, a conditioned association may be formed that enables the drug to evoke the positive affect originally produced by the social milieu. This theory is somewhat tangential to our discussion of discriminative drug effects, as it treats the drug as a conditioned stimulus that evokes a reinforcing affect rather than as a discriminative stimulus (Cook *et al.*, 1960; Harris, Claghorn, & Schoolar, 1968).

4. Recall of Consequences Is Prevented by SDL

Only the initial effects of drug during the onset of drug action are easily recalled. These effects, which take place before a substantial change in state has occurred, are positively reinforcing. A dissociative barrier prevents recall of the subsequent negative consequences of drug use. Evidence supporting this proposal has been obtained by Tamerin, Weiner, and Mendelson (1970), who showed that alcoholics selectively fail to remember much of the dysphoric content of their drinking episodes.

5. Dissociation Increases after Repeated Drinking

Storm and Smart (1965) postulated that the penetrability of the dissociative barrier between the no-drug and the alcohol state might decrease with prolonged drinking. They attributed various aspects of alcoholism to this increase in dissociation. Although this theory was apparently responsible for many of the early SDL studies in human subjects, it has not been consistently supported by the resulting data in either human or animal subjects (Goodwin *et al.*, 1974; McKim, 1976; Overton, 1972a; Weingartner & Faillace, 1971).

HYPOTHESIZED CONSEQUENCES OF SDL

The impact of SDL on everyday life has been the subject of much speculation. We have already mentioned several possible consequences of SDL in man, including the influence of SDL on dream recall, the possible relationship of discriminability and abuse, the cueing properties of affects, and the proposed role of time-of-day as a retrieval cue. Several other important possibilities will be considered in the following paragraphs.

Dissociation of Psychotherapeutic Gains

The beneficial effects of psychoactive drugs often do not persist after drug treatment is discontinued (Dollard & Miller, 1950). Heistad (1957) postulated that changes in drug stimuli and the resulting generalization decrements could cause such relapses by causing the patient to forget the new behaviors that he had acquired via psychotherapy while drugged. The assertion has been repeated by Otis (1965), Strobel (1967), and others from time to time. In view of the importance of the issue, there are surprisingly few studies directed at evaluating the significance of this effect. With the major tranquilizers, only one study has been reported (Kurland, Cassell, & Goldberg, 1968). There are no studies with antidepressant drugs, butyrophenones, lithium, or with minor tranquilizers at subintoxicating doses. The only adequately studied instance involves the possibility that stimulant drugs used with hyperactive children may produce SDL, and here the results are mixed, although mainly negative (Bustamante, Jordan, Vila, Gonzalez, & Insua, 1970; Hurst, Radlow, Chubb, & Bagley, 1969; Roffman, Marshall, Silverstein, Karkalas, Smith, & Lal, 1972; Sprague, 1972; Swanson & Kinsbourne, 1976).

If we can extrapolate from research with animals, it appears unlikely that the generalization decrements postulated by Heistad are strong enough to constitute a major side effect of most normal therapeutic drug usage. Among the commonly used psychoactive drugs, only the minor tranquilizers and sedatives have produced strong SDL effects in animal subjects, and these effects have occurred only at intoxicating doses (Overton, 1973, 1976). Similarly, all published SDL studies in humans (except those using stimulants) have used intoxicating doses to obtain even the modest SDL effects that are usually observed. Since the vast majority of psychoactive prescriptions involve subintoxicating doses, it appears at this time unlikely that this type of drug usage involves important SDL side-effects.

It should be noted, however, that some types of psychoactive drug usage do involve intoxicating dosages; for example, high doses of minor tranquilizers are used to relieve anxiety during flooding and desensitization procedures. SDL might be an important effect in this instance. Also, the animal data may not have good predictive value as regards the degree of SDL produced by various types of psychoactive drugs, and at least a few clinical studies would be very reassuring in this regard.

Dissociated Learning under Surgical Anesthesia

Dissociated learning under surgical anesthesia has been proposed, and the animal data do predict that dissociated learning should occur at some level of anesthesia (presumably at about Stage 2). Experimental evidence regarding this possibility is contradictory, and too few experiments have been performed to allow confident prediction (Adam, Castro, & Clark, 1974; Cheek, 1959; Levinson, 1965; Osborn, Bunker, Cooper, Frank, & Hilgard, 1967). An obvious consequence of SDL during surgical anesthesia would be to reduce casual conversation in the operating room, as such talk might be misinterpreted and remembered by the patient.

State Boundedness

Fischer has proposed a system of altered states of consciousness induced by drugs and other factors (Fischer, 1971a,b,c, 1975; Fischer & Landon, 1972). In his system, a continuum of arousal levels is postulated (called the ergotropic-trophotropic continuum) and a variety of drugs are assumed to move the subject to various points along this continuum. Learning that occurs at a particular point on the continuum is "state bound," and the degree of memory generalization between states is determined by the similarity or dissimilarity of the learning and retrieval conditions (as in SDL). Although Fischer has developed this system without very much explicit reference to the drug discrimination literature, the strongest critique of his proposal does come from that literature. Specifically, the geometry of Fischer's proposed system of states and the resulting predictions regarding generalization between states conflict with experimental findings. All stimulants and hallucinogenics produce ergotropic effects in Fischer's system. However, animal studies have shown that these drugs induce multidimensional shifts in state rather than shifts along a one-dimensional continuum. For example, several different types of hallucinogens have been identified with definite state boundaries, distinguishing the effects of these compounds (Kuhn, White, & Appel, 1977; Winter, 1975). These experimental findings conflict directly with Fischer's proposed structure for altered states of consciousness.

Multiple Systems of Logic

In a rather provocative book, Tart has proposed that altered states of consciousness are discrete and are "stabilized" by various mechanisms (Tart, 1975). Additionally, he has argued that autonomous and separate systems of logic and reality can be developed in various drug-induced states with each such system dissociated from the others (Tart, 1972). One important feature of Tart's proposal is that drug-altered perceptions of reality have a validity equal to and sometimes superior to that of our normal nondrug perceptions. If so, it is surprising that more instances of drug self-administration have not developed among various animal species, due to the improved survival value that would presumably result from pharmacologically improved perceptions of reality.

Tart considers that experimental drug SDL studies add little to our knowledge of altered states because such studies accept simple drug administration as an adequate state-inducing procedure without performing any manipulation check to determine what state has actually been induced. This methodological point has some merit, and Weingartner has found that subjective reports of the strength of drug effects do improve his ability to predict the size of observed SDL effects (Weingartner, this volume; Weingartner & Murphy, 1977). However, the drug-discrimination literature shows that Tart's argument is generally wrong in animal subjects and that repeated injections of a compound do reliably reinstate a uniformly altered state.

Acquired Drug Effects

One important consequence of drug discriminability may be that it enables drugs to essentially "acquire" new effects as the result of a behavioral conditioning history (Harris & Balster, 1970; Overton, 1972a). In the usual drug-discrimination paradigm, it appears somewhat stilted to argue that "pressing the left bar" becomes a new drug effect as the result of conditioning. However, using a somewhat different paradigm, Overton (1972b) was able to make conditioned fear contingent upon the presence of phenobarbital in the Estes and Skinner CER task (CER = conditioned emotional reaction). Thus, discriminative conditioning caused drug to acquire a fear-increasing effect, whereas it previously had exhibited a fear-decreasing effect in the same task. We can at least say that the discriminatively conditioned response to drug overrode and reversed the pre-existing intrinsic drug effect. Perhaps it is reasonable to say that the drug acquired a new effect that it did not previously possess.

It appears certain that a variety of other drug effects could be behaviorally modified in further analogous experiments. More crucial issues concern the extent to which such newly acquired drug effects may generalize to tasks other than those in which they were learned and the resistance to extinction of the acquired drug effects. No data are available regarding these questions.

Might such acquired drug effects possibly occur to a significant degree in a naturalistic situation? For example, if a person repeatedly used a minor tranquilizer *only* on those occasions when he was exposed to anxiety-provoking stimuli, would the drug gradually acquire discriminative fear-eliciting properties by virtue of this repeated association? The proposal appears to be feasible from the point of view of dosage since drug discriminations can be established with dosages below toxic levels and probably within the therapeutic dose range. However, the regularity with which drug was paired with anxiety would be rather poorly controlled in most naturalistic situations, and this might interfere with efficient conditioning. Obviously, effects like this do not universally occur—or else all intermittently applied drugs would eventually lose their effectiveness due to repeated pairings with the disease state they were intended to eliminate. Nontheless, since it has been possible to reverse the effects of one drug in one situation, it will no doubt be possible to alter some effects of other drugs in different situations (e.g., Siegel, 1975), and it appears very worthwhile to test the generality and strength of this phenomenon. Laboratory experiments will be very useful in investigating this phenomenon. However, valuable data could also be obtained in clinical drug-use situations; these would allow a test of the degree to which drugs actually do alter their effects in naturalistic drug-use situations and take on new discriminative effects as the result of differential reinforcement. Hopefully, this question will be investigated in the near future.

SUMMARY

This chapter has summarized several theories regarding the causes of state-dependent learning, and a variety of proposals regarding its possible scope and significance in everyday life. Included here are many very good ideas—

condensations of the best available knowledge and thought regarding the causes and properties of SDL and drug discriminations. Also included are several less reasonable proposals that conflict with current data: a sort of conceptual trash that has accumulated over the years due to intellectual creativity and our poor understanding of the phenomenon.

As a spectator and participant in this arena of ideas for several years, the author has been especially gratified by recent attempts to study affect-specific recall. It has always appeared to me that the major hazard in ascribing SDL to interoceptive stimuli (such as fear or arousal) lay in the fact that so little was known about these stimuli. By this time, the psychophysics of discriminable drug effects is almost certainly better understood than that of other interoceptive stimuli, excepting those that originate in defined sensory receptors and enter the CNS via known afferent sensory pathways. Hence, investigations of the cue properties of affect are most welcome, as they may partially remedy our ignorance about "feelings" and other interoceptive stimuli.

An amazingly high percentage of SDL and drug-discrimination studies have been nontheoretical and not obviously designed to prove or disprove any particular postulate regarding the nature of the phenomenon. Probably as a result, we have made considerable advances in our understanding of the phenomenology of SDL and drug discriminations without too much concurrent progress in the adequacy of our theories regarding their causes. It appears that more rapid progress toward an understanding of the causes of state dependency would occur if more theory-testing experiments were performed.

Sometimes during the process of theory development and theory testing it helps to take a broad perspective on all the various proposals and options available rather than to focus myopically on any particular proposal. The present paper aims to assist in this effort by gathering together in one place almost all of the major available theoretical proposals and speculations regarding the causes of state dependent learning. No new theories have been proposed here, although some old ones may have been restated more clearly than before. The paper has not attempted to be especially critical nor to focus exclusively on testable hypotheses. Hopefully, this concise presentation of theory may assist in developing new and testable formulations regarding the nature of state dependent learning.

ACKNOWLEDGMENTS

The author is grateful to Dr. Jeanne Marecek, Derri Shtasel, Evi Eskin, and Rhoda Porter for extensive editorial and technical assistance.

REFERENCES

Abse, D. W. *The diagnosis of hysteria.* Bristol: John Wright, 1950.
Abse, D. W. *Hysteria and related mental disorders.* Bristol: John Wright, 1966.

Adam, N., Castro, A. D., & Clark, D. L. State dependent learning with a general anesthetic (Isoflurane) in man. *T-I-T J. Life Sci.*, 1974, *4*, 125–134.

Alvarez-Buylla, R., & Carrasco-Zanini, J. A conditioned reflex which reproduces the hypoglycemic effect of insulin. *Acta Physiol. Latinoam.*, 1960, *10*, 153–158.

Aman, M. G., & Sprague, R. L. The dissociative effects of methylphenidate and dextroamphetamine. *J. Nervous Mental Disease*, 1974, *158*, 268–279.

Ashby, W. R. *Design for a brain.* New York: Wiley, 1954.

Bailey, C. J. The effectiveness of drives as cues. *J. Comp. Physiol. Psychol.*, 1955, *48*, 183–187.

Balster, R. L. The effectiveness of external and drug produced internal stimuli in the discriminative control of operant behavior. Unpublished doctoral thesis, Dept. Psychol., Univ. of Houston, Houston, Texas, 1970.

Barry, H., III. Classification of drugs according to their discriminable effects in rats. *Federation Proc.*, 1974, *33*, 1814–1824.

Barry, H., III, Miller, N. E., & Tidd, G. E. Control for stimulus change while testing effects of amobarbital on conflict. *J. Comp. Physiol. Psychol.*, 1962, *55*, 1071–1074. (a)

Barry, H., III, Wagner, A. R., & Miller, N. E. Effects of alcohol and amobarbital on performance inhibited by experimental extinction. *J. Comp. Physiol. Psychol.*, 1962, *55*, 464–468. (b)

Belleville, R. E. Control of behavior by drug-produced internal stimuli. *Psychopharmacologia* 1964, *5*, 95–105.

Berger, B., & Stein, L. Asymmetrical dissociation of learning between scopolamine and Wy 4036, a new benzodiazepine tranquilizer. *Psychopharmacologia*, 1969, *14*, 351–358.

Bilodeau, I. McD., & Schlosberg, H. Similarity in stimulating conditions as a variable in retroactive inhibition. *J. Exp. Psychol.*, 1951, *41*, 199–204.

Bindra, D., & Reichert, H. Dissociation of movement initiation without dissociation of response choice. *Psychon. Sci.*, 1966, *4*, 95–96.

Bindra, D., & Reichert, H. The nature of dissociation: Effects of transitions between normal and barbiturate-induced states on reversal learning and habituation. *Psychopharmacologia*, 1967, *10*, 330–344.

Bliss, D. K. Dissociated learning and state-dependent retention induced by pentobarbital in rhesus monkeys. *J. Comp. Physiol. Psychol.*, 1972, *84*, 149–161.

Bliss, D. K. Theoretical explanations of drug-dissociated behaviors. *Federation Proc.*, 1974, *33*, 1787–1796.

Bustamante, J. A., Jordan, A., Vila, M., Gonzalez, A., & Insua, A. State dependent learning in humans. *Physiol. Behav.*, 1970, *5*, 793–796.

Butcher, R. E., & Stutz, R. M. Discriminability of rewarding sub-cortical brain shock. *Physiol. Behav.*, 1969, *4*, 885–887.

Bykov, K. M. *The cerebral cortex and the internal organs.* New York: Chemical Pub. Co., 1957.

Cameron, O. G., & Appel, J. B. Conditioned suppression of bar-pressing by stimuli associated with drugs. *J. Exp. Anal. Behav.*, 1972, *17*, 127–137.

Cameron, O. G., & Appel, J. B. A behavioral and pharmacological analysis of some discriminable properties of d-LSD in rats. *Psychopharmacologia*, 1973, *33*, 117–134.

Case, T. J., & Funderbunk, W. H. An effect of curare on the central nervous system. *Trans. Am. Neurol. Assoc.*, 1947, *72*, 195–196.

Cheek, D. B. Unconscious perception of meaningful sounds during surgical anesthesia as revealed under hypnosis. *Am. J. Clin. Hypnosis*, 1959, *1*, 101–113.

Cherkashin, A. N., & Azarashvili, A. A. Dissociated learning in animals. *Soviet Psychol.*, 1972, *10*, 303–314.

Chodoff, P., & Lyons, H. Hysteria, the hysterical personality and "hysterical" conversion. *Am. J. Psychiat.*, 1957, *114*, 734–740.

Chute, D. L., & Wright, D. C. Retrograde state-dependent learning. *Science*, 1973, *180*, 878–880.

Colpaert, F. C., Niemegeers, C. J. E., & Janssen, P. A. J. Theoretical and methodological considerations on drug discrimination learning. *Psychopharmacologia*, 1976, *46*, 169–177.

Colpaert, F. C., Niemegeers, C. J. E., Kuyps, J. J. M. D., & Janssen, P. A. J. Apomorphine as a discriminative stimulus and its antagonism by haloperidol. *European J. Pharmacol.*, 1975, *32*, 383–386.

Conger, J. J. The effects of alcohol on conflict behavior in the albino rat. *Quart. J. Studies Alc.* 1951, *12*, 1–29.

Connelly, J. F., Connelly, J. M., & Epps, J. O. Disruption of dissociated learning in a discrimination paradigm by emotionally-important stimuli. *Psychopharmacologia*, 1973, *30*, 275–282.

Connelly, J. F., Connelly, J. M., & Phifer, R. Disruption of state-dependent learning (memory retrieval) by emotionally-important stimuli. *Psychopharmacologia*, 1975, *41*, 139–143.

Cook, L., Davidson, A., Davis, D. J., & Kelleher, R. T. Epinephrine, norepinephrine and acetylcholine as conditioned stimuli for avoidance behavior. *Science*, 1960, *131*, 990–991.

Crisler, G. Salivation is unnecessary for the establishment of the salivary conditioned reflex induced under morphine. *Am. J. Physiol.*, 1930, *94*, 553–556.

Crowder, W. F., Smith, S. G., Davis, W. M., Noel, J. T., & Coussens, W. R. Effect of morphine dose size on the conditioned reinforcing potency of stimuli paired with morphine. *Psychol. Record.*, 1972, *22*, 441–448.

Dallett, K., & Wilcox, S. G. Contextual stimuli and proactive inhibition. *J. Exp. Psychol.*, 1968, *78*, 475–480.

Dollard, J., & Miller, N. E. *Personality and Psychotherapy.* New York: McGraw-Hill, 1950.

Dougherty, J. Stimulus control and multiple personalities., Unpublished manuscript, 1977.

Duncan, P. M., & Copeland, M. A state-dependent effect from temporary septal area dysfunction in rats. *J. Comp. Physiol. Psychol.*, 1975, *89*, 537–545.

Eich, J. E., Weingartner, H., Stillman, R. C., & Gillin, J. C. State-dependent accessibility of retrieval cues in the retention of a categorized list. *J. Verb. Learn. Verb. Behav.*, 1975, *14*, 408–417.

Ellen, P., & Powell, E. W. Differential conditioning of septum and hippocampus. *Exp. Neurol.*, 1966, *16*, 162–171.

Ellinwood, E. H. "Accidental conditioning" with chronic methamphetamine intoxication: Implications for a theory of drug habituation. *Psychopharmacologia*, 1971, *21*, 131–138.

Evans, F. J. Hypnosis and sleep: Techniques for exploring cognitive activity during sleep. In E. Fromm & R. E. Shor (Eds.), *Hypnosis: Research developments and prespectives.* Chicago: Aldine, 1972. Pp. 43–83.

Evans, F. J., Gustafson, L. A., O'Connell, D. N., Orne, M. T., & Shor, R. E. Response during sleep with intervening waking amnesia. *Science*, 1966, *152*, 666–667.

Evans, F. J., Gustafson, L. A., O'Connell, D. N., Orne, M. T., & Shor, R. E. Sleep-induced behavioral response. *J. Nervous Mental Disease*, 1969, *148*, 467–476.

Evans, F. J., Gustafson, L. A., O'Connell, D. N., Orne, M. T., & Shor, R. E. Verbally induced behavioral responses during sleep. *J. Nervous Mental Disease*, 1970, *150*, 171–187.

Feldman, R. S. The mechanism of fixation prevention and "dissociation" learning with chlordiazepoxide. *Psychopharmacologia*, 1968, *12*, 384–399.

Fischer, R. Arousal-statebound recall of experience. *Diseases Nervous System*, 1971, *32*, 373–382. (a)

Fischer, R. The "flashback": Arousal-statebound recall of experience. *J. Psychedelic Drugs*, 1971, *3*, 31–39. (b)

Fischer, R. A cartography of the ecstatic and meditative states. *Science*, 1971, *174*, 897–904. (c)

Fischer, R. Cartography of inner space. In Drug Abuse Council (Ed.), *Altered states of consciousness: Current views and research problems.* Washington, D.C.: Drug Abuse Council, 1975. Pp. 1–58.

Fischer, R., & Landon, G. M. On the arousal state-dependent recall of 'subconscious' experience: Stateboundness. *Brit. J. Psychiat.*, 1972, *120*, 159–172.

Franks, C. M. Alcohol, alcoholism and conditioning: A review of the literature and some theoretical considerations. *J. Mental Sci.*, 1958, *104*, 14–33.

Freedman, N. L. Recurrent behavioral recovery during spreading depression. *J. Comp. Physiol. Psychol.*, 1969, *68*, 210–214.

Freedman, N. L., Pote, R., Butcher, R., & Suboski, M. D. Learning and motor activity under spreading depression depending on EEG amplitude. *Physiol. Behav.*, 1968, *3*, 373–376.

Gardner, E. L., Glick, S. D., & Jarvik, M. E. ECS dissociation of learning and one-way cross-dissociation with physostigmine and scopolamine. *Physiol. Behav.*, 1972, *8*, 11–15.

Gardner, L. A. An experimental re-examination of the dissociative effects of curareform drugs. Masters thesis, Dept. Psychol., Oberlin College, Oberlin, Ohio, 1961.

Gardner, L. A., & McCullough, C. A reinvestigation of the dissociative effect of curareform drugs. *Am. Psychologist*, 1962, *17*, 398.

Gazzaniga, M. S., & Hillyard, S. A. Language and speech capacity of the right hemisphere. *Neuropsychol.*, 1971, *9*, 273–280.

Girden, E. Cerebral mechanisms in conditioning under curare. *Am. J. Psychol.*, 1940, *53*, 397–406.

Girden, E. Generalized conditioned responses under curare and erythroidine. *J. Exp. Psychol.*, 1942, *31*, 105–119. (a)

Girden, E. The dissociation of blood pressure conditioned responses under erythroidine. *J. Exp. Psychol.*, 1942, *31*, 219–231. (b)

Girden, E. The dissociation of pupillary conditioned reflexes under erythroidine and curare. *J. Exp. Psychol.*, 1942, *31*, 322–332. (c)

Girden, E. Conditioned responses in curarized monkeys. *Amer. J. Psychol.*, 1947, *60*, 571–587.

Girden, E., & Culler, E. A. Conditioned responses in curarized striate muscle in dogs. *J. Comp. Psychol.*, 1937, *23*, 261–274.

Godden, D. R., & Baddeley, A. D. Context-dependent memory in two natural environments: On land and underwater. *Brit. J. Psychol.*, 1975, *66*, 325–331.

Goodwin, D. W. Two species of alcoholic blackout. *Am. J. Psychiat.*, 1971, *127*, 1665–1670.

Goodwin, D. W. Alcoholic blackout and state-dependent learning. *Federation Proc.*, 1974, *33*, 1833–1835.

Goodwin, D. W., & Hill, S. Y. Short-term-memory and alcoholic blackout. *Ann. N.Y. Acad. Sci.*, 1973, *215*, 195–199.

Goodwin, D. W., Powell, B., Bremer, D., Hoine, H., & Stern, J. Alcohol and recall: State-dependent effects in man. *Science*, 1969, *163*, 1358–1360.

Goodwin, D. W., Powell, B., Hill, S. Y., Lieberman, W., & Viamonte, J. Effect of alcohol on dissociated learning in alcoholics. *J. Nervous Mental Disease*, 1974, *158*, 198–201.

Greenspoon, J., & Ranyard, R. Stimulus conditions and retroactive inhibition. *J. Exp. Psychol.*, 1957, *33*, 55–59.

Grossman, S. P., & Miller, N. E. Control for stimulus-change in the evaluation of alcohol and chlorpromazine as fear-reducing drugs. *Psychopharmacologia*, 1961, *2*, 342–351.

Harris, R. T., & Balster, R. L. An analysis of psychological dependence. In R. T. Harris, W. M. McIsaac, & C. R. Schuster, Jr. (Eds.), *Advances in mental science II: Drug dependence.* Austin, Texas: Univ. of Texas Press, 1970. Pp. 214–216.

Harris, R. T., Claghorn, J. L., & Schoolar, J. C. Self administration of minor tranquilizers as a function of conditioning. *Psychopharmacologia*, 1968, *13*, 81–88.

Hart, B. The conception of dissociation. *Brit. J. Med. Psychol.*, 1926, *6*, 241–263.

Hebb, D. O. *The organization of behavior, a neuropsychological theory.* New York: Wiley, 1949.

Hecht, T., Baumann, R., & Hecht, K. The somatic and vegetative-regulatory behavior of the healthy organism during conditioning of the insulin effect. *Conditional Reflex*, 1967, *2*, 96–112.

Heistad, G. T. A bio-psychological approach to somatic treatments in psychiatry. *Amer. J. Psychiat.*, 1957, *114*, 540–545.

Heistad, G. T., & Torres, A. A. A mechanism for the effect of a tranquilizing drug on learned emotional responses. *Univ. Minnesota Med. Bull.*, 1959, *30*, 518–527.

Holloway, F. A., & Wansley, R. A. Multiple retention deficits at periodic intervals after active and passive avoidance learning. *Behav. Biol.*, 1973, *9*, 1–14. (a)

Holloway, F. A., & Wansley, R. A. Multiphasic retention deficits at periodic intervals after passive-avoidance learning. *Science,* 1973, *180,* 208–210. (b)

Holmgren, B. Conditional avoidance reflex under pentobarbital. *Boletin del Instituto de estudios medicos y biologicos* (*Mexico*), 1964, *22,* 21–37.

Huang, J. T. Amphetamine and pentobarbital effects on the discriminative response control by deprivation of rapid eye movement sleep (REMS). *Federation Proc.,* 1973, *32,* 786.

Huang, J. T., & Ho, B. T. Discriminative stimulus properties of d-amphetamine and related compounds in rats. *Pharmacol. Biochem. Behav.,* 1974, *2,* 669–673.

Hunt, H. F. Some effects of drugs on classical (type S) conditioning. *Ann. N.Y. Acad. Sci.,* 1956, *65,* 258–267.

Hurst, P. M., Radlow, R., Chubb, N. C., & Bagley, S. K. Effects of d-amphetamine on acquisition, persistence, and recall. *Am. J. Physiol.,* 1969, *82,* 307–319.

Iwahara, S., & Matsushita, K. Effects of drug-state changes upon black-white discrimination learning in rats. *Psychopharmacologia,* 1971, *19,* 347–358.

Iwahara, S., & Noguchi, S. Drug-state dependency as a function of overtraining in rats. *Jap. Psychol. Research,* 1972, *14,* 141–144.

Iwahara, S., & Noguchi, S. Effects of overtraining upon drug-state dependency in discrimination learning in white rats. *Jap. Psychol. Research,* 1974, *16,* 59–64.

Iwahara, S., & Sakama, E. Effects of chlordiazepoxide upon habituation of open-field behavior in white rats. *Psychopharmacologia,* 1972, *27,* 285–292.

Iwahara, S., & Sugimura, T. Effects of chlordiazepoxide on black-white discrimination acquisition and reversal in white rats. *Jap. J. Psychol.,* 1970, *41,* 142–150.

Janet, P. *The major symptoms of hysteria.* New York: Hafner, 1965.

John, E. R. *Mechanisms of memory.* New York: Academic Press, 1967.

Jones, C. Central versus peripheral mechanisms in discriminative response control by d-amphetamine and related compounds. Unpublished masters thesis, Dept. Psychol., Univ. of Houston, Texas, 1972.

Joy, R. M., & Prinz, P. N. The effect of sleep altering environments upon the acquisition and retention of a conditioned avoidance response in the rat. *Physiol. Behav.,* 1969, *4,* 809–814.

Kamano, D. K. Effects of stimulus associated with amobarbital administration on avoidance behavior. *Physiol. Psychol.,* 1973, *1,* 321–323.

Keane, T. M., & Lisman, S. A. Multiple task disruption of alcohol state-dependent retention. Paper presented at the meeting of the Eastern Psychological Association, New York City, 1976.

Kilbey, M. M., Harris, R. T., & Aigner, T. G. Establishment of equivalent external and internal stimulus control of an operant behavior and its reversal. *Proc. Am. Psychol. Assoc.,* 1971, *6,* 767–768.

Klein, S. B. Adrenal-pituitary influence in reactivation of avoidance-learning memory in the rat after intermediate intervals. *J. Comp. Physiol. Psychol.,* 1972, *79,* 341–354.

Klein, S. B., & Spear, N. E. Forgetting by the rat after intermediate intervals ("Kamin Effect") as retrieval failure. *J. Comp. Physiol. Psychol.,* 1970, *71,* 165–170.

Korol, B., Sletten, I. W., & Brown, M. L. Conditioned physiological adaptation to anticholinergic drugs. *Am. J. Physiol.,* 1966, *211,* 911–914.

Kruglikov, R. I., Aleksandrovskaya, M. M., & Dish, T. N. Disturbance of conditioned activity and morphological changes in rats' brain resulting from paradoxical sleep deprivation (Rs). *Zhurnal Uyshei Nervoi Deistelnosti Imeni I. P. Pavolva (Moskva),* 1975, *25,* 471–476.

Kuhn, D. M., White, F. J., & Appel, J. B. Discriminative stimulus properties of hallucinogens: Behavioral assay of drug action. In H. Lal (Ed.), *Research applications of drug-induced discriminable stimuli.* New York: Plenum Pub. Corp., 1977, 137–154.

Kurland, H. D., Cassell, S., & Goldberg, E. M. The effects of chlorpromazine on the transferability of learning in humans. Paper presented at the meeting of the Society of Biological Psychiatry, Washington, 1968.

Lal, H. *Research applications of drug-induced discriminable stimuli.* New York: Plenum Pub. Corp., 1977.

Lal, H., & Drawbaugh, B. S. Effect of naloxone, haloperidol, mecamylamine, phenoxybenzamine, or propranolol on the reversal of morphine abstinence (hypothermic) by morphine or conditional stimulus associated with morphine. *Proceedings of Committee on Drug Dependence, Nat. Acad. Sci.*, 1973, 382–385.

Lal, H., Reddy, C., & Rossman, M. Control of morphine-withdrawal hypothermia by conditional stimuli. *Proceedings of Committee on Drug Dependence, Nat. Acad. Sci.*, 1972, 258–264.

Lang, W. J., Brown, M. L., Gershon, S., & Korol, B. Classical and physiologic adaptive conditioned responses to anticholinergic drugs in conscious dogs. *Inter. J. Neuropharmacol.*, 1966, *5*, 311–315.

Langford, A., Freedman, N., & Whitman, D. Further determinants of interhemispheric transfer under spreading depression. *Physiol. Behav.*, 1971, *7*, 65–71.

Levinson, B. States of awareness during general anaesthesia. *Brit. J. Anaesthesia*, 1965, *37*, 544–546.

Ley, P., Jain, V. K., Swinson, R. P., Eaves, D., Bradshaw, P. W., Kincey, J. A., Crowder, R., & Abbiss, S. A state-dependent learning effect produced by amylobarbitone sodium. *Brit. J. Psychiat.*, 1972, *120*, 511–515.

Ludwig, A. M., Brandsma, J. M., Wilbur, C. B., Bendfeldt, F., & Jameson, D. H. The objective study of a multiple personality. *Arch. Gen. Psychiat.*, 1972, *26*, 298–310.

Macht, M. L. The effects of manipulations of affective state on free recall and performance. Paper presented at the meeting of the Eastern Psychological Association, New York, 1976.

Mayse, J. F., & DeVietti, T. L. A comparison of state-dependent learning induced by electroconvulsive shock and pentobarbital. *Physiol. Behav.*, 1971, *7*, 717–721.

McIntyre, D. C., & Reichert, H. State-dependent learning in rats induced by kindled convulsions. *Physiol. Behav.*, 1971, *7*, 15–20.

McKim, W. A. The effects of pre-exposure to scopolamine on subsequent drug state discrimination. *Psychopharmacologia*, 1976, *47*, 153–155.

Miller, N. E., & Barry, H. Motivational effects of drugs: Methods which illustrate some general problems in psychopharmacology. *Psychopharmacologia*, 1960, *1*, 169–199.

Miller, R. R., Malinowski, B., Puk, G., & Springer, A. D. State-dependent models of ECS-induced amnesia in rats. *J. Comp. Physiol. Psychol.*, 1972, *81*, 533–540.

Moffett, A., & Ettlinger, G. Opposite responding to position in the light and dark. *Neuropsychologia*, 1967, *5*, 59–62.

Morgan, C. T., & Stellar, E. *Physiological psychology*. New York: McGraw-Hill, 1950.

Nagge, J. W. An experimental test of the theory of associative interference. *J. Exp. Psychol.*, 1935, *18*, 663–682.

Nielson, H. C. Evidence that electroconvulsive shock alters memory retrieval rather than memory consolidation. *Exp. Neurol.*, 1968, *20*, 3–20.

Nielson, H. C., Justesen, D. R., & Porter, P. B. Effects of anticonvulsant drugs upon patterns of seizure discharge and brain thresholds: Some implications for memory mechanisms. *Psychol. Rep.*, 1968, *23*, 843–850.

O'Brien, C. P., O'Brien, T. J., Mintz, J., & Brady, J. P. Conditioning of narcotic abstinence symptoms in human subjects. *Drug and Alcoh. Depend.*, 1975/6, *1*, 115–123.

Osborn, A. G., Bunker, J. P., Cooper, L. F., Frank, G. S., & Hilgard, E. R. Effects of thiopental sedation on learning and memory. *Science*, 1967, *157*, 574–576.

Otis, L. S. Drive conditioning: Fear as a response to biogenic drive stimuli previously associated with painful stimulation. Unpublished doctoral thesis, Dept. Psychol., Univ. of Chicago, Chicago, Illinois, 1956.

Otis, L. S. Dissociation and recovery of a response learned under the influence of chlorpromazine or saline. *Science*, 1964, *143*, 1347–1348.

Otis, L. S. Discussion. In D. Bente & P. B. Bradley (Eds.), *Neuro-psychopharmacology*. Vol. 4. Amsterdam: Elsevier, 1965. Pp. 120–122.

Overton, D. A. Discriminative behavior based on the presence or absence of drug effects. *Am. J. Psychol.*, 1961, *16*, 453–454.

Overton, D. A. State-dependent or "dissociated" learning produced with pentobarbital. *J. Comp. Physiol. Psychol.*, 1964, *57*, 3–12.

Overton, D. A. State-dependent learning produced by depressant and atropine-like drugs. *Psychopharmacologia*, 1966, *10*, 6–31.

Overton, D. A. Dissociated learning in drug states (state-dependent learning). In D. H. Efron, J. O. Cole, J. Levine & R. Wittenborn (Eds.), *Psychopharmacology, a review of progress, 1957–1967*. U.S. Pub. Health Serv. Pub. No. 1836. Washington, D.C.: U.S. Govt. Printing Office, 1968, Pp. 918–930. (a)

Overton, D. A. Visual cues and shock sensitivity in the control of T-maze choice by drug conditions. *J. Comp. Physiol. Psychol.*, 1968, *66*, 216–219. (b)

Overton, D. A. Control of T-maze choice by nicotinic, antinicotinic, and antimuscarinic drugs. *Proc. 77th Annual Convention, Amer. Psychol. Assoc.* 1969, *4*, 869–870.

Overton, D. A. Discriminative control of behavior by drug states. In T. Thompson & R. Pickens (Eds.), *Stimulus properties of drugs*. New York: Appleton, 1971. Pp. 87–110.

Overton, D. A. State-dependent learning produced by alcohol and its relevance to alcoholism. In B. Kissen & H. Begleiter (Eds.), *The biology of alcoholism*. Vol. II. *Physiology and behavior*. New York: Plenum, 1972. Pp. 193–217. (a)

Overton, D. A. Making the CER conditional on an arbitrary drug state. Paper presented at the meeting of the Eastern Psychological Association, Boston, 1972. (b)

Overton, D. A. State-dependent learning produced by addicting drugs. In S. Fisher & A. M. Freedman (Eds.), *Opiate addiction: Origins and treatment*. Washington, D.C.: V. H. Winston, 1973. Pp. 61–75.

Overton, D. A. Experimental methods for the study of state-dependent learning. *Federation Proc.*, 1974, *33*, 1800–1813.

Overton, D. A. Comparative efficacy of various drugs in a T-maze drug discrimination task. *Neurosci. Abs.*, 1975, *1*, 335.

Overton, D. A. Drug state dependent learning. In M. Jarvik (Ed.), *Psychopharmacology in the practice of medicine*. New York: Appleton, 1976, Pp. 73–79.

Overton, D. A., & Batta, S. K. Relationship between abuse liability of drugs and their degree of discriminability in the rat. In T. Thompson & K. Unna (Eds.), *Predicting Dependence Liability of Stimulant and Depressant* Drugs, Baltimore: University Park Press, 1977 (in press).

Overton, D. A., Ercole, M. A., & Dutta, P. Discriminability of the postictal state produced by electroconvulsive shock in rats. *Physiol. Psychol.*, 1976, *4*, 207–212.

Peck, J. H., & Ader, R. Illness-induced taste aversion under states of deprivation and satiation. *Animal Learn. Behav.*, 1974, *2*, 6–8.

Perez-Cruet, J. Drug conditioning and drug effects on cardiovascular conditional functions. In T. Thompson & R. Pickens (Eds.), *Stimulus properties of drugs*. New York: Appleton, 1971. Pp. 15–38.

Perez-Cruet, J. Conditional reflex changes in dopamine metabolism induced by methadone as an unconditional stimulus. In J. M. Singh & H. Lal (Eds.), *Drug addiction*. Vol. 3. New York: Stratton, 1974. Pp. 249–250.

Pert, A., & Avis, H. H. Dissociation between scopolamine and mecamylamine during fear conditioning in rats. *Physiol. Psychol.*, 1974, *2*, 111–116.

Petersen, R. C. Isolation of processes involved in state-dependent recall in man. *Federation Proc.*, 1974, *33*, 550.

Pickens, R., & Dougherty, J. A. Conditioning of the activity effects of drugs. In T. Thompson & R. Pickens (Eds.), *Stimulus properties of drugs*. New York: Appleton, 1971. Pp. 39–50.

Powell, B. J., Goodwin, D. W., Janes, C. L., & Hoine, H. State-dependent effects of alcohol on autonomic orienting responses. *Psychon. Sci.*, 1971, *25*, 305–306.

Pusakulich, R. L. Analysis of cue use in state dependent learning. *Dissertation Abs. Int.* (*Part B*), 1974, *34*, 4095.

Pusakulich, R. L., & Nielson, H. C. Neural thresholds and state-dependent learning. *Exp. Neurol.*, 1972, *34*, 33–44.

Razran, G. Conditioned responses in animals other than dogs. A behavioral and quantitative critical review of experimental studies. *Psychol. Bull.*, 1933, *30*, 261–326.

Reed, V. G., & Trowill, J. A. Stimulus control value of spreading depression demonstrated without shifting depressed hemispheres. *J. Comp. Physiol. Psychol.*, 1969, *69*, 40–43.

Reichert, H. Exploration of the drug-induced dissociation mechanism by intracranial injections and post-trial drug states. Masters thesis, Dept. Psychol., Univ. of Waterloo, Waterloo, Ontario, 1968.

Reichert, H. Evidence for a memory consolidation interpretation of state dependent learning. *Dissertation Abs. Int.* (*Part B*) 1971, *32*, 1255.

Rickles, W. H., Jr., Cohen, M. J., Whitaker, C. A., & McIntyre, K. E. Marijuana induced state-dependent verbal learning. *Psychopharmacologia*, 1973, *30*, 349–354.

Roffman, M., Marshall, P., Silverstein, A., Karkalas, J., Smith, N., & Lal, H. Failure to demonstrate "amphetamine state" controlling learned behavior in humans. In J. Singh, L. Miller, & H. Lal (Eds.), *Drug addiction.* Vol. 3. *Clinical and socio-legal aspects.* Mt. Kisco, New York: Futura, 1972, Pp. 53–57.

Rosecrans, J. A., Goodloe, M. H., Bennett, G. J., and Hirschhorn, I. D. Morphine as a discriminative cue: Effects of amine depletors and naloxone. *European J. Pharmacol.*, 1973, *21*, 252–256.

Russek, M., & Pina, S. Conditioning of adrenalin anorexia. *Nature*, 1962, *193*, 1296–1297.

Sachs, E. The role of brain electrolytes in learning and retention. *Federation Proc.*, 1961, *20*, 339.

Sachs, E. The effects of central injections of potassium and calcium on avoidance learning in the cat. Unpublished doctoral thesis, Univ. of Rochester, 1962.

Sachs, E. Dissociation of learning in rats and its similarities to dissociative states in man. In J. Zubin & H. Hunt (Eds.), *Comparative psychopathology.* New York: Grune & Stratton, 1967. Pp. 249–304.

Sachs, E., Weingarten, M., & Klein, N. W., Jr. Effects of chlordiazepoxide on the acquisition of avoidance learning and its transfer to the normal state and other drug conditions. *Psychopharmacologia*, 1966, *9*, 17–30.

Schechter, M. D., & Rosecrans, J. A. C.N.S. effect of nicotine as the discriminative stimulus for the rat in a T-maze. *Life Sci.* (Part I) 1971, *10*, 821–832.

Schechter, M. D., & Rosecrans, J. A. Nicotine as a discriminative cue in rats: Inability of related drugs to produce a nicotine-like cueing effect. *Psychopharmacologia*, 1972, *27*, 379–387.

Schneider, A. M. Retention under spreading depression: A generalization decrement phenomenon. *J. Comp. Physiol. Psychol.*, 1966, *62*, 317–319.

Schneider, A. M. Control of memory by spreading cortical depression: A case for stimulus control. *Psychol. Rev.*, 1967, *74*, 201–215.

Schneider, A. M. Stimulus control and spreading cortical depression: Some problems reconsidered. *Psychol. Rev.*, 1968, *75*, 353–358.

Schneider, A. M., & Ebbesen, E. Interhemispheric transfer of lever pressing as stimulus generalization of the effects of spreading depression. *J. Exp. Anal. Behav.*, 1967, *10*, 193–197.

Schneider, A. M., & Hamburg, M. Interhemispheric transfer with spreading depression: A memory transfer or stimulus generalization phenomenon. *J. Comp. Physiol. Psychol.*, 1966, *62*, 133–136.

Schneider, A. M., & Kay, H. Spreading depression as a discriminative stimulus for lever pressing. *J. Comp. Physiol. Psychol.*, 1968, *65*, 149–151.

Schuster, C. R., & Balster, R. L. The discriminative stimulus properties of drugs. In T. Thompson

& P. B. Dews (Eds.), *Advances in behavioral pharmacology.* Vol. 1. New York: Academic Press, 1977, 85–138.

Settle, R. G. A comparison of state dependent learning produced by pre- and post-acquisition intrathoracic administration of sodium pentobarbital. Unpublished Masters thesis, Dept. Psychol., Univ. of Missouri of Columbia, 1973.

Shannon, H. E., & Holtzman, S. G. A pharmacological analysis of the discriminative effects of morphine in the rat. *Problems of drug dependence, 1975.* Proceedings of the 37th Annual Scientific Meeting Committee on Problems of Drug Dependence. Washington, D.C.: National Academy of Sciences, 1975. Pp. 698–719.

Siegel, S. Conditioning of insulin-induced glycemia. *J. Comp. Physiol. Psychol.*, 1972, *78,* 233–241.

Siegel, S. Evidence from rats that morphine tolerance is a learned response. *J. Comp. Physiol. Psychol.*, 1975, *89,* 498–506.

Spear, N. E. Forgetting as retrieval failure. In W. K. Honig & H. James (Eds.), *Animal memory.* New York: Academic Press, 1970. Pp. 45–109.

Spear, N. E., Klein, S. B., & Riley, E. P. The Kamin effect as "State-dependent learning": Memory-retrieval failure in the rat. *J. Comp. Physiol. Psychol.*, 1971, *74,* 416–425.

Sprague, R. L. Psychopharmacology and learning disabilities. *J. Operational Psychiat.*, 1972, *3,* 56–67.

Stillman, R. C., Weingartner, H., Wyatt, R. J., Gillin, J. C., & Eich, J. State-dependent (Dissociative) effects of marijuana on human memory. *Arch. Gen. Psychiat.*, 1974, *31,* 81–85.

Storm, T., & Caird, W. K. The effects of alcohol on serial verbal learning in chronic alcoholics. *Psychon. Sci.*, 1967, *9,* 43–44.

Storm, T., & Smart, R. G. Dissociation: A possible explanation of some features of chronic alcoholism and implications for treatment. *Quart. J. Studies Alc.*, 1965, *26,* 111–115.

Stroebel, C. F. Biochemical, behavioral and clinical models of drug interactions. In H. Brill (Ed.), *Neuro-psycho-pharmacology.* New York: Excerpta Medica Foundation, 1967. Pp. 453–458.

Stutz, R. M. Stimulus generalization within the limbic system. *J. Comp. Physiol. Psychol.*, 1968, *65,* 79–82.

Stutz, R. M., Butcher, R. E., & Rossi, R. Stimulus properties of reinforcing brain shock. *Science,* 1969, *163,* 1081–1082.

Swanson, J. M., & Kinsbourne, M. Stimulant-related state-dependent learning in hyperactive children. *Science,* 1976, *192,* 1354–1357.

Takahashi, R., Nagayama, H., Kido, A., & Morita, T. An animal model of depression. *Biol. Psychiat.*, 1974, *9,* 191–204.

Tamerin, J. S., Weiner, S., & Mendelson, J. H. Alcoholics' expectancies and recall of experiences during intoxication. *Am. J. Psychiat.*, 1970, *126,* 1697–1704.

Tart, C. T. States of consciousness and state specific sciences. *Science,* 1972, *176,* 1203–1210.

Tart, C. T. *States of consciousness.* New York: Dutton, 1975.

Thigpen, C. H., & Cleckley, H. M. *The three faces of Eve.* New York: McGraw-Hill, 1957.

Thompson, C. I., & Grossman, L. B. Loss and recovery of long-term memories after ECS in rats: Evidence for state-dependent recall. *J. Comp. Physiol. Psychol.*, 1972, *78,* 248–254.

Thompson, C. I., & Neely, J. E. Dissociated learning in rats produced by electroconvulsive shock. *Physiol. Behav.*, 1970, *5,* 783–786.

Thompson, T., & Pickens, R. (Eds.) *Stimulus properties of drugs.* New York: Appleton, 1971.

Tilson, H. A., & Rech, R. H. Conditioned drug effects and absence of tolerance to d-amphetamine induced motor activity. *Pharmacol. Biochem. Behav.*, 1973, *1,* 149–153.

Tulving, E., & Madigan, S. A. Memory and verbal learning. *Annual Rev. Psychol.*, 1970, *21,* 437–484.

Wann, P. D. Amnestic and dissociative effects of kindled convulsions in rats. Unpublished Masters thesis, Dept. Psychol., Carleton Univ., Ottawa, Canada, 1971.

Wansley, R. A., & Holloway, F. A. Multiple retention deficits following one-trial appetitive training. *Behav. Biol.*, 1975, *14,* 135–149.

Webb, W. B. Drive stimuli as cues. *Psychol. Reports,* 1955, *1,* 287–298.

Weingartner, H., & Faillace, L. A. Alcohol state-dependent learning in man. *J. Nervous Mental Disease,* 1971, *153,* 395–406.

Weingartner, H., Miller, H., & Murphy, D. L. Mood-state-dependent retrieval of verbal associations. *J. Abnorm. Psychol.,* 1977, *86,* 276–284.

Weingartner, H., & Murphy, D. Effects of drug states and clinical phenomena on the storage and retrieval of information in man. *Psychopharmacology* Bulletin, 1977, 13, 66–67.

Winter, J. C. The effects of 3,4-dimethoxyphenylethylamine in rats trained with mescaline as a discriminative stimulus. *J. Pharmacol. Exp. Therap.,* 1974, *189,* 741–747.

Winter, J. C. The effects of 2,5-dimethoxy-4-methylamphetamine (DOM), 2,5-dimethoxy-4-ethylamphetamine (DOET), d-amphetamine, and cocaine in rats trained with mescaline as a discriminative stimulus. *Psychopharmacologia,* 1975, *44,* 29–32.

Wright, D. C. Differentiating stimulus and storage hypotheses of state-dependent learning. *Federation Proc.,* 1974, *33,* 1797–1799.

Wright, D. C., & Chute, D. L. State dependent learning produced by post trial intrathoracic administration of sodium pentobarbital. *Psychopharmacologia,* 1973, *31,* 91–94.

Wright, D. C., Chute, D. L., & McCollum, G. C. Reversible sodium pentobarbital amnesia in one trial discrimination learning. *Pharmacol. Biochem. Behav.,* 1974, *2,* 603–606.

Zentall, T. R. Effects of context change on forgetting in rats. *J. Exp. Psychol.,* 1970, *86,* 440–448.

18

State Dependent Retrieval Based on Time of Day[1]

Frank A. Holloway

Department of Psychiatry and Behavioral Sciences
University of Oklahoma Health Sciences Center

INTRODUCTION

Although the majority of state dependent learning (SDL) studies have focused on alteration of state by drugs, several studies have found state dependent dissociative phenomena accompanying "natural" changes in the state of the organism. For example, the early studies of Hull (1933), Leeper (1935), and Kendler (1946) suggested that organismic states induced by deprivation of food or water provided a sufficient basis for differential responding by rats in a T-maze. The drive stimuli (S_D) resulting from the deprivation manipulation were said to enter into an association with a response (Bolles, 1975), an interpretation not dissimilar from "drug stimuli" explanations for SDL. (See Bliss, 1974.) With the advent of additional S_D studies in the 1950s, however, (e.g., Bolles, 1958; Kendler & Levine, 1951), it became apparent that the incentive values for hunger and thirst motivation were not sufficiently equated in most investigations. In one study in which drive level was documented (Yamaguchi, 1952), the effects predicted by the associational interpretation were present but quite small, leading Bolles (1975, p. 276) to suggest that "if drive stimuli enter into associative control of behavior at all, they do so only very weakly." Similarly, Webb (1955, p. 296) concludes: "The data support a notion that the conditioning of drive stimuli is not easy, and, where external cues are readily available, drive stimuli are likely to play only limited roles in mediating behavior."

[1] Supported in part by the following grants: National Science Foundation Grant No. GB-41722, National Institute of Mental Health Biological Sciences Training Grant No. C-1216500 (supporting graduate student Richard A. Wansley), and Public Health Service Grant No. 14702.

The consensus that hunger- or thirst-drive states provide cues that are difficult to condition is similar to the conclusions from Soviet studies (Bykov, 1959), which indicate that conditioning of interoceptive responses proceeds at a slow rate. Overton (1964) also noted that drugs that have primarily peripheral effects on the autonomic nervous system or muscles have few, if any, state dependent or discriminative properties. While hunger, thirst, and other interoceptive events may not be adequate discriminative stimuli for behavioral conditioning, evidence remains that some aspects of central nervous system (CNS) state changes may provide a basis for state dependent phenomena. For example, SDL or retrieval effects have been reported for the following nonpharmacologically induced state changes: electrically induced seizure states (DeVietti & Larson, 1971; Nielson, 1968); rapid eye movement (REM) sleep deprivation (Joy & Prinz, 1969); sleep or hypnotic states (Evans, 1972); affective or mood states (Weingartner, 1973); and states associated with different times of day and/or different phases of biological rhythms (Holloway & Wansley, 1973a,b; Stroebel, 1967). Since most of these studies employed partial or complete experimental designs similar to the drug-transfer paradigm (retention testing in drug state, which is the same or different from that present during training), the studies are vulnerable to the criticism that performance factors confound any possible state dependent outcomes of such experimental paradigms (see Overton, 1974).

The most systematic data supporting a state dependent retrieval effect that results from naturally (or at least nonpharmacologically) induced changes in organismic state are those related directly or indirectly to biological rhythms. The sleep studies of Evans (1972) demonstrated state dependent retrieval in humans by the introduction of information and sampling for recall in different portions of the REM–nonREM sleep cycle and/or the sleep–wake cycle. The REM cycle is an example of an ultradian rhythm (i.e., having a period of less than 24 hr), while the sleep–wake cycle is generally circadian (i.e., having a period of approximately 24 hr). The studies of Stroebel (1967) and Holloway and Wansley (1973a,b) also suggest that circadian and/or ultradian rhythmic processes may produce state dependent oscillations in the retention performance of learned behaviors in animals. Finally, Bolles (1975) has shown in extensive investigations that animals appear able to utilize nondeprivation-related internal events to anticipate food or water presented once each 24-hr period.

The next section of this chapter characterizes briefly the empirical data that indicate state dependent retrieval effects based on periodically occurring changes in the state of the organism. The third section examines evidence that addresses the question of whether animals can use time-of-day information from internal sources to anticipate daily food or water or to learn differential responses or response patterns. The fourth section examines some of the biological parameters involved in time-of-day state dependent phenomena. The final section considers models of state dependent retrieval that may account for the presented data and includes a brief discussion of the relevance of biological rhythmic phenomena to more traditional studies of drug discrimination or of learning and memory.

MULTIPHASIC RETENTION PHENOMENA RELATED TO TIME OF DAY

Based on research done in the mid-1960s, Stroebel (1967) reported several interesting sets of data from animals maintained under a 12-hr light–12-hr dark regimen: (*1*) Animals trained in a conditional emotional response (CER) task at a specific time of day were more resistant to extinction when the test was given a week or so later at the same time of day as training than if the extinction tests were given at different times; (*2*) animals learned the CER task more rapidly when all training trials were given at one time of day than when the training trials were given throughout the day. Stroebel interpreted these data as indicating that an animal could utilize internal rhythmic processes acting like a circadian clock to assist in the recall of the learned task. While such phenomena were found in both rats and primates, no control studies were presented to clarify the nature of the phenomena, that is, motivational, associative, encoding effects, and so on.

Multiple Retention Deficits as a Function of Training-Testing Interval

Avoidance Conditioning Studies

Holloway and Wansley (1973a,b) reported results similar to those found by Stroebel. In all our investigations, male albino rats were maintained on *ad lib.* food and water and were adapted to 12-hr light (0800–2000)–12-hr dark (2000–0800) housing conditions for at least 2 weeks prior to training. The animals were given either a single training trial or were trained to a moderately weak criterion. Independent groups of rats (Sasco-Holtzman or Sprague–Dawley) were then tested at various intervals after the single training session. In the first experiment (Holloway & Wansley, 1973a), retention performance of rats was examined at successive 6-hr intervals after a single training trial in a step-through passive-avoidance task, in which the animals received shock upon entering a darkened chamber. As can be seen in Figure 18.1, retention (indicated by longer step-through latencies) was best immediately after training or at successive 12-hr intervals from the occasion of training but was poor 6 hr after training or at multiples of 12 hr from this 6-hr post-training point.

Two aspects of these data deserve comment. (*1*) Two periodicities were evident in the latencies it took animals to enter the chamber in which they were previously shocked, one at 12 hr and one at 24 hr. *Performance was best when testing occurred at the same time of day as training.* (*2*) The multiple retention decrements appeared to wane at longer intervals (i.e., 66 hr). We sug-

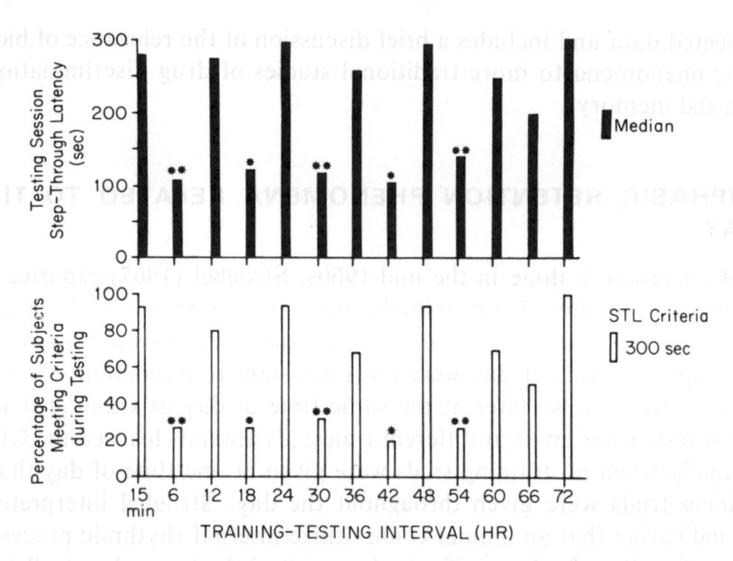

Figure 18.1. Median step-through latencies (STL) at testing sessions (top) and percentages of subjects meeting the 300 second STL criterion at testing session (bottom). Significant successive paired comparisons for each measure are indicated: * $p < .05$; ** $p < .01$. Redrawn from Holloway, F. A., and Wansley, R. A. Multiple retention deficits at periodic intervals after passive avoidance learning. *Science*, 1973, *180*, pp. 208–210. Copyright © by the American Association for the Advancement of Science. Reproduced by permission.

gested that some undetermined biological rhythmic process was affecting retention performance either by the direct interaction of biorhythms with performance or by the production of periodic state dependent retrieval deficits as a result of fluctuations or shifts in organismic state at certain times after training.

A second study (Holloway & Wansley, 1973b) examined whether the oscillations in retention performance reflected either a strict interaction of circadian or ultradian factors (e.g., corticosteroid or activity levels) with performance or state dependent retrieval deficits caused by endogenous or induced fluctuations in organismic state. Retention performance was examined at various intervals subsequent to training at 1 of 4 times during the day (0300–0600, 0900–1200, 1500–1800, 2100–2400) with either the one-trial passive avoidance task or a multitrial, 1-way active (step-up) avoidance procedure. We reasoned that the state dependent explanation would be favored over a performance explanation if the patterns of retention as a function of the training-testing interval (TTI) were similar for both avoidance procedures and for training at each of the four equally spaced periods throughout the day. The active-avoidance data displayed in Figure 18.2 show that the 15-min and 24-hr TTI groups had better retention scores than the 6-, 18-, or 30-hr TTI groups across all training times. (The 12-hr TTI group was intermediate.) On the basis of these data, we hypothesized that some undetermined constellation of periodically fluctuating internal events defined the state of the organism at the occasion of training and that reinstatement of this training state became a relevant condition for optimal retrieval of the original conditioning. More succinctly, certain aspects of naturally fluctuating states of the organism may become state-specific factors in the retrieval of learned responses.

Appetitive Conditioning Studies

In our initial reports (Holloway & Wansley, 1973a,b), we suggested that the multiple retention deficit phenomenon may be related to the so-called Kamin effect, a deficit frequently found at intermediate intervals after training. (See Brush, 1971.) Gabriel (1968) made a similar suggestion. Many of the more recent Kamin effect studies (Anisman & Waller, 1971; Barrett, Leith & Ray, 1971; Pinel & Mucha, 1973) interpreted the intermediate interval avoidance decrement as a performance effect that resulted from behavioral inhibition or reactivity induced by the footshock used during training. In an attempt to avoid possible confounding by any footshock-induced factors and to test the generality of the multiple retention deficit phenomenon, we examined retention performance in an appetitive task (Wansley & Holloway, 1975b). Independent groups of albino rats, tested at various intervals following a single training trial, ran down a simple runway to a water-filled tube. All animals were water deprived, but time since last drink was controlled for both training and testing trials. As indicated in Figure 18.3, the test-trial latencies to lick from the drink tube at the end of the runway oscillated in a periodic 12-hr cycle similar to that seen for the avoidance-retention data. In a control experiment, the rat's exposure to water subsequent to running the alley was delayed and given in a different apparatus. Under these conditions, the 12-hr oscillatory pattern in

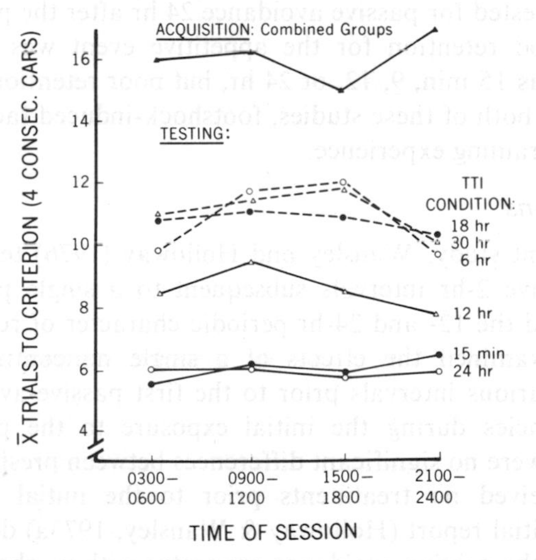

Figure 18.2. One-way active-avoidance training and testing performance (trials to four consecutive conditioned avoidance responses—CAR) as a function of training time and training testing interval (N = 10/subgroup.) (Redrawn from Holloway & Wansley, 1973b).

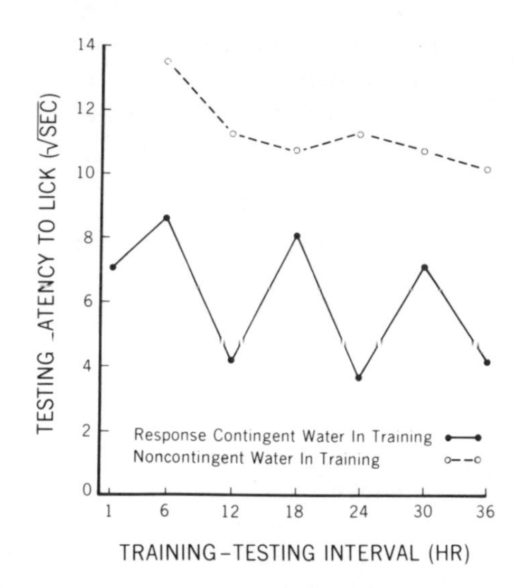

Figure 18.3. Testing latencies to make appetitive response (licking) at various TTIs subsequent to a single training trial involving response-contingent reinforcement or delayed, non-contingent water (N = 10/group). (Redrawn from Wansley & Holloway, 1975b.)

runway latencies was absent. Hunsicker (1974) found similar results when he pre-exposed rats (Hooded) to sugared milk in the shock compartment of a passive-avoidance task, subsequently (at various intervals) giving independent groups a single passive-avoidance trial. Memory for the appetitive experience should have interfered with development of the passive-avoidance behavior. When Hunsicker tested for passive avoidance 24 hr after the passive-avoidance training trial, good retention for the appetitive event was found when the critical interval was 15 min, 9, 12, or 24 hr, but poor retention was noted at 3, 6, 15, or 18 hr. In both of these studies, footshock-induced factors were absent during the initial training experience.

Other Investigations

In another recent study, Wansley and Holloway (1976) tested independent groups at successive 2-hr intervals subsequent to a single passive-avoidance trial and replicated the 12- and 24-hr periodic character of retention performance. We also examined the effects of a single noncontingent footshock, administered at various intervals prior to the first passive-avoidance trial, on step-through latencies during the initial exposure to the passive-avoidance apparatus. There were no significant differences between preshock animals and animals that received no treatments prior to the initial exposure to the apparatus. Our initial report (Holloway & Wansley, 1973a) demonstrated that mere exposure to the passive-avoidance apparatus with no shock present in the shock compartment had no effect on subsequent step-through latencies in the apparatus. Again, such control experiments suggest that the periodic deficits in retention performance are not easily explained by stress or shock-induced performance variables.

Thus far, only one instrumental conditioning study has specifically examined retention performance at similar TTIs. It failed to find a multiphasic effect. Caul, Barrett, Thune, and Osborne (1974) examined retention performance of Fisher-bred rats in an automated Y-maze in which light signaled the side on which the animal could avoid or escape shock. They found a single decrement in number of avoidances at 1 hr post-training but no decrements at 6, 18, or 30 hr. They also reported no decrements at any of the TTIs for their discrimination index.

There are several possible reasons for the failure to find multiphasic effects. There were several differences in strain, task requirements, and environmental lighting conditions between the Caul study and ours. All of the avoidance or appetitive procedures that have yielded multiphasic results, however, have been of the go–no-go variety, in which retention performance was indicated by the withholding or initiation of some response. The state dependent interpretation of the multiple retention deficits is based on the saliency and relevancy of the organism's state at the occasion of training for retrieval of that learning. In tasks in which specific external stimuli signal both the occasion of responding and the type of response required, internal-state factors may be masked or obscured. Finally, it is possible that the intermediate-interval deficits reported by Caul *et al.* and the multiphasic deficits may represent entirely different processes.

Other Characteristics of Multiphasic Retention Phenomena

As mentioned earlier, there are at least two plausible explanations for the multiphasic retention performance data of Holloway and Wansley: (*1*) Some endogenous physiological rhythm defines the state of the organism at the time of training such that shifts away from this state at certain training-testing intervals constitute shifts away from internal conditions present at training and produce a state dependent retrieval failure, or (*2*) the training procedure triggers or resets some rhythmic processes that facilitate or interfere with the performance. A shift in internal state induced by the training procedure may also produce a state dependent retrieval deficit. (See Spear, 1973.) We have performed two experiments to address the issue of whether such a shift in internal state could be based on some endogenous rhythmic process or whether it could be induced by the training procedure. In the first study, we employed an extinction procedure in which the acquisition-extinction interval was held constant. In the second study, we examined the effects of prior shock and interpolated shock on acquisition and relearning, respectively.

Transfer of Information between Consecutive Trials

The experimental design of the first study (Holloway & Sturgis, 1976) was derived from Capaldi's (1967) sequential theory of the partial reinforcement effect (PREE). In this experiment, 5 independent groups of rats were trained to escape shock in a straight alley on a schedule of partial reinforcement and were extinguished (20 trials with ITI equal to 30 sec) in the alley 24 hr after the last training trial. Differential resistance to extinction as a function of the intertrial interval between nonreinforced (N) and the following reinforced trials was measured. The sequence of acquisition trials was RRRN–RRRRN–RRRRN–R. (N-R ITIs were variable, while all other ITIs equaled 30 sec.) With this procedure, we intended to minimize strict performance factors per se by separating the critical retention interval and the interval between the exposure to the training procedure and subsequent testing.

Capaldi's analysis of the PREE involves the conditioning of the memory of nonreinforcement (S^n) to the subsequent reinforced instrumental response (R_I). Resistance to extinction is said to be a function of the strength of the latter S^n-R_I association. If the state dependent hypothesis of the multiple retention deficit is correct, then the state of the rat at the time of an N-trial should be a relevant condition for the retrieval of the S^n. Thus, if the intertrial interval between N-trials and following R-trials were systematically varied, the availability of S^n, hence the strength of the S^n-R_I association, should be a cyclic function of this ITI with a period of approximately 12 hr. Groups of rats trained with ITIs between the N and R trials of 15 min, 12 hr, or 24 hr were more resistant to extinction than rats for which this interval had the

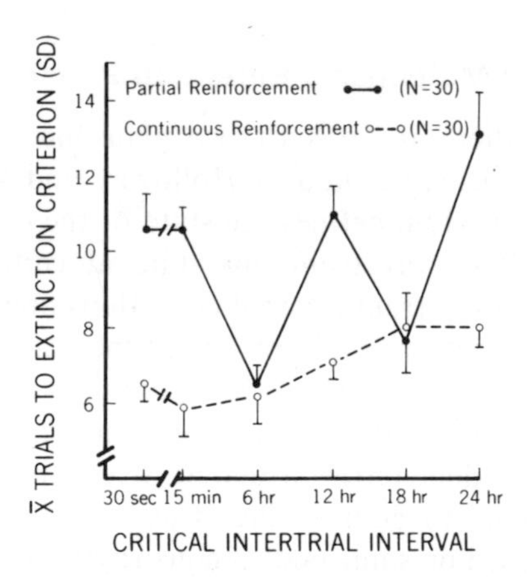

Figure 18.4. Mean trials to extinction criterion (three successive trials with shock escape latencies < 60 sec.) for groups given partial reinforcement during acquisition (the sequence RRRN–RRRRN–RRRRN–R) or continuous reinforcement during acquisition (the sequence RRRR–RRRRR–RRRRR–R). The critical ITI is indicated by the dashes in the latter sequences; all other ITIs were 30 sec. Extinction was 20 nonreinforced trials with 30-sec ITIs.

intermediate values of 6 or 18 hr. A subsequent experiment, which examined the existence of a PREE by comparing rats under continuous and partial reinforcement schedules, replicated these findings (see Figure 18.4), demonstrating a partial reinforcement effect when the critical intertrial interval between the N and R trials of the partial reinforcement group was 15 min, 12 hr, and 24 hr but not when the N-R interval was 6 or 18 hr.

These extinction data support the hypothesis that some endogenous, cyclically determined state of the organism modulates the accessibility of memory of the nonreward on reinforced trials such that the association between the memory of nonreward and the subsequent reinforced escape response may or may not become conditioned, depending on the interval between the N and R trials. The data further suggest: (*1*) The pattern of differences in extinction scores is not easily explained by a direct influence on extinction performance by endogenous factors since the pattern of extinction across the N-R ITIs was similar during the extinction given at different times during the day. (*2*) Any relevant state changes are likely to be endogenous changes rather than some shock-induced rhythmic process since in this experiment rats were shocked on both R and N trials.

Temporal Relationship of Pre- and Post-Training Shock Effects on Acquisition or Retention

In another experiment (Holloway, unpublished study), we trained 10 groups (N = 10/group) of rats in the step-up active-avoidance task. Five groups were tested 24 hr later (a TTI that in the past produced superior relearning performance), and five groups were tested 30 hr later (a TTI that usually yielded inferior relearning performance). Subgroups of animals were given either noncontingent shock in an apparatus dissimilar to the training apparatus or shock in the training apparatus at 1 of 5 intervals following the training session: 15 min, 6, 12, 18, or 23.75 hr. If the training shock experience indeed induced any physiological oscillation, then a second shock event might again set up such oscillations, with the result that the 24-hr TTI animals that received interpolated shock 6 or 18 hr after training would be impaired relative to the groups that received interpolated shock 15 min, 12 hr, or 23.75 hr after training. It was further anticipated that the usual deficit in thc 30-hr TTI groups might be mitigated in those groups that received shock 6 or 18 hr after training.

If, on the other hand, the state dependent retrieval effects previously reported were based on a pretraining endogenous rhythmic process, then one would expect the interpolated shock to interact with retention performance only in those groups that received the interpolated shock at post-training times in which superior retention performance was usually found (i.e., 15 min, 12 hr, or 23.75 hr). The upper portion of Figure 18.5, depicting the retention performance of the 24-hr TTI groups, shows that animals that received interpolated shock at those times when retention performance was usually good (the 15-min, 12-hr, and 23.75-hr group) were impaired relative to the perfor-

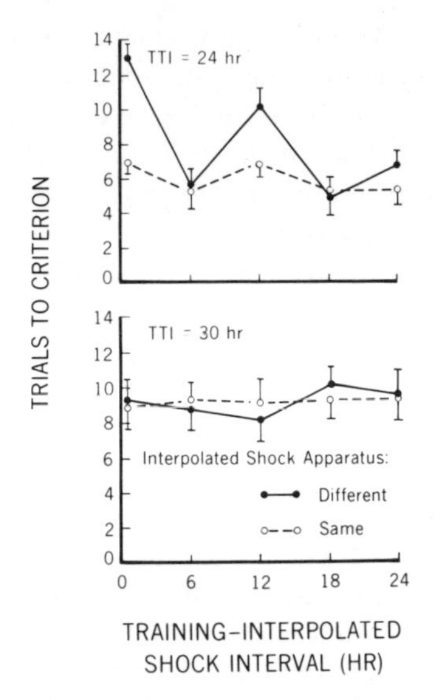

Figure 18.5. Active-avoidance testing performance (trials to four successive avoidances) with TTIs of 24 hr (top; N = 10/group) and 30 hr (bottom; N = 10/group) with interpolated shock at various intervals post-training in the training apparatus or a different chamber.

mance of those groups that received interpolated shock at 6 or 18 hr post-training. Retention performance in the 30-hr TTI groups was uniformly poor (lower portion of Figure 18.5). We suggested that those TTIs that usually yield good retention performance are post-training times when the state of the organism is similar to that present at training. They also represent, however, conditions in which an interpolated treatment can modify or interact with the memory of the training experience, even producing a subsequent retention performance deficit.

The results of this experiment suggest that post-training interpolated shock did interact with the training experience in the 15-min, 12-hr, and 24-hr groups but appeared to have virtually no effect whatsoever in the 6- or 18-hr groups. In essence, 6- and 18-hr groups behaved as if they had received no interpolated shock treatment. A second experiment showed that preshock given in the same or a different apparatus 24 hr, 18 hr, 12 hr, 6 hr, or 15 min prior to training in the active-avoidance task had no effect on active-avoidance acquisition performance except to weakly impair acquisition when presented in the same apparatus 15 min before training. We concluded that the results of this experiment again supported the notion that some pretraining rhythmic process became associated with original training and became part of what was learned on that occasion or a determining matrix for future access to the learning experience.

DIFFERENTIAL RESPONDING BASED ON TIME OF DAY

While there have been no systematic studies of whether an animal can learn to make response A at one time of day and response B at another time of day, at least three sets of data provide tentative answers to this issue. It will be recalled from the studies of Holloway and Wansley (1973a,b) that retention performance was always best when testing occurred at the same time of day as training had been given.

Circadian Anticipation of Food

An early report by Richter (1927) indicated that a large proportion of a rat's daily activity occurred during the hours just preceding regularly scheduled feeding. Birch, Burnstein, and Clark (1958) found that rats fed during the same 2-hr period daily began, by the thirty fifth day of the regimen, to develop a temporal discrimination indicated by anticipatory approaches to the food trough. Although Birch and his colleagues had sought evidence for hunger-produced stimuli, their data clearly indicated that during the course of a 48-hr period of food deprivation, food-trough responses increased just prior to the 24-hr interval, subsided, and increased again prior to 48 hr. Subsequently, Bolles (1961) suggested that some internal or external cues related to time of day rather than drive stimuli might best account for the data. He trained rats to press a lever for food either during a 1-hr period each 24 hr or for a 1-hr period each 29 hr. Essentially, his animals were able to anticipate the 24-hr circadian feeding but not the 29-hr feeding. Furthermore, as in Stroebel's results (1967), Bolles's 24-hr animals extinguished more slowly at the 24-hr time than at 19- or 29-hr periods.

In another study, Bolles and Stokes (1965) demonstrated that rats could not readily learn to anticipate 19- or 29-hr feeding cycles but could easily learn to anticipate 24-hr feedings regardless of whether food was regularly provided during the middle of the dark period or during the middle of the light period. A subsequent study (Bolles, 1968) also indicated that rats may be able to acquire circadian anticipation of water. Since the water anticipation effect was smaller than that for food, Bolles, Riley, Cantor, and Duncan (1974) suggested that the water-anticipation effect may be mediated by eating, which frequently follows water intake. Bolles *et al.* (1974) found that rats were unable to develop an anticipatory conditioned emotional reaction to electric shock regularly scheduled once a day. The authors suggested that circadian anticipatory responding may in fact "be a characteristic of and be limited to the rat's food regulatory system."

Thus, animals appear able to utilize some aspect of their internal state to anticipate food presented regularly every 24 hr. Such anticipation seems more related to processes governed by some biological clock than to cues induced by deprivation or external feeding schedules.

Differential Classical Conditioning Based on Time of Day

Asratyan (1961, 1965) reported that constant features of an environment, both internal and external, may become conditioned background stimuli that evoke tonic reflexes. The studies of his Russian colleagues that support this hypothesis are termed "switching" experiments and involve classical conditioning of different types of responses to the same (conditional) stimulus (CS) but in different contexts. In one such experiment (by Sakhyulina, as reported by Asratyan, 1965), an auditory stimulus was paired with shock to the left leg in the morning and to the right leg in the afternoon. As expected time-of-day dependent conditioned leg flexions (left leg flexion in morning and right leg flexion in the afternoon) could be elicited by the same CS. Asratyan interpreted these data to mean that the particular tonic reflex evoked by the constant stimuli that characterized a particular experimental context determined the type of conditional response elicited by the phasic CS.

Although the time-of-day facet of this set of data is interesting and supportive of the notion that time of day may be an important contextual factor in conditioning, controls for the role of internal versus external parameters were not mentioned.

Differential Instrumental Conditioning Based on Time of Day

There are several reports that animals may be able to utilize some information about time of day, that is, the light–dark cycle or the phase of some internal biological clock, and that such time-of-day discriminative processes may have important adaptive significance. Koltermann (1971) reports that bees can remember scents or colors at the same time of day in which the sensory training initially took place but not as well at other times of day. Pittendrigh (1974) cites evidence suggesting that both the natural light–dark cycle and internal circadian clocks may play a critical role in the ability of birds to find their way to home roosts after foraging.

There have been no systematic studies in rodents of differential response conditioning based on time of day. Overton did report that rats were unable to learn to escape shock by running to the right or left in a T-maze when the differential parameter signifying the appropriate response was morning versus evening (a 12-hr time-of-day difference; Overton, 1964) or morning versus afternoon (a 6-hr time-of-day difference; Overton, personal communication). Stavnes (1971), in a somewhat different task, also failed to find left–right differential responding by mice when the differential condition was time of day. As suggested earlier in this chapter, an animal's internal state may play a lesser role in influencing retrieval operations in tasks that involve highly salient external cues, which under some circumstances may even mask internal state changes. I addressed the latter problem by using a task in which external cues presumably played a minimal role (Holloway, 1975). I attempted to train animals to different patterns of operant behavior based on FR or DRL schedules of reinforcement depending on the time of day each reinforcement contingency was in effect. The differential reinforcement of low rates of responding (DRL) schedule should promote less responding than the fixed ratio (FR) schedule of reinforcement.

In this investigation, all animals were maintained on a 12–12 light–dark schedule and were shaped to bar-press for Noyes food pellets. Four Long-Evans male rats received alternating, daily half-hour training sessions in the food-reinforced bar-press task in which a DRL 15 sec schedule of reinforcement was given in the morning (0900–1100) or afternoon (1500–1700) and a FR 15 reinforcement schedule was given on the alternate day at the opposite time. Two other animals received daily alternating DRL and FR sessions exclusively in the morning hours, and 2 final animals received all of their sessions in the afternoon. During the initial 10 min of each session, no reinforcement was available. Responding during this initial extinction period served as a measure for whether the animal could respond with the pattern of behavior (low or high rates of responding) appropriate for that day. Those animals that received DRL/FR schedules at different times of day should perform better than animals that received the DRL/FR schedules at the same time each day. All animals received at least 20 daily sessions.

The results of this experiment, shown in Figure 18.6, are presented in terms of mean reinforcement scores (across sessions) for the initial 10-min test

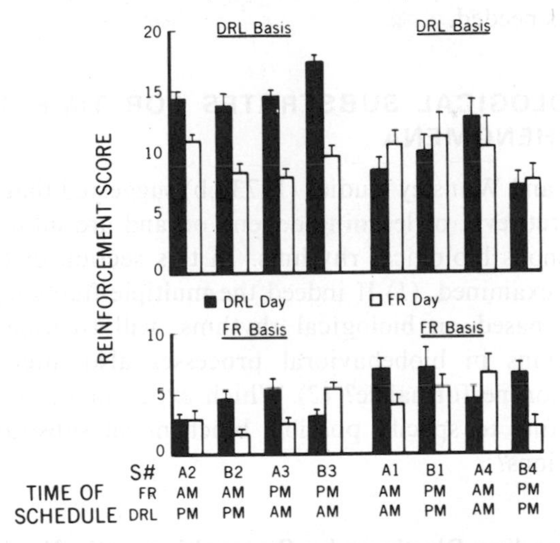

Figure 18.6. Food reinforced bar-press performance during initial 10-min extinction period of daily operant task in animals trained under DRL and FR schedules at the same time each day (the 4 animals on the right) or at different times each day (the 4 animals on the left). A significant DRL–FR day difference was noted for DRL scores in animals trained at different time (upper left quadrant; $t(3) = 6.58$, $p < .01$) but not for animals trained at the same time ($t = .09$). A significant difference was also found among animals trained at the same versus different times for DRL scores on DRL days ($t(6) = 3.13$, $p < .05$).

(extinction) period. This score was derived by counting the number of reinforcements the animal would have received on the basis of a FR or DRL contingency, for example, FR scores were the total number of responses divided by 15, and DRL scores were the number of responses following a response-free interval of at least 15 sec. The top portion of Figure 18.6 illustrates DRL scores, the bottom portion, FR scores. The graphs on the right illustrate performance of animals receiving DRL/FR sessions at different times of the day, while the two on the left depict animals receiving DRL/FR sessions at the same time each day. The dark bars represent DRL sessions, and the open bars represent the FR sessions. The only systematic effect of the experimental variable was seen for reinforcement scores calculated on the DRL basis. Animals trained with FR and DRL schedules at different times of the day had higher DRL reinforcement scores on those days when the DRL contingency was in effect than on those days when the FR contingency was in force. No such differences were apparent in the animals that received DRL and FR sessions at the same time each day. Further, on days when the DRL schedule was in force, the "same-time" animals had significantly lower DRL reinforcement scores than the "different-time" animals.

These data only partially support the hypothesis that differential responding can be based on time-of-day factors. Differential responding was apparent only during the initial extinction period and then only if reinforcement scores were derived on a DRL basis. Clearly, however, these are pilot data, and much additional work is needed.

POSSIBLE BIOLOGICAL SUBSTRATES FOR TIME-OF-DAY RETENTION PHENOMENA

The Holloway and Wansley studies (1973a,b) suggested that some aspects of learning and/or retrieval of learning depend on and are influenced by certain unknown endogenous biological rhythms. In this section of the chapter, two questions will be examined. (*1*) If indeed the multiple fluctuations in retention performance are based on biological rhythms, will treatments that disrupt periodic fluctuations in biobehavioral processes also alter the fluctuating pattern of retention performance? (*2*) Which experimental strategies may be useful in attempting to specify possible biochemical substrates for multiple retention fluctuations?

Abolition of Circadian Rhythms by Suprachiasmatic Nucleus Lesions

Wansley attempted to investigate the involvement of biological rhythms in the multiphasic retention phenomena by making discrete radio frequency lesions in a portion of the hypothalamus called the suprachiasmatic nucleus (Wansley & Holloway, 1975a). Recent research by Moore (1974) shows that this hypothalamic area receives direct input from the primary visual pathway via the retinal hypothalamic tract. Research by Moore and by Zucker and his colleagues (Moore, 1974; Rusak & Zucker, 1975) indicates that this nucleus is essential for 24-hr fluctuations of such behaviors as eating and drinking and such neuroendocrine substrates as corticosteroids or the essential enzyme required for the conversion of pineal serotonin to melatonin, *N*-acetyltransferase. Lesions of this nucleus not only abolish rhythmic fluctuations in these latter substrates when animals are entrained by external light–dark cycles but also abolish free-running circadian rhythms normally present under constant dark conditions. It is unclear, however, whether this small area of the hypothalamus acts as a master pacemaker for all rhythmically fluctuating biological substrates or as a master coupler for the appearance of circadian fluctuations in various physiological processes.

Based on these latter findings, we examined eating, drinking, and spontaneous activity at different times of the day before and after lesions of the suprachiasmatic nucleus (Holloway, 1975). Before the lesion, animals consumed 70% of their daily food intake and 85% of their daily water intake during the dark period of the 12-hour light–12-hour dark cycle. After the lesion, approximately equal amounts of food and water were consumed in the light and dark. The nocturnal peak and diurnal trough in spontaneous activity levels noted prior to surgery were absent after surgery. Clearly, the lesion treatment influenced the 24-hr rhythmicity of both consumatory behaviors and spontaneous activity. Approximately 1½ weeks after surgery, retention performance subsequent to passive- or active-avoidance training was examined at various post-training intervals. Figure 18.7 shows the retention performance after 1 trial passive-avoidance training for nonoperated controls that received

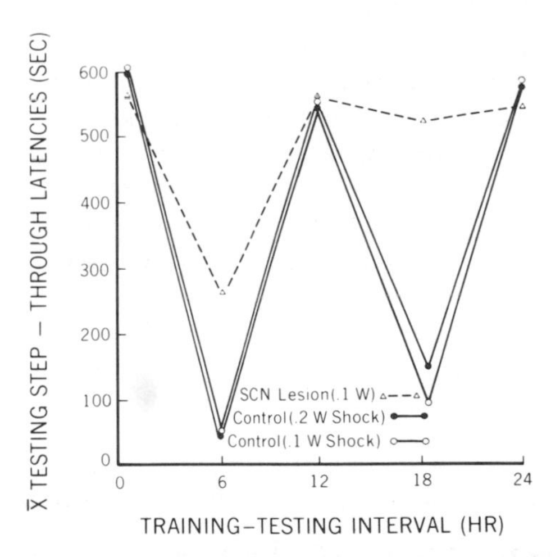

Figure 18.7. Mean step-through latencies in the 1-trial passive-avoidance task at various TTIs in controls trained under 2 shock intensities ($N = 8$/group) and in animals with bilateral lesions of the suprachiasmatic nucleus ($N = 5$/group). A significant ($p < .05$) 6 versus 18-hr TTI difference was noted in the lesion group but not in control animals.

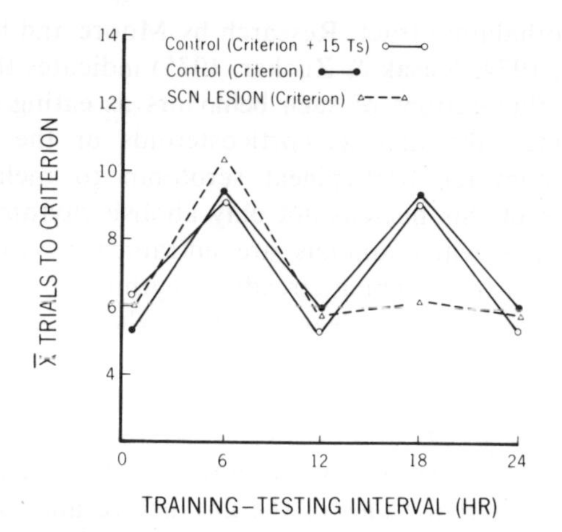

Figure 18.8. Mean trial to active-avoidance criterion (four successive avoidances) in control animals (N = 8/group) trained to criterion or to criterion plus 15 additional trials and in animals (N = 9/group) with bilateral lesions of the suprachiasmatic nucleus. A significant ($p < .05$) difference between the 6- and 18-hr TTI was noted for lesion but not control animals.

a moderate or high-level shock and for animals with bilateral lesions of the suprachiasmatic nucleus. Figure 18.8 shows retention performance of animals trained to four consecutive conditioned avoidance responses (CARs) in the step-up active-avoidance task, for control animals that received a low- and high-level shock and for animals that had bilateral lesions of the suprachiasmatic nucleus. The results clearly suggest that the multiple character of retention performance seen in control groups is absent in the lesioned animals. These data also suggest that the initial deficits frequently seen by others following training (i.e., the Kamin effect), may be related to processes other than those mediating our multiphasic retention performance effects.

These results support our suggestion that multiphasic retention phenomena are based on processes acting like biological rhythms. Other treatments, like constant light, that are known to disrupt rhythmic fluctuations in biobehavioral processes may also influence the degree to which multiphasic effects are seen. Miller (unpublished study) reports the absence of multiple fluctuations for retention of passive-avoidance conditioning in animals housed under constant-light conditions.

Possible Biochemical Substrates Underlying Multiphasic Effects

While we now have tentative data that indicate the possible dependence of certain aspects of the multiphasic fluctuations in retention performance on biological rhythmic processes, we do not know which rhythmically fluctuating processes or biochemical substrates are involved. There are several severe methodological difficulties involved in a search for possible biological oscillators acting as state markers. If one were to examine biochemical correlates of retention performance, one of the primary interpretive problems would be to identify which oscillators are pertinent for the behavior in question. Specifically, if a behavior is phased by some constantly fluctuating internal process (the internal clock), one must then distinguish between this relevant substrate and other irrelevant processes that also are constantly fluctuating. It is also evident that no single biochemical substrate is likely to account for the observed data, that is, 12-hr and 24-hr periodic fluctuations in retention performance that are independent of the time of day during which training occurs or during which testing is given. Figure 18.9 illustrates certain hypothetical rhythmic physiological processes. The top portion of the figure illustrates why no single circadian process could account for the multiphasic effects presented by Holloway and Wansley. For example, the state changes illustrated in the top portion of the figure for training at points A and C clearly would not yield a similar TTI retention performance curve. A similar analysis holds for a 12-hr periodic process. The bottom portion of the figure illustrates an arbitrary situation in which two 12-hr processes 3 hr out of phase could producc the same values for the two arbitrary physiological states X and Y every 12 hr. The reinstatement of similar states every 12 hr would be inde-

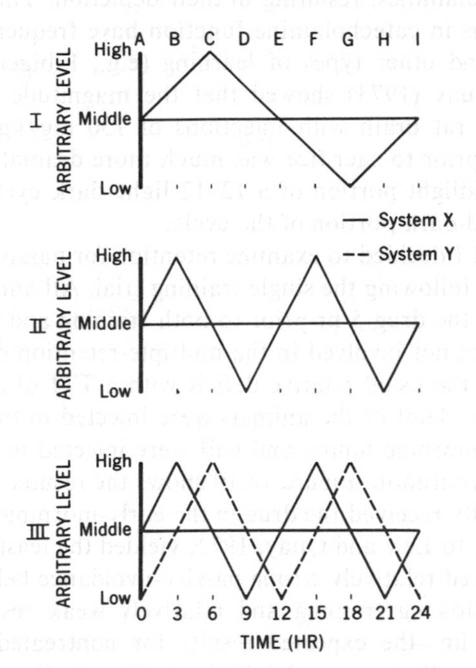

Figure 18.9. Hypothetical physiological processes having periodic properties: (I) Variable with a 24-hr periodicity; (II) variable with a 12-hr periodicity; (III) 2 variables with 12-hr periodicities, 3 hr out of phase.

pendent of when training occurred in the cycle. Clearly, if at least two oscillators are required to account for the data, and there is no adequate criterion for distinguishing relevant from irrelevant oscillators accompanying retention performance, the search for simple biochemical correlates is likely to prove futile.

Possible Catecholamine Mechanisms

Our initial attempts to address the issue of possible biochemical correlates involves the following rationale: If the level of some substrate, for example, catecholamines, were pharmacologically held constant in the training and testing sessions, and if retention fluctuations were in part dependent on such biochemical processes, then normally present deficits in retention performance would not be seen. However, this approach involves two profound problems in experimental design: (*1*) The pharmacological treatment itself may produce state-dependent events that mask any other endogenous changes; (*2*) the sensitivity of the organism to pharmacological treatments is not constant throughout the day.

In our first study (Holloway, 1975), we hoped to take advantage of the circadian sensitivity curve for the psychopharmacological potency of many drugs. In this study, we used alpha-methyl-para-tyrosine, which blocks the synthesis of catecholamines, resulting in their depletion. This drug was chosen because alternations in catecholamine function have frequently been found to affect avoidance and other types of learning (e.g., Fibiger, Phillips, & Zis, 1974). Lew and Quay (1973) showed that the magnitude of norepinephrine depletion in whole rat brain with injections of 150 mg/kg of alpha-methyl-para-tyrosine 4 hr prior to sacrifice was much more dramatic with drug injections during the midlight portion of a 12–12 light–dark cycle than with injections during the mid-dark portion of the cycle.

Walter Tapp and I decided to examine retention for passive-avoidance training 24, 30, or 36 hr following the single training trial. All animals were injected with 200 mg/kg of the drug 5 hr prior to both training and testing sessions. If catecholamines were not involved in the multiple-retention deficit phenomena, we expected to see the usual relative deficit with a TTI of 30 hr but not with TTIs of 24 or 36 hr. Half of the animals were injected in the late-dark period and trained in the morning hours, and half were injected in the midlight cycle and trained in the afternoon. Figure 18.10 shows the results of this experiment. Animals that initially received the drug in the early-morning hours (the time of day that, according to Lew and Quay, 1973, yielded the least depletion of catecholamines) displayed relatively strong passive-avoidance behavior when tested at 24 or 36 hr following training and relatively weak response suppression when tested at 30 hr—the expected results for nontreated control animals. However, the groups that received initial injection in the midmorning hours (the time of day yielding more drastic reduction in catecholamine levels) displayed no retention performance decrement 30 hr after training. Further

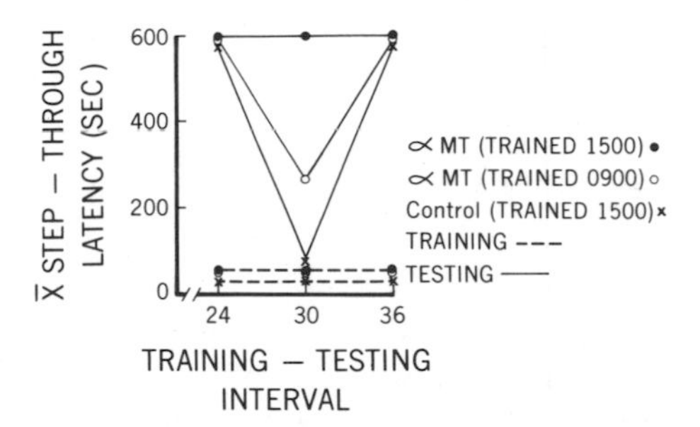

Figure 18.10. Training and testing step-through latencies in a 1-trial passive-avoidance task for animals injected with alpha-methyl-para-tyrosine during the hours 0400–0600 or 1000–1200 (N = 5/group) and for noninjected controls (N = 10/group). The only time of injection difference noted was at the 30-hr TTI condition between the two drug groups ($p < .028$).

work is required to determine whether the absence of the 30-hr deficit in the latter case is because of an abolition of the fluctuation in catecholamines or whether the biological impact of the drug on the organism is greatest when it is injected at this time of day, leading to a drug-induced state dependency that merely masks the expected deficit.

In a second experiment, we examined retention of one-way active-avoidance conditioning at intervals of 15 min, 6 hr, 12 hr, 18 hr, or 24 hr following training. All animals in this study were given three injections of 50 mg/kg of alpha-methyl-para-tyrosine 4 hr apart with the last injection 4 hr prior to training. This drug dose is known to deplete whole brain dopamine and norepinephrine for periods of 24–48 hr. All training occurred between 3 and 6 P.M. As can be seen in Figure 18.11, acquisition performance was impaired, but no fluctuations in retention performance were noted.

The interpretation of these data remains ambiguous at this time. Some aspects of rhythmic catecholamine processes may, in fact, partially mediate fluctuations in multiple-retention performance. However, the earlier analysis of possible rhythmic substrates (see Figure 18.9 and discussion in text) indicated that at least two rhythmic substrates were required to account for the fact that the pattern of multiple-retention deficits is independent of time of training.

Several alternative explanations for our drug data are possible. First, catecholamines may be a substrate underlying all biorhythmic phenomena. Pilot studies in my laboratory indicate that alpha-methyl-para-tyrosine injections at least do not alter circadian patterns in drinking or activity. Second, elimination of one substrate, like the catecholamines, may produce a situation in which some other substrate continues to influence retention performance, but only when training occurs at certain critical times during the day (i.e., removal of the Y substrate in the bottom portion of Figure 18.9). The passive-avoidance data, in fact, indicate that an animal with drug treatment and train-

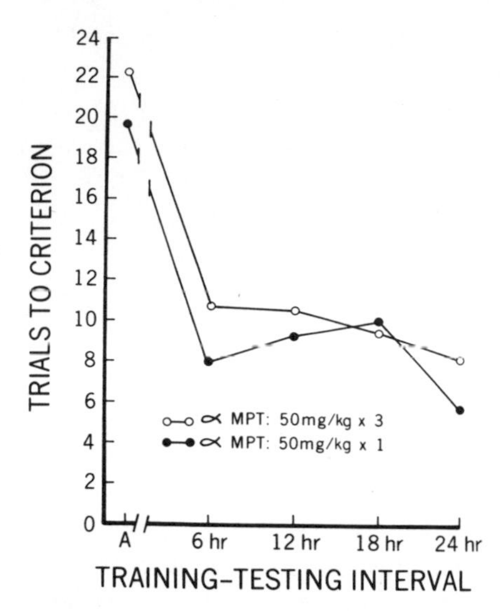

Figure 18.11. Mean trials to 4 successive avoidances during training and testing for animals injected with alpha-methyl-para-tyrosine (αMT; 50 mg/kg × 3, 4 hr apart and 4 hr prior to training and 50 mg/kg 4 hr prior to training).

ing during the morning displayed a retention decrement at 30 hr, but drug treatment and training at a later time did not. This experimental strategy, however, does not control for the differential sensitivity of the organism to alpha-methyl-para-tyrosine throughout a 24-hr period. According to a third interpretation, the drug injections in the early-morning hours may simply reflect ineffective dose levels for that time of day.

Other Possible Neurochemical Substrates

If some class of neurotransmitter or neuroendocrine substrates were mediating the time-of-day retention fluctuations via some state-dependent retrieval mechanism, then such a substrate should have at least three properties: (*1*) Any pharmacological alteration of substrate activity should produce changes in the magnitude of post-training retention performance decrements. (*2*) Any pharmacological treatment that blocked or facilitated the activity of the substrate also should produce SDL effects and (*3*) The substrate should display circadian and/or ultradian changes in level or activity. At least three classes of chemical substrates have been found to affect and/or vary with retention performance at intermediate intervals after training: ACTH (e.g., Klein, 1972), catecholamines (Underhill, Rucker, McDiarmid, & Sparber, 1973), and acetylcholine (Thompson & Nielson, 1971).

According to Overton's (1971) summary of the relative potency of drugs with discriminative properties, epinephrine, norepinephrine, nicotinic agents, and acetylcholine have moderate discriminative properties, as do most CNS anticholinergic compounds like scopolamine and atropine. Antiadrenergic compounds, however, tend to have very weak discriminative properties. Recently, several studies indicated that ACTH may produce state-dependent effects (e.g., Gray, 1975). There is even evidence that alpha-methyl-para-tyrosine, the drug we used in our TTI studies, may produce SDL effects (Zornetzer, Gold, & Hendrickson, 1974). Finally, all of these 3 classes of substrates have been shown to display circadian and/or ultradian fluctuations throughout successive 24-hr periods (e.g., Friedman & Walker, 1972; Ganong, 1974; Reis, Wurtman, Weinbein, & Corbelli, 1968; Scheving, Harrison, Gordon, & Pauly, 1968).

Another neurochemical class could, in fact, be added to the field of candidate neural substrates for the time-of-day retention phenomena, namely, the indole amines. Both serotonin and melatonin display circadian fluctuations in rat brain (Klein, 1974). Furthermore, the suprachiasmatic nucleus of the hypothalamus, an area intimately involved both in circadian phenomena and the multiphasic retention effect, is also rich in serotonin terminals (Fuxe, 1965). While there have been to my knowledge no state dependent studies with serotonin per se, several hallucinogins (e.g., LSD) with analogous chemical structures are known to have state-dependent properties and to block brain serotonin activity (see Barry, 1974).

MODELS FOR STATE DEPENDENT PHENOMENA BASED ON TIME OF DAY

On the basis of studies reported in this chapter, it would appear that natural changes in the state of the organism resulting from food or water deprivation and other interoceptive events do not provide a strong basis on which differential responding can develop. The data supporting state dependent phenomena based on time of day do suggest, however, that some undetermined, rhythmically modulated aspect of the state of the organism at the time of training may become a relevant condition for the retrieval of all or some aspects of that training experience on subsequent occasions. It seems unlikely that the basis for such a time-of-day dependency can be adequately handled by a "state as stimulus" model. Such a model implies that the state of the organism would actively elicit the response pattern or response that was learned during the original training period. Bliss (1974), Weingartner, Adefris, Eich, and Murphy (1976), and Wright (1974) discussed models for drug-state dependency in which the drug state acts to change certain organizational properties of the brain related to encoding and retrieval of information. Such a brain organization or storage model for state dependent retrieval phenomena implies a passive process. That is, if the state of the organism at the time of training determines in some sense the manner or mode in which responses or information are encoded, then such a state would have to be reinstituted in order for environmental cues or conditioned stimuli to elicit the learned behavior on a subsequent occasion. This passive model of state dependency suggests that the state of the organism is a relevant factor in retrieval but not in an associational sense. The state would not elicit what was originally learned in the sense of classical conditioning but would gate or control accessibility to previously acquired responses.

I am not aware of systematic studies that address the validity of the latter encoding/retrieval specificity model for drug state dependency. There certainly are no studies on such a model supporting time-of-day state dependent retrieval phenomena. It is clear, however, that the rate at which an animal learns or acquires new responses varies systematically throughout the day. (See Davies, Navaratman, & Redfern, 1973; Holloway & Wansley, 1973b.) Furthermore, Stephens, McGaugh, and Alpern (1967) show that the degree to which electroconvulsive shock subsequent to the training trial in mice produces amnesia for the training experience varies significantly throughout the day.

In summary, the involvement of biological rhythms in learning, memory, and performance is an exciting although frequently neglected area in psychobiological studies. The data reviewed in this chapter suggest that there are situations in which such rhythmic parameters play important roles in retention, sometimes producing state dependent retrieval effects. It is becoming increasingly apparent that failure to control for time-of-day factors in biobehavioral experiments over the last 10 years has produced a multitude of conflicting reports in the literature. Margules, Lewis, Dragovich, and Margules (1972), for example, appear to have clarified certain discrepancies in the literature concerning the effects of intracranial injections of norepinephrine on food intake by examination of different portions of the light–dark cycle. Time-of-day parameters most likely play a particularly critical role in all psychopharmacological studies, including drug SDL investigations. Virtually every drug that has been studied in a dissociative or a drug discrimination paradigm is known to display a marked fluctuation in its potency as a function of time of day. Even controlling for time-of-day effects by always training and/or testing at the same time each day may be insufficient since substantial variance may result from within-group variations in the timing of experimental sessions. Many of the time-dependent parameters of learning and retention (e.g., incubation, consolidation) were derived from studies that at least tacitly assumed stable baselines or steady states across time in untreated animals. The impact of biological rhythms on various biobehavioral processes makes the latter assumption increasingly untenable.

ACKNOWLEDGMENTS

The author wishes to express his appreciation to Drs. Joan Holloway, Raymond Russin, and Harold L. Williams for their critical reading of the manuscript, to my research assistants Alan Lincoln and Robert Sturgis, and to my graduate student Walter Tapp. Special appreciation is due my graduate student Richard Wansley, who has collaborated with me on this project for several years.

REFERENCES

Anisman, H., & Waller, T. G. Effects of conflicting response requirements and shock compartment confinement on the Kamin effect in rats. *J. Comp. Physiol. Psychol.*, 1971, *77*, 240–244.

Asratyan, E. A. Some aspects of the elaboration of conditioned connections and formation of their properties. In J. F. Delafresnaye (Ed.), *Brain mechanisms and learning.* Oxford: Blackwell, 1961. Pp. 95–113.

Asratyan, E. A. *Conditioned reflex and compensatory mechanisms.* London: Pergamon Press, 1965.

Barrett, R. J., Leith, N. J., & Ray, O. S. Kamin effect in rats: index of memory or shock induced inhibition. *J. Comp. Physiol. Psychol.*, 1971, *77*, 234–239.

Barry, H. Classification of drugs according to their discriminable effects in rats. *Federation Proc.*, 1974, *33*, 1814–1824.

Birch, D., Burnstein, E., & Clark, R. A. Response strength as a function of hours of food deprivation under a controlled maintenance schedule. *J. Comp. Physiol. Psychol.*, 1958, *51*, 350–354.

Bliss, D. K. Theoretical explanations of drug-dissociated behaviors. *Federation Proc.*, 1974, *33*, 1787–1795.

Bolles, R. C. A replication and further analysis of a study on position reversal learning in hungry and thirsty rats. *J. Comp. Physiol. Psychol.*, 1958, *51*, 349.

Bolles, R. C. The generalization of deprivation-produced stimuli. *Psychol. Rep.*, 1961, *9*, 623–626.

Bolles, R. C. Anticipatory general activity in thirsty rats. *J. Comp. Physiol. Psychol.*, 1968, *65*, 511–513.

Bolles, R. C. *Theory of motivation.* (2nd ed.) New York: Harper, 1975.

Bolles, R. C., Riley, A. L., Cantor, M. B., & Duncan, P. M. The rat's failure to anticipate regularly scheduled daily shock. *Behav. Biol.*, 1974, *11*, 365–372.

Bolles, R. C., & Stokes, L. W. Rats' anticipation of diurnal and a-diurnal feeding. *J. Comp. Physiol. Psychol.*, 1965, *60*, 290–294.

Brush, F. R. Retention of aversively motivated behavior. In F. R. Brush (Ed.), *Aversive learning and conditioning.* New York: Academic Press, 1971. Pp. 401–465.

Bykov, K. M. *The cerebral cortex and the internal organs.* Moscow: Foreign Languages, 1959.

Capaldi, E. J. A sequential hypothesis of instrumental learning. In K. W. Spence and J. T. Spence (Eds.), *The psychology of learning and motivation: Advances in research and theory.* Vol.1. New York: Academic Press, 1967. Pp. 67–156.

Caul, W. F., Barrett, R. J., Thune, G. E., & Osborne, G. L. Avoidance decrement as a function of training-testing interval: Single cycle or multiphasic? *Behav. Biol.*, 1974, *11*, 409–414.

Davies, J. A., Navaratnam, V., & Redfern, P. H. A 24-hour rhythm in passive avoidance behavior in rats. *Psychopharmacologia*, 1973, *32*, 211–214.

DeVietti, T. L. & Larson, R. C. ECS effects: evidence for state-dependent learning in rats. *J. Comp. Physiol. Psychol.*, 1971, *74*, 407–415.

Evans, F. J. Hypnosis and sleep: Techniques for exploring cognitive activity during sleep. In E. Fromm & R. E. Shor (Eds.), *Hypnosis: Research and developments.* Chicago: Aldin-Atherton, 1972. Pp. 43–83.

Fibiger, H. C., Phillips, A. G., & Zis, A. P. Deficits in instrumental responding after 6-hydroxydopamine lesions of the nigro-neostriatal dopaminergic projection. *Pharmacol. Biochem. & Behav.*, 1974, *2*, 87–96.

Friedman, A. H. & Walker, C. A. The acute toxicity of drugs acting at cholinoceptive sites, and twenty-four rhythms in brain acetylcholine. *Arch. Toxicol.*, 1972, *29*, 39–49.

Fuxe, K. Evidence for the existence of monoamine nerve terminals in the central nervous system. IV. Distribution of monoamine terminals in the central nervous system. *Acta Physiol. Scand.*, 1965, *64*, Suppl. 247, 41–85.

Gabriel, M. Effects of intersession delay and training level on avoidance extinction and intertrial behavior. *J. Comp. Physiol. Psychol.*, 1968, *66*, 412–416.

Ganong, W. F. Brain mechanisms regulating the secretion of the pituitary gland. In F. O. Schmitt and F. G. Wordon (Eds.), *The neurosciences, third study program.* Cambridge, Massachusetts: MIT Press, 1974. Pp. 549–564.

Gray, P. Effect of adrenocoxtico-tropic hormone on conditioned avoidance in rats interpreted as state-dependent learning. *J. Comp. Physiol. Psychol.*, 1975, *88*, 281–284.

Holloway, F. A. State-dependent retrieval based on "natural" states. Symposium paper presented at meeting of the Southwestern Psychological Association, Houston, Texas, 1975.

Holloway, F. A. Non-monotonic temporal gradients reflecting susceptibility of learned responses to interference. Unpublished study.

Holloway, F. A., & Sturgis, R. D. Periodic decrements in retrieval of the memory of non-reinforcement reflected in resistance to extinction. *J. Exp. Psychol. Animal Behavior Processes*, 1976, *2*, 335–341.

Holloway, F. A., & Wansley, R. A. Multiphasic retention deficits at periodic intervals after passive avoidance learning. *Science*, 1973, *180*, 208–210. (a)

Holloway, F. A., & Wansley, R. A. Multiple retention deficits at periodic intervals after active and passive avoidance learning. *Behav. Biol.*, 1973, *2*, 1–14. (b)

Hull, C. L. Differential habituation to internal stimuli in the rat. *J. Comp. Psychol.*, 1933, *16*, 255–273.

Hunsicker, J. R. The retention function for appetitive responding in rats. Unpublished masters thesis, Univ. of Oklahoma, Norman, Oklahoma, 1974.

Joy, R. M., & Prinz, P. N. The effect of sleep altering environments upon the acquisition and retention of a conditioned avoidance response in the rat. *Physiol. Behavior*, 1969, *4*, 809–814.

Kendler, A. A., & Levine, S. Studies of the effect of change of drive: I. from hunger to thirst in a T-maze. *J. Exp. Psychol.*, 1951, *41*, 429–436.

Kendler, H. H. The influence of simultaneous hunger and thirst drives upon the learning of two opposed spacial responses of the white rat. *J. Exp. Psychol.*, 1946, *36*, 212–220.

Klein, D. C. Circadian rhythms in indole metabolism in the pineal gland. In F. O. Schmitt & F. G. Worden (Eds.) *The neurosciences, third study program.* Cambridge, Massachusetts: MIT Press, 1974. Pp. 509–516.

Klein, S. B. Adrenal-pituitary influence in reactivation of avoidance learning memory in the rat after intermediate intervals. *J. Comp. Physiol. Psychol.*, 1972, *79*, 341–359.

Koltermann, R. Circadian memory rhythm after scent and colour training with honey bees. *Z. vergl. Physiologie*, 1971, *75*, 49–68.

Leeper, R. The role of motivation in learning: a study of the phenomenon of differential control of the utilization of habits. *J. Genet. Psychol.*, 1935, *46*, 3–40.

Lew, G. M. & Quay, W. B. The mechanism of circadian rhythms in brain and organ contents of norepinephrine: Circadian changes in the effects of methyltyrosine and 6-hydroxydopamine. *Comp. Gen. Pharmacol.*, 1973, *4*, 375–381.

Margules, D. L., Lewis, M. J., Dragovich, J. A. & Margules, A. S. Hypothalamic norepinephrine: Circadian rhythms and the control of feeding and drinking behavior. *Science*, 1972, *178*, 640–643.

Moore, R. Y. Visual pathways and the central nervous system control of diurnal rhythms. In F. O. Schmitt & F. G. Worden (Eds.), *The neurosciences, third study program*, Cambridge, Massachusetts: MIT Press, 1974. Pp. 537–546.

Nielson, H. C. Evidence that electroconvulsive shock alters memory retrieval rather than memory consolidation. *Exp. Neurol.*, 1968, *20*, 3–20.

Overton, D. A. State-dependent or "dissociated" learning produced with pentobarbital. *J. Comp. Physiol. Psychol.*, 1964, *57*, 3–12.

Overton, D. A. Discriminative control by drug states. In T. Thompson & R. Pickens (Eds.), *Stimulus properties of drugs.* New York: Appleton, 1971. Pp. 87–110.

Overton, D. A. Experimental methods for the study of state-dependent learning. *Federation Proc.*, 1974, *33*, 1800–1813.

Pinel, J. D. J., & Mucha, R. F. Incubation and Kamin effects in the rat: changes in activity and reactivity. *J. Comp. Physiol. Psychol.*, 1973, *84*, 661–668.

Pittendrigh, C. S. Circadian oscillations in cells and the circadian organization of multi-cellular systems. In F. O. Schmitt & F. G. Worden (Eds.), *The neurosciences, third study program*. Cambridge, Massachusetts: MIT Press, 1974. Pp. 437–458.

Reis, D. J., Wurtman, R. J., Weinbein, M., & Corbelli, A. A circadian rhythm of norepinephrine regionally in cat brain: Its relationship to environmental lighting and to regional diurnal variations in brain serotonin. *J. Pharmacol.*, 1968, *164*, 135–145.

Richter, C. P. Animal behavior and internal drives. *Quart. Rev. Biol.*, 1927, *2*, 307–343.

Rusak, B., & Zucker, I. Biological rhythms and animal behavior. *Ann. Rev. Psychol.*, 1975, *26*, 137–171.

Scheving, L. E., Harrison, W. H., Gordon, P., & Pauly, J. E. Daily fluctuation (circadian and ultradian) in biogenic amines of the rat brain. *Am. J. Physiol.*, 1968, *214*, 166–173.

Spear, N. E. Retrieval of memory in animals. *Psychol. Rev.*, 1973, *80*, 163–194.

Stavnes, K. L. State-dependent learning: its relation to biogenic amines and natural states. Unpublished doctoral dissertation, Univ. of Colorado, Boulder, Colorado, 1971.

Stephens, G., McGaugh, J. L., & Alpern, H. P. Periodicity and memory in mice. *Psychon. Sci.*, 1967, *8*, 201–202.

Stroebel, C. F. Behavioral effects of circadian rhythms. In J. Zubin and H. F. Hunt (Eds.), *Comparative psychopathology*. New York: Grune & Stratton, 1967. Pp. 158–172.

Thompson, R. W. & Nielson, C. The effect of scopalamine on the Kamin effect: a test of the parasympathetic over-reaction hypothesis. *Psychon. Sci.*, 1971, *23*, 41–42.

Underhill, W. R., Rucker, W. B., McDiarmid, C. G., & Sparber, S. B. Time factors in behavioral control of brain norepinephrine and plasma corticosterone levels. *Proc. 81st Ann. Conv. APA*, 1973, 1021–1022.

Wansley, R. A. & Holloway, F. A. Lesions of the suprachiasmatic nucleus (hypothalamus) alter the normal retention performance in the rat. *Neuroscience Abstracts*, 1975, *1*, 810. (a)

Wansley, R. A. & Holloway, F. A. Multiple retention deficits following one-trial appetitive training. *Behav. Biol.*, 1975, *14*, 135–149. (b)

Wansley, R. A. & Holloway, F. A. Oscillations in retention performance after passive avoidance training. *Learn. & Motiv.*, 1976, *1*, 296–302.

Webb, W. B. Drives as cues. *Psychol. Rep.*, 1955, *1*, 287–298.

Weingartner, H. Studies of affect-specific recall in manic-depressives. Paper presented at Amer. Psychol. Assoc. Convention, Montreal, 1973.

Weingartner, H., Adefris, W., Eich, J. E., & Murphy, D. L. Encoding-imagery specificity in alcohol state-dependent learning. *J. Exp. Psychol.*, 1976, *3*, 83–87.

Wright, D. C. Differentiating stimulus and storage hypotheses of state-dependent learning. *Federation Proc.*, 1974, *33*, 1797–1799.

Yamaguchi, H. G. Gradients of drive stimuli (S_D) intensity generalization. *J. Exp. Psychol.*, 1952, *43*, 298–304.

Zornetzer, S. F., Gold, M. S., & Hendrickson, J. Alpha-Methyl-p-tyrosine and memory: State dependency and memory failure. *Behav. Biol.*, 1974, *12*, 135–141.

19

The Engram: Lost and Found

Douglas L. Chute
Department of Psychology
University of Houston

STATE DEPENDENT LEARNING

The ability of an organism to retrieve stored memory traces seems to depend in part on the condition or "state" of the central nervous system (CNS). Researchers investigating state dependent effects usually begin with the observations of Girden and Culler (1937) and Girden (1940, 1947). Using curare or dihydro-b-erythroidine HBr, they conditioned autonomic and skeletal motor responses in a number of species. When the animals were subsequently tested in a nondrugged state, no evidence of conditioning was found. If the drug was present again, redintegrating the state, the conditioned responses returned. Girden termed the phenomenon "dissociation of learning."

Discrimination and Transfer Strategies

Two lines of research, represented in this book, have evolved. Overton (1972) identifies the two research strategies as "drug-discrimination" procedures and "transfer" procedures. Drug discrimination is the more sensitive of the two, in which subjects are rapidly brought under "stimulus" control, more rapidly usually than in the analogous sensory-discrimination paradigms. The major findings and applications that have evolved from experimental designs using the discrimination techniques are reviewed in a number of sources, including Overton (1968, 1972), Thompson and Pickens (1971), Schuster and Balster, (1977), and the preceding chapters.

Transfer procedures evaluate changes in performance of the same response in differing states. The most common design is the 2 × 2 factorial illustrated in Table 19.1. The organism is trained either drugged (D) or nondrugged (ND)

Table 19.1
The 2 × 2 Design[a]

Acquisition state	Retention test state	Transfer
ND	ND	+
ND	D	0
D	ND	0
D	D	+

[a] ND = nondrugged; D = drugged; + = transfer occurred; 0 = little or no transfer occurred.

and tested either drugged or nondrugged. When training and testing states match (D–D or ND–ND), transfer (+) occurs; when states do not match (ND–D or D–ND), little or no transfer (0) is observed. Variations of the 2 × 2 design for studying transfer effects include repeated measures procedures and "savings" determinations in which subjects serve as their own controls. Overton (1974) provides a comprehensive critique of the methods employed in studying state-dependent learning.

The basic discrimination and transfer strategies are mentioned here because they provide the data base for two distinct theoretical approaches, each of which may be concerned with separate mechanisms. Implicit in much of the literature on drug discrimination and state dependent learning (SDL), however, is the assumption of an underlying commonality or unitary general mechanism of action. Although different pharmacological and other "state-producing" agents are known to affect separate pathways, transmitter substances, or central loci, the notion seems to exist, on a molar level, that discrimination and transfer paradigms address a single conceptual process, albeit from differing aspects. These conceptual issues are occasionally obscured by problems of nomenclature in which drug discrimination and state-dependent learning are used interchangeably, as if these terms were synonymous.

STATE-AS-STIMULUS MODELS

Perhaps the strongest and most comprehensive advocacy of a unitary mechanism is provided by Schuster and Balster (1977), although they emphasize that the understanding of mechanism is not "a necessary precondition to a functional analysis on the behavioral level" of stimulus control by drugs. Schuster and Balster propose that a "state as stimulus" model can explain the role of drugs in both the discrimination and transfer procedures.

Drugs are stimuli (Catania, 1971) and can serve as conditioned, unconditioned, reinforcing, or discriminative stimuli. As such, drugs and presumably many other state-producing agents are observed to parallel the principles of operant and classical conditioning. A powerful description and prediction of drug-behavior interactions is therefore available. Certainly for discrimination procedures, the descriptive validity of behavioral analysis is high; however, results from studies using transfer strategies seem, in many cases, incompatible with such an analysis. Five lines of evidence have been proposed that provide varying degrees of difficulty for a stimulus model: failure of transfer, asymmetrical dissociation, central versus peripheral effects, postacquisition treatments, and endogenous biorhythm state dependency.

DEFICIENCIES OF STATE-AS-STIMULUS MODELS

Failure of Transfer

John (1967) suggested that the complete lack of response sometimes reported in state-change conditions was unlike normal exteroceptive stimulus control. Schuster and Balster find this argument weak and suggest that the superior stimulus control by drugs (i.e., a steeper generalization gradient) does not require the postulation of a fundamentally different process.

Asymmetrical Dissociation

In the transfer procedure, change across states, either D–ND or ND–D, often results in differential or asymmetrical levels of response. Usually, animals in the ND–D group perform at levels superior to those in the D–ND group. Since the shift from D–ND and ND–D could be presumed equivalent in magnitude, any generalization decrement should be symmetrical. As Schuster and Balster point out, asymmetry does not require the postulation of another mechanism. Rather the effect could be a complex drug interacton with learning and performance or a simple consequence of experimental design in which discrimination procedures, unlike transfer procedures, are not amenable to analysis for asymmetry.

Central versus Peripheral Effects

Centrally acting drugs are much more efficacious in producing discrimination than compounds that do not cross the blood–brain barrier. If centrally acting agents only could come under stimulus control, it would be detrimental to a state-as-stimulus model. As Schuster and Balster note, however, central activity seems to be neither a necessary nor sufficient condition for discrimination.

The problems presented by the issues surrounding failure of transfer, asymmetrical dissociation, and central versus peripheral effects may be coped with in a unitary mechanism implied by the stimulus model. The following two areas, however, provide much greater difficulty in interpretation from the state-as-stimulus point of view.

Table 19.2
Stimulus versus Retrieval[a]

Post acquisition state	Retention-test state	Stimulus hypothesis	Retrieval hypothesis
ND	ND	+	+
ND	D	0	0
D	ND	+	0
D	D	0	+

[a] ND = nondrugged; D = drugged; + = transfer occurred; 0 = little or no transfer occurred.

Postacquisition Treatments

In the 2 × 2 design of the transfer procedure, drug treatment has typically been administered prior to acquisition and again before retention testing. Drug effects on the stimulus-sampling interval are therefore indistinguishable from effects on the information storage/retrieval interval following acquisition. Chute and Wright (1973) used 12.5 mg/kg sodium pentobarbital to drug rats immediately after acquisition in a 1-trial passive-avoidance task. In such a postacquisition design, the drug can be presumed effective during the storage/retrieval interval but not present during the stimulus-sampling interval. Table 19.2 illustrates the predicted results for a stage-as-stimulus hypothesis and for a competing state-specific retrieval hypothesis. Retrograde SDL was demonstrated by Chute and Wright (1973) and Wright and Chute (1973a), supporting a retrieval mechanism for state dependency (Figure 19.1).

Holloway (1973) reported asymmetrical dissociation in an active avoidance task for a retrograde state dependent design using ethanol. Unless backward

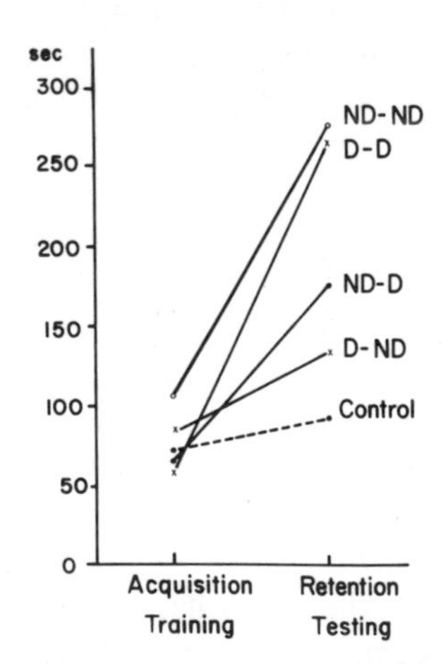

Figure 19.1. The longer retention test latencies of groups ND–ND and D–D show transfer and support a retrieval hypothesis (D = drugged; ND = nondrugged). Redrawn from Chute, D. L., and Wright, D. C. Retrograde state dependent learning. *Science,* 1973, *180,* p. 878. Copyright © 1973 by the American Association for the Advancement of Science. Reproduced by permission.

associations are assumed possible, the existence of state dependent effects with postacquisition treatment obviates a unitary mechanism model based on a stimulus hypothesis.

Endogenous Biorhythm State Dependency

State dependency phenomena based on endogenous events like the time of day are reviewed by Holloway in another chapter. A state-as-stimulus model like that proposed by Schuster and Balster would be hard put to explain why endogenous state cues do not bombard the organism with uncalled memory, given, in the case of Holloway's experiments, a rhythmic ultradian fluctuation.

STATE DEPENDENT RETRIEVAL MODELS

A number of alternatives to the stimulus model have been proposed (e.g., Overton, 1975) and have the advantage of high descriptive and predictive power in those areas identified above as problematical for a state-as-stimulus position. The models center around state-specific storage or retrieval mechanisms and have generally been derived from experiments employing a transfer strategy. As Overton (1968) notes, however, most CNS models make rather arbitrary assumptions about the effects of drugs on brain function. Recent proposals (e.g., Bliss, 1974, Chute & Wright, 1973) have an implicit notion of state dependency as a passive gating mechanism, such that "state matching" during information storage and information retrieval is necessary for performance. With overtraining, the necessity for state matching may decrease, that is, performance becomes state independent. The passive model differs somewhat from some previously hypothesized CNS mechanisms (e.g., Girden & Culler, 1937; Holmgren, 1965; Overton, 1968) in that the engram is not "rerouted" to different CNS structures. Instead, the presence or absence of a drug may specify distinct spatiotemporal firing patterns in the same representational system. Incidentally, any model other than passive gating is hard put to include endogenous state dependent results.

Electrophysiological Evidence in State Dependent Retrieval

E. Roy John has reported that consistent changes in electrophysiological activity develop with learning. The changes in the evoked response are independent of afferent stimulation and may be considered as representations of the released activity associated with a specific memory, which John terms the "readout component" (John, 1972). Extrapolation from the results in John's laboratory led to the hypothesis that performance decrements produced by state change may occur by the passive gating out of the "readout component."

The "readout component" identified in the following experiment is not strictly identical with John's interpretation. The term is used advisedly here to describe the differences between evoked potentials from correctly responding and incorrectly responding trials in the early development of the conditioned avoidance response. An analogous change in evoked response is reviewed by Ilyutchenok (1975) and termed a "delayed evoked response." Although a good deal of variability exists in the late components of the evoked responses in the study described below, differences were easily identified and readily distinguished from control evoked potentials in which the stimulus was not a significant cue. If SDL results from the failure to retrieve the stored memory trace, then it might be expected that evoked potentials from trained animals will or will not show a "readout component"/"delayed evoked response" depending on their state.

Subjects were 56 naive, male, albino rats (Holtzman) weighting 250 to 400 gm. 16 animals were excluded from the study when, after 5 days of training, they showed little or no learning. The remaining 40 animals were assigned 5 per group to the eight experimental groups indicated in Table 19.3.

Two animals from each group had epidural cortical skull screws implanted over the visual cortex and frontal sinus for monopolar recording of evoked responses. Six days prior to training, these animals were anesthetized with 50 mg/kg sodium pentobarbital 20 min following a .15 cc injection of .4 mg/cc atropine. After the area was shaved and washed, a midline incision was made in the scalp and tissue retracted by gentle periosteal elevation. The temporal muscles were retracted, and cotton packing to the dorsolateral flexures prevented seepage. The skull was prepared; 2 electrode holes and one anchor

Table 19.3
Avoidance Performance by Groups[a]

	Avoidance responses		
Group	Criterion Day 1	Criterion Day 2	Criterion Day 3
ND–ND–ND	68a	65a	68a
ND–ND–D	62a	65a	3d
ND–D–ND	65a	2d	54a
ND–D–D	68a	4d	12c, d
D–D–D	64a	63a	64a
D–D–ND	63a	59a	25b, c
D–ND–D	63a	17b, c, d	53a
D–ND–ND	63a	9d	32b

[a] The first nondrugged (ND) or drugged (D) in the group designation indicates a nondrugged (saline) or drugged (sodium pentobarbital) state during training to criterion. The second ND or D indicates state during the first day of training following criterion, and the third ND or D indicates state on the second day following criterion. Following each avoidance response entry is a letter indicating the results of a Duncan's Multiple Range Test. Any entry with the same letter is not significantly different from any other with that letter. Entries with unlike letters are significantly different at least $p < .05$. Some entries have more than one letter, and they do not differ from any other entry which has one or more of these letters.

hole were drilled and packed with Gelfoam. Electrodes with pin connectors and an anchor were made from .80 stainless-steel screws. The indifferent electrode was placed 2 mm to the right of the sagital suture, 4 mm rostral to bregma. The active electrode was placed 3 mm to the right of the sagital suture, 3 mm rostral to lambda. The anchor screw was positioned 4 mm to the left of the sagital suture, midway between bregma and lambda. A dental acrylic base secured the screws, connectors, and a connector-stabilizing anchor post. Before closure, a topical antiseptic (Neosporin) was applied. Clean procedure was indicated throughout, with electrodes and instruments soaked in a 1:750 tincture of benzalkonium chloride solution. The animals were killed 120 hr after the last day of training, frozen, and later necropsied. Minimal evidence of infection was found, and less than 20% of the electrode sites showed significant gliosis where some electrodes pierced the dura.

The behavioral apparatus consisted of a brightly polished metal box 25 × 25 × 50 with a grid floor and a long, low lever (12 gm activation pressure) to the right of a door. The door and top were clear Plexiglas. The experimental room was dimly lit with a shaded 100 W lamp. The conditional stimulus (CS) was a 10-sec train of 2-Hz flashes delivered by a Grass PS_2 photo stimulator (intensity 16). The reinforcer, footshock (FS), was a 65 V, 200 Hz, 80% duty cycle (4 msec on, 1 msec off), train of monophasic square waves applied to the grid floor.

All animals were given 2 sessions, 24 hours apart, to habituate to the box with only the CS presented. For 1 session of habituation, the animals were drugged (D) with an i.p. injection of 15 mg/kg sodium pentobarbital, and on the other day they were nondrugged (ND) with an equal volume of normal saline. These dosages were maintained throughout the experiment and were injected 15 min prior to the start of a session. The first day of D or ND was counterbalanced. As in later training, a habituation session consisted of 20 presentations of the CS on a variable interval schedule (with contraints) of 40 sec. The CS had a normal duration of 10 sec unless the animal pressed the bar, which terminated the CS and stopped a response latency clock.

Electrophysiological recordings from the prepared animals and a calibration signal were FM tape recorded, and the first four evoked potentials (EPs) from each of the 20 trials were averaged by a computer of average transients in 250 msec epochs. Because the CS and shock had not yet been paired, the EPs from the habituation session provided control data where the CS had not gained any "significance" as a cue.

Following habituation, the animals received daily training sessions in either D or ND states to a criterion of 60% shock avoidance. Onset of the CS was followed 4 sec later by the onset of footshock for the remaining 6 sec the CS was presented, or until a response was made. An avoidance response had to occur within the first 4 sec of a trial to prevent the onset of shock for that trial. The session in which the animal first reached the avoidance criterion was designated Criterion Day 1. Two days of additional training followed Criterion

Day 1, and these were designated Criterion Day 2 and Criterion Day 3. Training continued as before, but the effective drug states D or ND were manipulated, depending on group designation. During train phases, bar-press response latencies were recorded, and evoked potentials were again taped. The first four EPs of each trial in a session were averaged, depending on the "correctness" of the response for that trial. If an animal correctly avoided shock, EPs were averaged with other responding trials. If an avoidance response was not made, EPs were averaged with other nonresponding trials in the session.

Evoked potentials were analyzed using two highly experienced independent judges as "pattern discriminators." The sorting was conducted "blind," after initial instructions to identify drugged and nondrugged patterns and "readout components" (Figure 19.2). The results are indicated in Figure 19.3.

Table 19.3 presents the number of avoidance responses per group for criterion day and the 2 successive days. A significant group effect ($F_{7,32}41.94$, $p < .01$), a significant training effect ($F_{2,32}145.12$, $p < .01$), and a significant interaction ($F_{14,64}34.19$, $p < .01$) were found using an $8 \times 5 \times 3$ analysis of variance with repeated measures.

Table 19.3 also shows the results of a Duncan Multiple Range Test. On Criterion Day 1 there were no significant differences between groups reflecting an equivalent ability to learn in either a D or ND state. For Criterion Day 2, the significantly poorer performance of state-change groups is characteristic of SDL. Although not significantly better, Groups D–ND–D and D–ND–ND have an observed performance superior to Groups ND–D–ND and ND–D–D indicative of asymmetrical dissociation. Asymmetrical dissociation is again observable on Criterion Day 3, in which Group D–D–ND is significantly superior to Group ND–ND–D. The asymmetrical effect suggests that transfer across drug states occurs less readily when an animal is trained ND and tested D. Since the asymmetry occurs in the nonusual direction, the possibility exists that drug side effects, like ataxia, could be responsible.

Multiple-state training could be operationally defined as 2 successive days of training in the alternate state following the attainment of the avoidance cri-

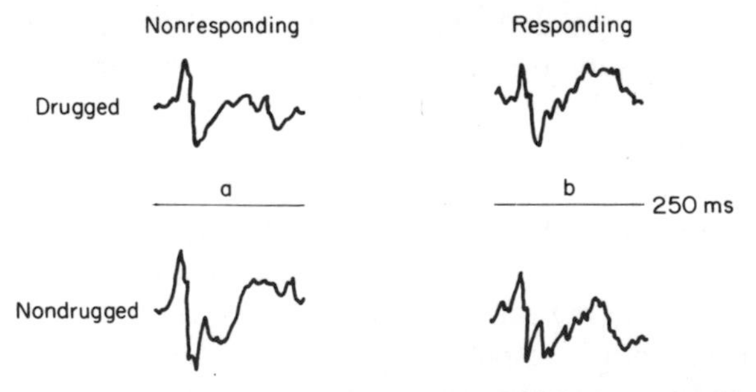

Figure 19.2. Illustration of nondrugged and drugged effects on evoked potentials and the "readout component"/"delayed evoked response." The examples are taken from the clearest records and do not represent a single animal.

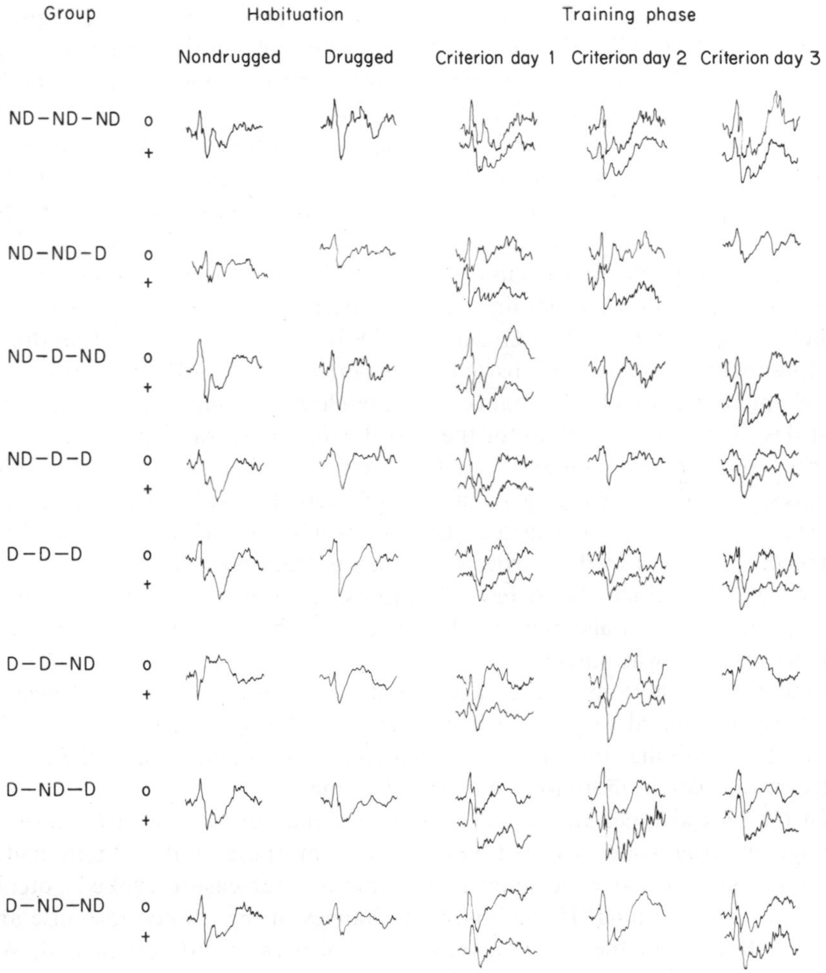

Figure 19.3. Evoked potentials (EPs) from 1 of the 2 animals per group of habituation days and final training phases. EPs from correctly responding trials are indicated with a "+," incorrectly responding trials with a "0." Judges identified correctly D or ND patterns in 100% of the 236 cases. "Readout component" like response patterns for correctly responding trials were identified in 90.18% of the cases. Incorrect responses, that is, response failures, from trials in which the state was the same as original training were judged to show the "readout component" like pattern in 59.38% of the cases. Incorrect responses from trials in which the drug state was different from original training were judged to have only 12.50% "readout components." "Readout components" were identified as occurring in habituation EPs 1.56% of the time. With a combined total of 236 discriminations, the judges disagreed on only 8, or 3.39%.

terion. Overtraining could similarly be defined as an additional day of training following criterion, prior to shifting states. In both cases, there is a weak effect exposed in the 2 × 2 × 2 design. The increase in performance of Groups ND–D–D and D–ND–ND indicates a facilitory effect of multiple-state training on Criterion Day 3. Again the significantly better performance of Group D–ND–ND compared to Group ND–D–D is an asymmetrical effect. Limited overtraining in the SDL phenomenon can be observed by noting that significantly less dissociation occurs on the first day of state change for Group D–D–ND, which had an extra day of state-same training, than occurs for Group D–D–ND. Although not significantly different, the comparison of Groups D–D–ND and D–ND–D is in the same direction. This result is also asymmetrical in that no differences were obtained for comparison of Group ND–ND–D with either Group ND–D–ND or Group ND–D–D. Clearly a change from drugged to nondrugged states is less disruptive of performance. SDL is not an "all-or-none" phenomenon. Performance of state-change groups was better for the first day of state change than for the first day of initial training.

The concomitant changes found in visual evoked responses as training progressed reflected learning in the rat of a conditioned avoidance response. Evoked responses from drugged (D) and nondrugged (ND) animals showed marked differences in the earlier components from 60 to 80 msec because of the drug effect alone. With few exceptions, a pattern could be identified in both D and ND animals about 90–130 msec into the epoch for trials in which a correct response was made.

Under the conditions of the experiment, two types of error could occur. In the first, an animal fails to respond correctly during a testing trial while in the original training state. For these incorrectly responding trials, the response pattern was present more than one half the time.

In the second type, an animal fails to respond correctly as a function of a change in drug state. Evoked potentials from these trials seldom had the response pattern and were judged the same as premeasure evoked potentials taken prior to training. If the observed changes in the evoked response are in fact a reflection of the engram, they were not permanently eradicated. When the original training state was redintegrated, or if multiple-state training or overtraining occurred, the evoked response pattern or engram was again "found."

State Dependent Retrieval and Retrograde Amnesia

Various treatments, like drugs, electroconvulsive shock (ECS), and spreading depression, when administered immediately after training, produce retrograde amnesia (RA). The treatments are thought to disrupt an hypothesized labile perseverative first-memory phase, thus blocking consolidation (e.g., McGaugh & Herz, 1972). A number of investigators have considered the possibility that memory disruption is an experimental design artifact in which

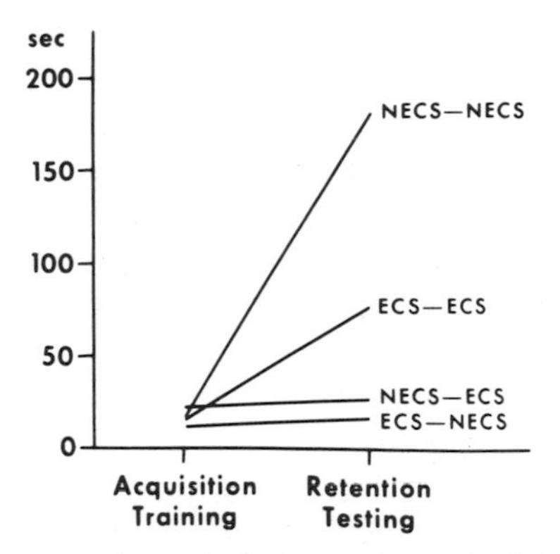

Figure 19.4. Retention test latencies show apparent state-dependent retrieval with electroconvulsive shock. (ECS = electroconvulsive shock; NECS = no electroconvulsive shock.)

the apparent amnesia represents a state dependent retrieval failure characteristic of a D–ND group. For example, Pearlman, Sharpless, and Jarvik (1961) report retrograde amnesia for sodium pentobarbital administered immediately after training. However, the retrograde SDL effect reported by Chute and Wright (1973) can be interpreted as reversing pentobarbital amnesia when the original postacquisition training state is redintegrated as in the complete 2 × 2 design. Wright, Chute, and McCollum (1974) also show amnesia reversal for sodium pentobarbital in an appetitive discrimination task measuring response choice rather than latency.

The state dependent transfer design has been applied to other amnestic treatments. For example, Thompson and Neely (1970) and Wright and Chute (1973b) report ECS-induced amnesia reversal in a state dependent context (Figure 19.4). Similarly, for spreading depression, Greenwood and Singer (1974) observed symmetrical dissociation.

DEFICIENCIES OF STATE DEPENDENT RETRIEVAL MODELS

A large part of the support for state dependent retrieval models comes from those experiments demonstrating retrograde state dependency and amnesia reversal. Unfortunately, replications are difficult to obtain, and a number of negative findings, although unpublished, exist. If the phenomena are so fragile that strain differences or minor procedural variations drastically affect the results, then generalizations of the model as a unitary explanatory concept of state effects is premature or unwarranted.

Reminder Effects

State dependent retrieval does not supply a satisfactory mechanism for the growing number of reports (e.g., Miller & Springer, 1973) of reversal of amnesia using "reminder" cues. Miller, Malinowski, Puk, and Springer (1972) evaluate the retrograde state-dependent model. They note that the strengths of the model are: prediction of amnesia time gradients, immediate post-treatment "memory," and some empirical data for amnesia reversal with state redintegration.

One weakness of the state dependent retrieval model identified by Miller and his students is not debilatory. In Experiment 2 of their report, Miller *et al.* (1972) note asymmetrical dissociation in which a D–D (i.e., footshock [FS] + electroconvulsive shock [ECS] – footshock [FS] + electroconvulsive shock (ECS) group performed as predicted, but the ND–D (i.e., NECS – FS + ECS) group showed no amnesia. Such asymmetry is not uncommon in the state-dependent literature and incidentally was not observed in the previosly reported study by Wright and Chute (1973b). The 2 × 2 design of Miller's Experiment 2 does not permit a distinction between a state dependent hypothesis and the competing "reminder" hypothesis. The position of Miller *et al.* (1972) is that the results of Experiment 2 can be adequately explained if FS + ECS prior to retention testing serves to "remind" the organism of the original training contingencies.

To distinguish between the state-dependent and reminder hypotheses, Miller conducted a third experiment. Here the approach resembled basically the 2 × 2 × 2 design in which the state dependent hypothesis would predict that a D–D–ND (i.e., FS + ECS – FS + ECS – NECS) group tested in the "state" different condition would show poor performance. The results were to the contrary and support the reminder position. Only a weak argument like multiple-state training or overtraining in the same state, mentioned previously, could account for the results of Experiment 3 in a state-dependent retrieval context.

The most important finding of Miller's group is that footshock alone is effective in reversing the effects of ECS, significantly more effective, in fact, than footshock plus ECS. A state dependent position would be hard put to claim that footshock alone provided better state matching for a FS + ECS initial training state. An only slightly more acceptable claim could be that the footshock may provide additional "training," thus, overtraining effects washout any state dependency. In fact, such an explanation has been implied (e.g., Gold, Haycock, & McGaugh, 1973; Haycock, Gold, Macri, & McGaugh, 1973), although from a different context. A strong reminder effect can be shown, however, in a design in which the reminder cue is unlikely to provide additional "training," and the general conclusions from Miller's laboratory seem to be supported. Shaw and Chute (1975) designed an experiment in which the reminder cue was kinesthetic and occurred at the time of retention testing. Since the cue did not mimic experimental treatments and did not precede in time retention testing, it is unlikely that the observed reminder effect could be because of an implicit additional "training" trial (Figure 19.5). Clearly, state matching alone is not as effective a means of reversing retro-

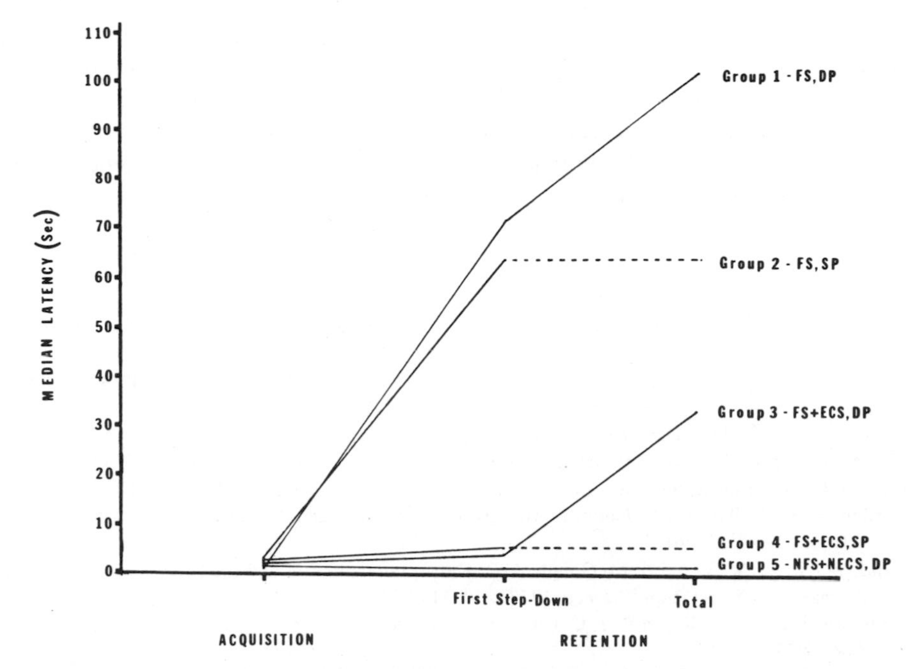

Figure 19.5. Kinesthetic reminder was provided at the time of retention testing by stepping from the top to the second level of a double platform (DP). Both normally trained animals (FS, DP) and presumed amnestic animals (FS–ECS, DP) showed significant improvement in performance. (FS = footshock; ECS = electroconvulsive shock; SP = single platform; DP = double platform.)

grade amnesia as the judicious application of some reminder cue. The comparative inability of state matching to consistently facilitate the retrieval of the engram suggests that state alone can not usually actively elicit a memory.

SUMMARY

Although a state-as-stimulus model has high descriptive and predictive power for many drug-behavior interactions, the model has difficulty encompassing all such interactions. Conversely, CNS or state dependent retrieval models, while handling these difficulties, do not provide parsimonious description or prediction of the functional characteristics of drugs as stimuli. As Overton (1968) suggests, there may well be separate mechanisms involved that obviate a unitary descriptive paradigm. The difference, then, between discrimination and transfer procedures becomes more than a simple divergence of perspective on the same conceptual problem.

The inability of state matching to reverse retrograde amnesia consistently and effectively, even when that "amnesia" is caused by highly dissociative agents, weakens state dependent retrieval models. State matching appears to be seldom a necessary or sufficient condition for amnesia reversal, with the implication that simple identity of state in acquisition and retention does not actively elicit the engram. A passive gating model of state-dependent retrieval does, however, provide a working mechanism that at best is not incompatible with retrograde state dependency, endogenous biorhythm state dependency, and reminder-effect results.

REFERENCES

Bliss, D. K. Theoretical explanations of drug-dissociated behaviors. *Federation Proc.*, 1974, *33*, 1787.

Catania, A. C. Discriminative stimulus functions of drugs. In T. Thompson & R. Pickens (Eds.), *Stimulus properties of drugs.* New York: Appleton, 1971.

Chute, D. L., & Wright, D. C. Retrograde state dependent learning. *Science*, 1973, *180*, 878.

Girden, E. Cerebral mechanisms in conditioning under curare. *Am. J. Psychol.*, 1940, *53*, 397.

Girden, E. Conditioned responses in curarized monkeys. *Am. J. Psychol.*, 1947, *60*, 571.

Girden, E. & Culler, E. A. Conditioned responses in curarized striate muscle in dogs. *Comp. Psychol.*, 1937, *23*, 261.

Gold, P., Haycock, J. W., & McGaugh, J. L. Retrograde amnesia and the "reminder affect": An alternative interpretation. *Science*, 1973, *180*, 1199.

Greenwood, P. M. & Singer, J. J. Cortical spreading depression induced state dependency. *Behav. Biol.*, 1973, *10*, 345.

Haycock, J. W., Gold, P. E., Macri, J., & McGaugh, J. L. Noncontingent footshock attentuation of retrograde amnesia: A generalization effect. *Physiol. Behav.*,1973, *11*, 99.

Holloway, F. A. Retrograde state dependent learning with ethanol. Paper presented at the annual meeting of the American Psychological Association, Montreal, 1973.

Holmgren, B. Drug dependent conditional reflexes. International Symposium on Cortical–Subcortical Relationships in Sensory Regulation, Havana, 1965.

Ilyutchenok, R. Yu. Pharmacological and neurochemical approaches to memory trace retrieval. *Current Developments in Psychopharmacology*, 1975, *1*, 109.

John, E. R. *Mechanism of Memory.* New York: Academic Press, 1967.

John, E. R. Switchboard versus statistical theories of learning and memory. *Science*, 1972, *177*, 850.

McGaugh, J. L., & Herz, M. J. *Memory consolidation.* San Francisco: Albion, 1972.

Miller, R. R., Malinowski, B., Puk, G., & Springer, A. D. State dependent models of ECS-induced amnesia in rats. *J. Comp. Physiol. Psychol.*, 1972, *81*, 533.

Miller, R. R. & Springer, A. D. Amnesia, consolidation, and retrieval. *Psychol. Rev.*, 1973, *80*, 69.

Overton, D. A. Dissociated learning in drug states (state dependent learning). In D. H. Efron, J. O. Cole, J. Levine, & J. R. Wittenborn (Eds.), *Psychopharmacology: A review of progress*, 1957–1967. U.S. Public Health Service Publication 1836, Washington, D.C., 1968. P. 918.

Overton, D. A. State dependent learning produced by alcohol and its relevance to alcoholism. In B. Kissen & H. Begleiter (Eds.), *The biology of alcoholism.* Vol. 2. *Physiology and behavior.* New York: Plenum Press, 1972.

Overton, D. A. Experimental methods for the study of state-dependent learning. Paper presented at a meeting of the Southwestern Psychological Association, Houston, 1975.

Pearlman, C. A., Jr., Sharpless, S. K., & Jarvik, M. E. Retrograde amnesia produced by anesthetic and convulsant agents. *J. Comp. Physiol. Psychol.*, 1961, *54*, 109.

Schuster, C. R., & Balster, R. L. The discriminative properties of drugs. In T. Thompson & P. B. Dews (Eds.), *Advances in behavioral pharmacology.* Vol. 1. New York: Academic Press, 1977.

Shaw, T. G., & Chute, D. L. Kinesthetic feedback and recovery from ECS-induced amnesia. Paper presented at a meeting of the Society for Neuroscience, New York, 1975.

Thompson, C. I., & Neely, J. E. Dissociated learning in rats produced by electroconvulsive shock. *Physiol. Behav.*, 1970, *5*, 783.

Thompson, T. & Pickens, R. (Eds.) *Stimulus properties of drugs.* New York: Appleton, 1971.

Wright, D. C., & Chute, D. L. State dependent learning produced by post-trial intrathoracic administration of sodium pentobarbital. *Psychopharmacologia*, 1973, *31*, 91. (a)

Wright, D. C., & Chute, D. L. Memory retrieval failures in passive avoidance following ECS. Paper presented at a meeting of the Midwestern Psychological Association, Chicago, 1973. (b)

Wright, D. C., Chute, D. L., & McCollum, G. C. Reversible sodium pentobarbital amnesia in one trial discrimination learning. *Pharmacol. Biochem. Behav.*, 1974, *2*, 603.

20

Human State Dependent Learning

Herbert Weingartner
Adult Psychiatry Branch
National Institute of Mental Health
and
University of Maryland

Research in human state dependent learning (SDL) can be seen as being quite new, but the topic has interested behavioral scientists for as long as there has been curiosity about the nature of stored experience as determined by conditions under which learning takes place. The SDL phenomenon can be conceptualized narrowly, in terms of a set of defining operations, or it can emerge as an all-encompassing arena for an enormously broad range of questions, theories, and areas of investigation. In this review, we have chosen a broad-stroked approach in describing some of the major methodological, theoretical, and applied issues that have emerged in the study of human state dependent learning. As such, it is not intended as an exhaustive or highly detailed account of research accomplished to date but as an orientation to the kinds of concepts and findings that make up research explicitly labeled SDL. The review is also intended as an overview for investigators who might want to venture into SDL research, not only psychopharmacologists who are most familiar with it, but researchers interested in human information processing.

At the outset, it should be pointed out that SDL paradigms have been rather alluring to many investigators. The extraordinary number of methodological and conceptual pitfalls which make research in this area very difficult, are by no means obvious. Some of these difficulties in study design and interpretation of findings will be discussed below, but it should be pointed out that these problems come in many forms and vary as a function of the specific focus of SDL research. Despite such warnings, it is also clear that the studies completed to date promise enormous potential leverage for future research that will attempt to delineate phenomena as different as the neurochemical effects of drugs in the storage and retrieval of experience and the use of SDL research strategies in engineering more viable models of memory. It is also becoming apparent that SDL studies may have substantial impact in providing research tools for better understanding such clinical phenomena as depression, drug-neurochemical mechanisms and specific characteristics of retrieval mechanisms in outputting memory, not merely because of the usefulness of the findings themselves but because an SDL approach invites inquiry across very different territories of thinking and investigations (Weingartner & Murphy, 1977). The first such translation and bridging of generally disparate areas of study have already occurred in moving from the SDL animal-learning paradigms, as they developed in the last decade (Overton, 1971), to the human studies of SDL completed during the last few years.

After attempting a brief statement of the basic SDL concept, we shall outline some of the major features of current and completed human SDL research. Some attempts will also be made to account for the apparent underlying foci of these efforts, which vary enormously. Following this, we shall present a sketch of the rapidly expanding SDL research that has been completed. Details of the methods used in these studies and their implications for demonstrating SDL will be considered separately. In this section, it will be clear that the very presence of SDL effects is often determined as much by the methods used to test the phenomenon as it is by the manipulations used to alter state. Following a discussion of methods, we shall present a summarizing statement of findings to date organized by problem areas. These findings have implications not only for theories of SDL, but they are potentially useful in generating new definitions of what constitutes a change in brain state with respect to a drug, neurochemical event, or disturbed behavioral states. The findings impinge also on theories of memory, defined through the joint use of biological and psychological concepts. Finally, we shall attempt to project or extrapolate these research endeavors into the near future by providing a research prospectus pointing out some possible points of application for SDL.

As it does in animal studies, the SDL phenomenon refers to both a set of operations and a set of findings. Although these operations have become increasingly complex (Overton, 1974), the basic design unit of approach and the germinal findings remain as follows. A set of behaviors or events is learned or stored in one state, defined either in behavioral, pharmacological, neurochemical, or inferred terms, and is later tested or retrieved in either the same or in some disparate state. For many researchers, this defining set of operations squarely sets SDL within the framework of generalization and discrimination learning. No further explanation of the mechanisms of SDL is required except for those that help to define these findings with respect to classic learning and performance notions in the context of an analysis of behavior. A subject learns (or stores, encodes, or inputs) some information while in a particular drug-defined state and later attempts to utilize (output) the information either while in that same (congruent) state or in a different (disparate) state. In general, information utilization is impaired when the subject's state is changed between the input and output phases of the experiment relative to the situation in which state is held constant for input and output. This pattern of results—or, more specifically, the interaction of drug state at input with drug state at output—defines the phenomenon of SDL.

The meaning of state disparate memory dissociation, with respect to its behavioral mechanism, is just now emerging from research described below. However, at the crudest level, learning, presumably stored in memory, cannot be easily elicited when tested in the disparate brain state. Even at this elementary descriptive level of the phenomenon, as will be shown, it is striking to what extent evidence for the very presence of SDL rests on methods used to test retrieval, the kinds of material learned, kinds of induced state change and their intensity.

A simple statement of apparent SDL has been made in dozens of animal studies, particularly in those using this paradigm to test for a drug discrimination and its consequent underlying neurochemical–behavioral mode of action. Until recently most of the animal SDL research has focused on issues that were not directed at studying learning mechanisms but emphasized issues directed at asking whether an animal is capable of a given discrimination, either one drug from another, one dose from another, or the drugged state from the undrugged state, all tested by generally simple learning–performance paradigms.

BRIEF OVERVIEW

The first human SDL studies were, in a sense, replications of animal studies, using such central nervous system (CNS) depressants as alcohol in the context of simple SDL experiments. Subjects were generally asked to learn a variety of different kinds of materials, usually verbal in nature, while either intoxicated or sober, for example—and were later tested on some retention task in either an intoxicated or sober state. The human studies seemed to resemble the structure of animal SDL in terms of an overall pattern of results, although these findings were by no means consistent in their fine detail. The initial studies also seemed to express an underlying theme that related SDL to some clinical issues such as the relevance of the phenomenon to issues of drug abuse (Overton, 1972, 1973). This was particularly true in the study of alcohol SDL and its relationship to the blackout state (Goodwin, 1974) and features of the drug state that might account for its apparently powerful reinforcing properties, particularly in the addicted individual (Weingartner & Faillace, 1971). These studies also suggested possible uses of SDL findings in the development of more effective therapeutic interventions in drug abuse.

Many of the initial SDL research efforts were crude, from the vantage point of contemporary neuropsychopharmacology or current notions of theory and method in human learning and memory. Nevertheless, the initial alcohol and occasional other drug-related SDL studies persuaded others, especially psychopharmacologists using human subjects, to explore other drug studies in which SDL paradigms were to be used to explore drug action and mechanism. Here, too, the research focus was not on the psychology or biology of memory and its mechanisms, as seen in the dissociative phenomenon, but on the drugs themselves.

Nevertheless, investigations of human SDL were expanding rapidly and attracted a growing number of different kinds of investigators, including, most recently, those interested in clinical phenomena other than drug abuse-related problems and investigators interested in the neurochemical correlates of disturbed behavior. The clinical syndromes studied for the first time with SDL techniques were the group of disorders characterized as the affective disturbances in man. This group of psychiatric disorders was particularly suited for study using SDL techniques, because of recent developments detailing neurochemical changes in relation to mood disturbances, as expressed in the discrete psychopharmacological models that have emerged both to treat effectively these mood disturbances and to clarify their biological underpinnings. In addition, anecdotal clinical observations seemed to preview what appeared to be an affect state-dependent learning and mood-related memory dissociation (Henry, Weingartner, & Murphy, 1973). Contemporary memory researchers have at last begun to pay attention to SDL findings. This recent interest has been triggered not by an interest in the pharmacological, neurochemical, or clinical SDL findings per se but by the suggestion that these studies might provide a testing ground for current conceptual models of memory. In many ways, this group of researchers find SDL a particularly foreign territory of study because of the history of isolation of human-memory research from investigations of animal learning on the one hand and developments in the biology of memory on the other. For most human-memory researchers, psychopharmacology and experimental clinical psychology exist yet farther out in the scientific galaxy. Nevertheless, some memory researchers, especially those interested in the presence of contextualism at the time of storage and retrieval of information from memory, have appreciated some of the parallelisms between their endeavors and those evolving in SDL research. This interest in pursuing the implication of SDL findings for theories of memory has in fact directly altered the form, focus, questions, and methods used in some current SDL studies (Eich, Weingartner, Stillman, & Gillin, 1975; Weingartner, Adefris, Eich, & Murphy, 1976).

METHODS OF APPROACH

At the level of empirical operations, human SDL has usually been investigated within the context of a 2 $\times$ 2 experimental design. Information is acquired during a drug or drug-free state and is outputted under drug or no-drug conditions. Given this format, four experimental conditions can be identified, two of which are termed congruent state conditions (input under drug–output drug, D–D, and input under no drug–output no drug, ND–ND), and two of which are denoted as disparate state conditions (input no drug–output drug, ND–D, and input drug–output no drug, D–ND).

Some of the many researchers who have used the 2 × 2 design in experiments on human SDL learning have found evidence of symmetrical state-dependent learning (e.g., Tarter, 1970). That is, the level of information utilization (or performance) attained under the ND–ND condition is higher than that achieved in the ND–D condition, and performance in D–D is superior to that in D–ND. Other investigators have adduced evidence of asymmetric SDL (e.g., Goodwin, Powell, Boemir, Hoine, & Stern, 1969). That is, D–D is superior to D–ND, but ND–ND and ND–D produce about equal performance levels. Until recently, the utility of the distinction between symmetric and asymmetric forms of SDL was unquestioned. This is not to say that researchers understood why SDL assumed different forms in different experiments. On the contrary, the nature of the mechanism underlying the symmetric versus asymmetric dichotomy was a mystery, but the existence of such a mechanism seemed irrefutable. This is not the case today. Overton (1974) has raised some serious questions about the adequacy of the 2 × 2 design as a tool for demonstrating SDL. The basic problem in Overton's view, is that centrally acting drugs typically produce a wide variety of cognitive-behavioral effects (e.g., impaired consolidation, retrieval deficits). Overton has shown that some of these effects may combine to mimic SDL effectively, either symmetrically or asymmetrically. Thus, in the absence of reliable information concerning the robustness of contributory cognitive-behavioral effects, investigators would be hard pressed to decide whether their empirical evidence of SDL is real or illusory. Overton acknowledges that there is no obvious methodological or analytical solution to this problem. It is clear, however, that Overton favors the method of drug discrimination (Overton, 1972, 1974) to the 2 × 2 experimental design, primarily on the grounds that drug discrimination designs seem particularly sensitive to "weak" dissociative effects.

We believe that Overton's arguments are focal not only in the conduct of future SDL research but also in interpreting and assessing the reliability of work completed to date. We have little that is new to add to the issue, except to say that we are less sanguine than Overton about the utility of drug discrimination methods in human SDL experimentation. It would seem that a necessary ingredient in establishing reliable drug discrimination is the repeated exposure of to-be-remembered information, and there is reason to believe that human SDL is attenuated by repeated learning or input trials. From our current perspective, the important point is that students of human state dependence should at least be aware of the serious methodological problems pointed out by Overton.

What are the methodological conditions that have seemed necessary for demonstrating human SDL? Research to date has isolated at least three factors that future studies may prove to be a bare minimum for demonstrating human dissociated learning. The first factor is drug type. Research on infrahuman SDL has shown that dissociation is not produced by agents that act exclusively outside the CNS. It is plausible to think that this conclusion may also hold for human SDL, but at present there are no reliable data for this assumption. Human SDL has been demonstrated with six pharmacologically diverse agents, namely, alcohol (Goodwin *et al.*, 1969), marihuana (Darley, Tinklenberg, Roth, & Atkinson, 1974), amphetamine (Bustamante, Rosello, Jordan, Pradere, & Insus, 1968), barbiturates (Bustamante, Jordan, Villa, Gonzolez, & Irsua, 1969), methylphenidate (Swanson & Kinsbourne, personal communication), and physostigmine (Weingartner & Murphy, 1977). Most published studies were conducted with the 2 × 2 design, as described earlier. Hence, the basic manipulation involved either the administration of a single dose of one of the drugs listed above or the administration of appropriate placebo material. Comparisons of different drugs or of different doses of the same drug have not been systematically studied despite their obvious inherent interest.

A second putatively necessary condition for demonstrating human SDL concerns drug dosage. It seems that the drug dosage most likely to evoke SDL is one that produces readily discernible cognitive-behavioral effects but does not impair cognitive functioning to the extent that even simple learning tasks cannot be successfully completed. Low drug doses do not seem to elicit discernible dissociative effects. The conclusion, at least for the time being, is that dissociated learning in man is most readily demonstrated when moderate rather than very high or very low doses of CNS agents are given.

The third necessary condition for demonstrating human SDL has to do in a broad sense with the concept of "information" as used in our description of the SDL phenomenon and in a more circumscribed sense with the concept of task-specific dissociated learning. Numerous investigators have found that human SDL effects are far more likely to obtain in some verbal learning-memory tasks than in others. However, few researchers have ventured any ideas as to why this is so. (Exceptions are, Hill, Schwin, Powell, & Goodwin, 1973; Stillman, Weingartner, Wyatt, Gillin, & Eich, 1974.) There is some reason to believe that experimental tasks that involve serial recall of to-be-retrieved events may be particularly sensitive to dissociative effects (Hill *et al.*, 1973). One type of task—or, more accurately, experimental paradigm—that fits the above description is serial recall of a list of highly meaningful nonsense syllables (Ley *et al.*, 1972). A second variant involves memory for the sequence in which a series of lights flash (Ley *et al.*, 1972). A third representative paradigm is standard free recall of a list composed of clusters of semantically related words, in which each cluster refers to a different concept (e.g. types of flowers, musical instruments, precious stones; see Eich *et al.*, 1975). With this last paradigm, we have consistently demonstrated that the sequential recall strategy used by many subjects in recollecting a conceptually clustered list is to recall the clusters in their original order of presentation. This occurs regardless of storage or recall state. In both normal brain states and drug-altered states, words within a given cluster tend to be recalled in an essentially all-or-none fashion (Tulving & Pearlstone, 1966), and the order in which within-cluster words are recalled does not typically mimic the order in which they were originally seen or heard. In other words, it appears that most subjects, when asked for free recall of a conceptually clustered word list, rely on serial information when searching for memory traces of the various clusters of information stored in memory. When a cluster is "found," its contents—conceptually related words—are read out in an all-or-none manner and without regard to their order of presentation. Once the contents of a cluster have been emptied, the search for clusters continues, and it is this phase of retrieval that seems to involve recollection of temporal (serial) relations among the to-be-retrieved clusters. A fuller treatment of this particular facet of the methodological implications of sequential search plans for understanding SDL phenomena is in preparation. The structure of this theme is simply this: under certain conditions, optimal performance in tests of free recall may depend on recollection of temporal (serial) relations among to-be-remembered events, and these may be disrupted when recall is attempted in some disparate state.

We have collected the results of 23 experiments that, from our point of view, used some paradigmatic variant of serial recall. Of these 23 cases, 19 (83%) yielded evidence of SDL. In contrast, 20 experiments were surveyed that, from our perspective, did not involve memory for temporal relations among to-be-retrieved events. Representative paradigms are item recognition and paired-associate learning. Only 3 of the 20 experiments (15%) yielded evidence of dissociated learning.

The conclusion that emerges from this survey of the human SDL literature is that the kind of mnemonic information most likely to be rendered inaccessible for retrieval because of a marked disparity between input and output drug states is information specifying temporal (or serial) relations among target events. Clearly, the veracity and generality of this conclusion remains to be formally tested and assessed. However, even in its present crude form, the idea of temporal-information dissociation invites some interesting speculation concerning the nature of the cognitive mechanism underlying SDL in man, and as we will see later, it is germane to issues concerning the applied clinical significance of the SDL phenomenon.

BRIEF SUMMARY OF PHARMACOLOGICALLY INDUCED SDL EFFECTS IN MAN

The initial SDL findings indicated that a partial asymmetrical learning-memory dissociation can be demonstrated by drug-induced changes in state using alcohol, barbiturates, and marihuana as tools to alter brain. The asymmetry of function appeared as a greater dissociation of information that was learned or stored in the drugged state and later retrieved or tested in the undrugged state than in the inverse set of conditions. Part of this asymmetrical effect seemed accounted for by the fact that, in these drug states, memory consolidation is markedly impaired (Weingartner and Faillace 1971), so that less information may have been effectively stored in memory when initial learning took place in the drug condition than in the undrugged state. As a result, far less information could be expected to be retrieved in the disparate, undrugged recall state. The key SDL element in this pattern of research findings was more complete retrieval of input if information stored drugged was retrieved in the drugged rather than in the undrugged state. Reinstituting the drug state at recall seems to reverse the disparate retrieval state dissociation. As has been pointed out, most of these SDL studies were constructed as 2 × 2 simple designs that included two disparate conditions: ND–D, D–ND; and two congruent conditions: ND–ND, D–D. Parenthetically, no early study, or for that matter recent study, has attempted literally to reverse a drug dissociation by introducing the drug immediately after a retrieval attempt in the undrugged disparate state. Such a design would contain the element train (drugged)—retrieve (undrugged and where a dissociation of memory store appears)—reverse retrieval state (reintroduce the drugged state present at the time of storage).

One issue that emerged from initial SDL studies, as was noted in the examination of methods, was that by no means all tasks or kinds of learning produced dissociative effects under disparate-state retrieval conditions. When state-disparate memory dissociations did appear, some tasks and stimuli elicited markedly stronger SDL effects than others. Furthermore, the same kind of task or stimulus was found not necessarily always to be weak or strong in articulating apparent SDL effects produced by a drug such as alcohol. Stimulus and task specificity has become an important area of study, particularly because of its implications for understanding the mechanism of this phenomenon.

In terms of the nature of the stimulus, it seems that, in general, difficult tasks are more likely to show drug dissociations than simple tasks. Highly overlearned material is also less likely to be dissociated in disparate state recall. These findings have prompted a number of studies of SDL in which both the drug and dosage used to induce dissociative effects could be assured and in which the focus of investigation was designed to elicit either stimulus-specific or task-specific differences in disparate-state retrieval. To accomplish this, first marihuana, then physostigmine were tested in two complete versions of a 2 × 2 design. In one, subjects learned categorized lists and later attempted free recall in either congruent or disparate retrieval conditions. Other subjects were studied under the same set of storage and retrieval conditions with respect to drug state and material to be learned. At the time of retrieval, however, subjects were provided with powerful retrieval cues in the form of the category names that represented the superordinate categories of the to-be-remembered events. By providing retrieval cues, dissociative effects produced by marihuana (Eich *et al.,* 1975) and physostigmine (Weingartner, Murphy,

Redmond and Eich, in preparation) could be effectively eliminated. The findings obtained with physostigmine to alter state under free and cued conditions are displayed in Figure 20.1; they are very similar to those obtained with marihuana to induce state change in an identical design. The effect of cuing as a means of erasing SDL effects has also been demonstrated in SDL studies with alcohol.

These findings suggest that dissociated information stored in memory in one state but irretrievable in some disparate state is merely inaccessible rather than unavailable in memory. That is, subjects "forget" previous learning because they are unable to generate appropriate retrieval strategies with which to query memory when retrieval is attempted in some disparate state. By providing subjects with powerful retrieval cues at the time memory is tested, events previously stored in some disparate state reappear in memory retrieval, and apparent dissociations can be erased.

The issue of what kind of information is inaccessible in disparate-state recall has also been directly addressed. In both the free versus cued recall studied using marihuana or physostigmine to induce state change, if a category of information was retrieved in disparate-state recall, then subjects were no less effective at recalling stored exemplars from that category compared to recall of category exemplars under congruent retrieval conditions. Subjects seemed to lose whole clusters (or higher-order memory units) when retrieval was attempted in some disparate state. It is also these higher-order memory units that, as clusters, are retrieved when subjects are provided with cues.

In another study (Weingartner *et al.*, 1976), the specific characteristics of individual to-be-remembered stimuli were systematically varied in a 2 × 2

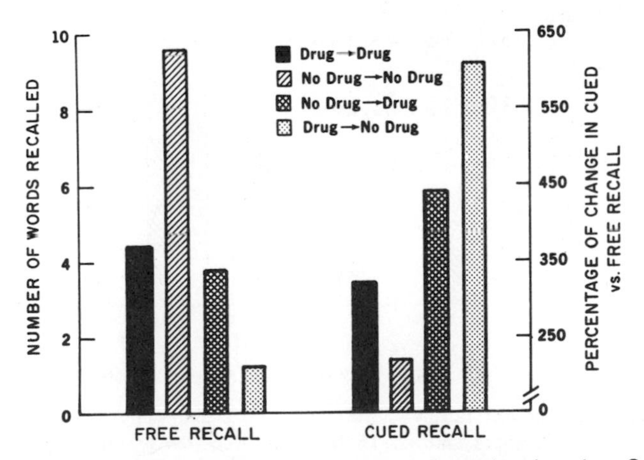

Figure 20.1. Cueing effects of physostigimine state dependent learning. Cued recall was always greater than free recall. The sequences of drug, no-drug treatments were fixed and as follows: no drug–no drug; no drug–physostigmine (1 mg); physostigmine–physostigmine; physostigmine–no drug; no drug–no drug. The time between learning and retrieval was 20 min.

SDL design. All of the stimuli were very common nouns, but some were highly imageable, for example, mountain, while others were less conducive to rapidly elicited, vivid images, for example, joy. The imagery variable relates to how easily stimuli are encoded, learned, and recalled in typical verbal learning studies. In this SDL study, strongly encoded, easily imageable input was less likely to be lost in recall when retrieval was attempted in some disparate state than the more fragilely stored events, such as words that are less concrete and imageable (Figure 20.2).

One other method of making a stimulus more likely to be powerfully encoded is to repeat the to-be-remembered event at the time it is stored or to reinforce the features of the stimulus that could alter how well it is learned. When this is in fact done in a modified SDL design, the weakly stored events are far more likely to be dissociated or forgotten in the disparate retrieval state than is saliently encoded input.

In summary, these findings indicate that SDL can be accounted for by a relative inaccessibility of information stored and potentially available in memory. Events stored in memory cannot be recalled because the retrieval strategies used to search memory are inappropriate when these are generated in the disparate recall state. However, highly encodable input is less likely to induce state-specific encoding and retrieval strategies and therefore is less susceptible to disparate recall state dissociation. Some additional findings link

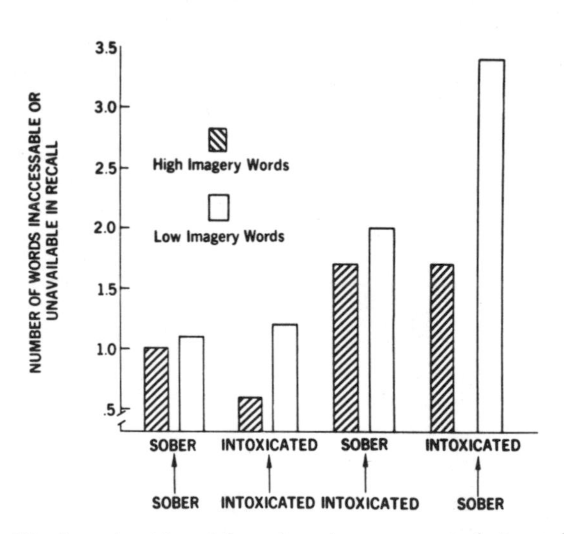

Figure 20.2. Words not retrieved from long-term memory but previously available in immediate memory. Subjects learned 15 high- and 15 low-imagery words either 20 min after drinking 6 oz of orange juice or consuming a cocktail of juice containing 3 oz of 190-proof alcohol. Recall was tested 6 hr later, either when subjects were sober or intoxicated. Each subject was studied under 4 conditions; learn sober–recall sober; learn intoxicated recall sober; learn intoxicated–recall intoxicated; learn sober–recall intoxicated. Sequencing of conditions was randomized across subjects.

these state-specific retrieval strategies with state-specific encoding strategies that operate at the time information is stored, transformed, and consolidated in memory (Weingartner, Snyder, Faillace, & Markley, 1970; Weingartner and Faillace, 1971).

NONPHARMACOLOGICALLY INDUCED SDL EFFECTS IN MAN

A few nonpharmacological SDL studies have been completed to date. These have included studies of disturbed mood states (Weingartner, Miller, & Murphy, 1977; Weingartner & Murphy, 1977), manipulations of mood through the use of noncontingent success and failure, and the use of sleep stages to define SDL brain state (Evans & Kihlstrom, 1973). The findings from these studies parallel those seen in drug-related SDL studies. For example, if patients with bipolar affective illness are systematically studied over a period of months, they are likely to exhibit some periods during which mood state is relatively stable or unchanging, including episodes of "normal," "depressed," and "manic" affect, while at other times they may move from one affective state to another, for example, from normal affect or hypomania into a manic episode. A group of these patients was repeatedly asked to self-generate sets of free associations and to retrieve them 4 days later. Sometimes patients generated and retrieved information in the same affective state (congruent conditions), while at other times they experienced a marked change in mood (disparate state conditions) during the four-day interval. When this occurred, information was far more difficult to retrieve from memory than under congruent-affect state conditions, even when such congruent conditions represented a constant period of acute mood disturbance with associated disturbances in learning and memory (Henry *et al.*, 1973). These findings are illustrated in Figure 20.3.

A recent study has also shown that when such patients, while depressed, are induced into a rapid lifting of mood through a sleep-deprivation experience or by manipulating noncontingent experiences of success in contrast to failure, dissociative memory effects are evident. The structure of these dissociative effects resembles those induced by drug-altered changes in brain state in that they involve information that, although inaccessible, is available and is associated with changes in encoding strategies as they relate to alterations in mood state. Furthermore, when mood-related dissociations appear, they seem to be more evident for those events that are not easily encoded when stored in memory (Weingartner, Post, Gerner, Murphy and Eich, in preparation).

Pharmacological studies of SDL also interrelate with mood-determined dissociative effects, as has been shown in at least one recent study (Weingartner *et al.*, in press). Here depressed patients were treated on separate occasions with either placebo or *d*- or *l*-amphetamine in rather large doses, 30 mg of *the base,* and the entire design was repeated after the patients were pretreated with lithium carbonate. Patients learned verbal material shortly after receiving placebo, or at the time of peak drug effect as measured by subjective and clinically observed state change (following amphetamine treatment). Many hours later, subjects were required to retrieve previously learned information when subjective drug response and drug levels had substantially returned to baseline levels. Amphetamines were shown to produce a weak but statistically significant SDL effect in these depressed patients. The effect was most marked in the absence of lithium carbonate. However, if individual differences in mood responses to amphetamine treatment are not ignored, measuring changes in affect, particularly with respect to activation, euphoria, and change in depression, made it possible to show that absolute change in mood along these relevant dimensions powerfully predicts ($r = .89$) the size of the dissociations in recall induced by these drug manipulations (Figure 20.4). The point of these findings is that drugs may induce SDL effects, but these can be attenuated or potentiated as a function of individual specific responses to the drug, as expressed in subjective experience or "reading" of the neurochemical events triggered by the introduction of the pharmacological agent. This individual specific modulation of a treatment that might alter mood and thereby induce mood-related SDL was also seen in the studies mentioned above, in which behavioral treatments such as manipulations of success–failure

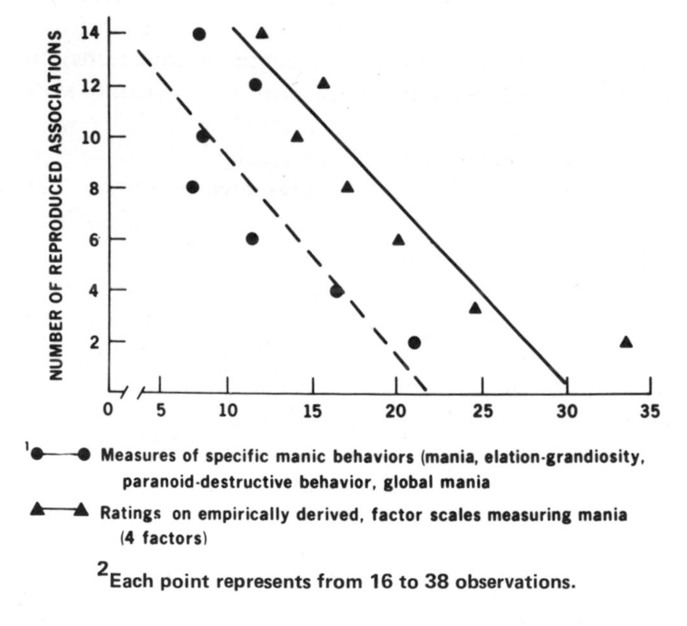

Figure 20.3. Absolute change in manic state-affect and its relationship to retrieval in self-generated learning. Nine subjects were studied twice in each treatment condition. The average subjective-state changes experienced just before learning information is plotted as a function of retrieval of stored information 6 hr later. Procedures used to study learning and measures of subjective state change are described in detail elsewhere (Henry *et al.*, 1973). Copyright 1973 by the American Psychological Association, Reprinted by permission.

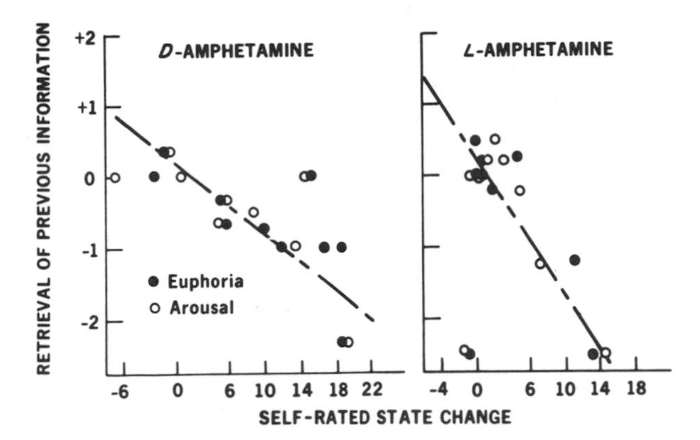

Figure 20.4. Self-rated change in state (arousal and euphoric) associated with *d*- or *l*-amphetamine and retrieval of previous information. Data based on 186 pairs of observations, obtained from 8 bipolar and patients. On each occasion, subjects generated 20 discrete associations to each of 2 standard noun stimuli and were asked to reproduce these associations 4 days later. At the same time, various components of patients clinical state were rated on empirically validated scales used to measure the intensity of affective state components. Retrieval of associations are described in relation to absolute change in effective state.

and sleep deprivation were used to alter mood. Here, too, the strongest dissociative effects were seen in subjects showing the most robust changes in mood in nominally the same experimental conditions.

The issue of individual differences in the expression of SDL effects is important for understanding how drugs act in a specific neurochemical milieu and in the context of a subject's behavioral history. It is clear that there are considerable individual differences in SDL response to a variety of drugs, and some of these have been directly interpretable (Rickles, Cohen, Whitaker & McIntyre, 1973; Weingartner & Faillace, 1971). The studies seem to show that marihuana–SDL effects are altered by the extent to which the subject is accustomed to using the drug. The same alcohol dose affects SDL in alcoholics more than it does normal controls. The size of amphetamine–SDL effects are predictable in terms of subjective drug response, but this has not been tested for other drugs to date. These findings suggest that the size of individual dissociative drug effects may provide some important insights into drug-behavioral history relationships.

THEORETICAL MODELS OF STATE DEPENDENT LEARNING

Some attempts have been made to formulate a theoretical framework in which to cast SDL findings so that they might be understood and integrated with other behavioral phenomena and theory (e.g., Wright, 1974). The simplest of these theoretical frameworks is based on some characteristics of operant theory, to account for SDL effects (Overton, 1971). In this system, a drug-defined state may function as a discriminative stimulus, and the elicitation of a given response depends on the discriminative stimuli present during training (Overton, 1971). Operationally, it is possible to teach an animal one response in a drug condition and a different, possibly incompatible, response in the undrugged state. These responses are maintained and are differentially elicited on the basis of a given state manipulation acting as a discriminative stimulus. Defining the drug state as a discriminative stimulus allows one to ask whether the organism can distinguish different intensities of state change, defined by differences in drug dose, and whether it is possible to discriminate two different drugs from each other. This model would describe the SDL phenomenon as one in which learning occurs under specific stimulus conditions, optimum performance being conditional on the reinstitution of the same internal state (stimulus) present at the time of learning. The issue of identifying the salient stimulus features from the larger set of internal and peripheral drug effects that might function as the discriminative stimuli is the key question to be answered within such a system.

It is also possible to view SDL as an example of a context manipulation that effectively alters the perception of a stimulus in a state-specific manner. In many ways, this view can be considered a simple extension of the situation. A given state provides a context of background in which information is uniquely perceived and encoded. This description of phenomena would be consistent with a good deal of classic perceptual theory, such as adaptation level theory, field dependence, and signal/noise theory. Brain states induced by drugs or behavior may produce unique contexts that alter figure-ground relations, change the relative salience of specific features of a stimulus in an information field, or alter inferred perceptual anchors that ultimately influence perceptual judgments of events to be later retrieved from memory. The perceptual field in which to-be-remembered events is perceived may also provide some cuing function at the time of recall. State dependent dissociation results when the context present at the time of storage is different from that present on retrieval of the trace event in some disparate context. Such a change concept may be seen as providing "inappropriate" anchors—for example, figure–ground relation—or other surrounding stimulus–cue characteristics that might alter the way target stimuli might be viewed or tested against a previously stored memory-trace pattern.

Another reasonable theoretical framework that may be a useful descriptive model of SDL emphasizes the role of unique or specific disruptions of normal brain function as a necessary condition for the production of memory dissociations. Most of the drugs that have been investigated in human SDL research do impair functioning, particularly at the higher doses commonly used. It may well be that when drugs induce SDL effects in man, they do so because parts of the brain are not functioning optimally at the time of storage or upon retrieval of information from memory. Some recent research in our laboratory suggests that alcohol may produce particularly disruptive effects on the storage of those verbal stimuli that do not easily elicit visual images (Weingartner *et al.*, 1976). The dissociative effects of alcohol are most marked for those stimuli that are also most disrupted at the time of storage. That is, stimuli without some pattern or image representation in memory are particularly susceptible to disparate state dissociation. It is also verbal rather than pattern information that is most likely to be disrupted in learning while intoxicated or in patients with focal brain lesions associated with long histories of alcoholism and poor diet. In fact, the alcohol-learning deficit resembles the learning associated with Korsakoff psychosis (Weingartner & Faillace, 1971), and these alcohol-induced memory changes are similar to those seen in asymmetrical, dominant temporal-lobe lesions in man. Similarly, sleep-deprivation induced SDL appears when learning and remembering verbal input but does not appear in disparate-state retrieval of pattern information (Weingartner, Post, Gerner, and Murphy, in preparation). It is possible that SDL effects are associated with nonhomogenous disruptions of the brain. That is, for drugs such as alcohol and marihuana, language-dominant hemisphere functions may be more disrupted than nondominant hemisphere-supported learning. Within this system, input may be seen as stored and retrieved not only less effectively but qualitatively differently than would be the case if the brain functioned normally.

Many of the current information-processing models postulate that information storage and retrieval from memory may be effectively described as occurring either in conceptually discrete stages or in terms of the strategies used to process input. The storage of experience, as input, may be pictured as moving from a sensory register, through an attentional system, into a short-term memory store, and finally into long-term memory storage. Some models describe the staging of information processing as occurring in parallel, while others view such processes as a serial or fixed sequence. In contrast, a level-of-processing model of information storage emphasizes the nature of the stored trace not with respect to where it might be in memory, for example, which discrete stage, but by the operations and encoding strategies used in processing input. Such a model of information processing masks many of the major controversies between different, highly specific memory theories advanced during the past few years. For example, some of these contrasting issues include the question of whether there are discrete or separate memory systems such as short- and long-term memory stores, what might be the distinguishing features of these different kinds of memory even if memory storage followed a discrete staging model, whether features of memory storage might be best defined by the way input has been processed and memory is searched, or whether it is even possible to theorize about the characteristics of the stored trace without considering retrieval condition. Some features of these current memory models, however, are particularly germane to the issue of human SDL, and these may emerge as important regardless of the relative merits of one theoretical position versus another. Likewise, some of the findings generated in psychopharmacological studies of cognitive function, particularly in SDL designs, have direct relevance to the nature of some integrative viable theory of memory storage and retrieval.

In the context of these broadly defined models of memory, the kinds and features of these models that might fit existing SDL and related findings are as follows: although some drugs seem to alter the trace of an experience at the level of some version of a sensory register (Weingartner *et al.*, 1970), it is unlikely that this facet of memory would have much of a role in determining SDL. Most of the dissociative effects seem to involve changes in the organization, meaning, or interpretation of events to be remembered and stored in memory. It seems, therefore, that programs and structures previously stored in long-term memory determine either state-specific scanning of information stored in some sensory register and/or state-specific encoding of information as it is consolidated in memory. Scanning strategies could easily be seen changing in a state-specific manner, in which elements in an informational field might be attended to differently in different brain states. For example, one could conceptualize state-specific programs that would operate on some to-be-processed input field, where such state-specific programs would determine the relative salience of the elements in a field of information. Such state-specific scanning strategies might also influence the sequence in which different elements within a perceptual field would become the focus of attention for further processing in memory. The most direct kinds of data that would support a notion of brain state-specific scan paths in processing input comes from findings like those emerging from eye-movement data in drug versus nondrug comparisons. Some preliminary data do in fact suggest that scan paths are reliably different in moving from one state to another, although stable across time within a given state (Stillman, Weingartner, & Eich, in preparation).

The role of organization in altering characteristics of the stimulus and effectively determining the nature of the transformation of input stimuli is at the heart of a great many information-processing models. Such a notion represents a particularly useful explanatory and descriptive tool for SDL. There are many views of what defines organization and of how or when it operates, either on input or at the time of retrieval. They all have a common focus on stimulus transformation as determined either by past experience stored in some organizational structures, contextual organization determinants, and/or organizational constraints at the time of retrieval. Some of the organizational notions that have been used include notions of schemata and their effects on distortions of input (Bartlett, 1932), gestalt psychological concepts, chunking theory (Miller, 1956), redundancy and uncertainty (Garner, 1962), associative and semantic organization (Deese, 1965), conceptual and semantic organization (Mandler, 1972), and encoding and retrieval specificity (Tulving & Thomson, 1973). These models emphasize systematic alterations in the encoding of input information at the time of storage in memory. They also stress the lack of congruence between the nominal to-be-remembered event and its functional trace in memory storage. In much the same way that stimulus transformations are determined by specific contextual or previously learned organizational constraints, the drug-altered state can also be seen as inducing specific strategies for organizing and encoding events. Such state-specific encoding strategies have been demonstrated (Weingartner *et al.*, 1970; Henry *et al.*, 1973). Information may therefore be dissociated in disparate-state recall because of a unique interpretation on transformation of input at the time of storage.

In the same way that organizational strategies may play a role in state-specific storage of information, it is also possible to see brain states associated with state-specific retrieval strategies used to search memory at the time of recall. Some studies have begun to document changes in retrieval strategies as a function of drug state (Darley *et al.*, 1974). These changes have generally been seen as quantitative shifts rather than a qualitative change in the way memory is searched at the time of recall. A notion of specific encoding strategies tied to specific retrieval strategies would predict that optimal recall levels would appear only under congruent storage conditions because of the unique retrieval strategies that would be used under some given retrieval condition. This theoretical position is both germane to SDL and supported by findings. For example, it has been shown that a given informational context induces a unique encoding of to-be-remembered events and that the retrieval of these events depends on the extent to which that informational context is reinstituted at the time of recall. Translated into the framework of SDL, the brain state present at the time of storage or learning and at the time of retrieval can be substituted for the information field or organizational set that transforms to-be-recalled target stimuli. The drug analog of encoding specificity would be that information stored while in a brain-altering drug state changes the interpretation of events. The stored event is associated with state-specific strategies for organizing, coding, and time-tagging the information. Optimal retrieval strategies for the recall of stored information depends on reinstituting that same drug state. If recall is attempted in a disparate state, such as the undrugged state, memory seems to fail. This would follow if the disparate state is associated with retrieval strategies that are qualitatively different (and therefore less appropriate) from those used in congruent recall states. In reinstituting the drug state present at the time of learning, subjects seem again to recall stored events more completely, since retrieval plans for searching memory are congruent with those associated with the state-specific storage of that same input. A theory of encoding specificity and retrieval specificity logically also stresses distinctions between availability of information in memory (Is the information there? Has it been stored?) and its accessibility (Can you get at it in a given test situation?). It is clear that subjects often know far more than they can demonstrate on tests of spontaneous or nominally noncued recall and that issues of what was stored or available in memory are as much a function of the specific retrieval task used as are the nature and strength of the stored memory trace.

This last theoretical view of memory fits many of the features of current SDL findings. Earlier we pointed out that when storage and retrieval states are altered by drugs (such as marihuana, physostigmine, and alcohol) or behaviorally (using sleep-deprivation to alter mood), information stored in these states is more effectively recalled under similar conditions. It has also been demonstrated that although each of these state changes induce SDL effects when memory is tested under free-recall conditions, these are erased under cued cued recall conditions. That is, by providing subjects with powerful and appropriate retrieval cues that presumably cannot be self-generated in the disparate recall state, subjects are able to access information that is available in memory but cannot be spontaneously recalled without such externally provided cues. SDL findings also consistently show that only certain kinds of information are lost in disparate-state recall. What is dissociated is not the fine detail of previously presented information but whole clusters of information (Eich *et al.*, 1975). It is these organized clusters, or higher-order memory units, that are recovered with cuing rather than the detailed information within a cluster. For example, when we presented subjects with many exemplars from sets of highly organized words such as vegetables, items of transportation or flowers and recall was attempted in a disparate state, whole categories of stimuli could not be recalled. Subjects forgot that flowers and transportation items had been presented earlier. If they recalled a presented category, they just as effectively regenerated and recalled the items within that category as in congruent state recall. In the disparate state, entire categories or higher-order memory units were dissociated or "forgotten." Providing the category names erased SDL dissociative effects for whole clusters of stored events. The implications of this model of memory and the parallel SDL findings suggest that the retrieval of a great deal of information stored in long-term memory merely awaits the formulation of the right kind of retrieval scheme or set of questions or, by implication, the reinstatement of some appropriate state.

OVERVIEW AND POSSIBLE NEW APPLICATIONS OF SDL MODELS

Although SDL research findings have not generated new models of memory, they have begun to provide a powerful and orthogonal framework for testing some of the notions of how information is stored and retrieved from memory. At the very least, it has provided some human-memory researchers with a relatively painless introduction to the biology of memory, an area almost totally ignored in traditional human-learning research. SDL research also provides a potential powerful interface between animal-learning concepts that ordinarily do not account for the findings and theories of human learning and memory phenomenona generated in the human-learning laboratory. Using drugs as tools for altering brain before, during, or after learning, or later when what has been learned is tested, is not quite the same as manipulating the informational context in which information is encoded, stored or retrieved from memory. However, there are enough common features and, more important, common questions, that can be raised with an SDL approach in order to provide the beginnings of a common forum for cross-fertilization between animal and human versions of what makes memory systems work. It is also apparent that SDL research paradigms invite some new appreciation of clinical phenomena and their underlying neurochemical events as seen in response to pharmacological manipulations. How depressed people interpret events, and how and what they remember, may tell us a great deal about the nature of underlying biological events associated with disturbances in mood, especially when the behavioral approach is coupled with systematic psychopharmacological investigation and simultaneous monitoring of CNS neurochemistry.

It is hard to resist speculating about where SDL research might lead, both in terms of its potential theoretical leverage and in possible applications of findings. We indulged in some speculation. For example, cognitive-developmental issues might be viewed and pursued in a somewhat new way using SDL concepts. Freud, Piaget, and other classical developmental theorists described changes in thinking as occurring in discrete stages or states, while others (e.g., Skinner, Bruner) stressed that such changes result from very different determinants and along a continuum. Such cognitive changes can alternatively be described as changes in the kinds of organizational or encoding strategies that systematically alter how input is transformed and what is necessary for effective retrieval. The 4-year-old child notices different facets of his environment and codes and associates events differently, in part as a function of a restrictive behavioral repertoire, specifically linguistic, and the kinds of organizations that tie previous learning together in long-term memory. The adult may find much of early learning lost in memory because it is inaccessible rather than unavailable. The memory of stored events can be retrieved, given some external cue that is not likely to be self-generated, such as hearing the sound of a stick along a picket fence in the town in which he grew up. The phenomenon is similar to external cuing factors erasing recall failures when storage has taken place in a disparate state. We could argue that the storage schemas and associated retrieval strategies used by children are not the same as those ordinarily used by adults. It might be quite interesting to approach developmental stages as a problem in brain state-specific storage and retrieval.

The nature of feeling or mood states in man may be productively studied with an SDL approach by integrating the behavioral as well as biological underpinnings that trigger and maintain such states. Amphetamine, for example, has been shown to induce activation in depressed individuals who also evidence SDL effects in response to this drug. The size of such effects is, however, directly related to the subjective antidepressant drug effects. On the other hand, one sees nondrug-related affect-determined SDL effects that are correlated with discrete neurochemical events. Jointly defining affect or mood states in man by combining data and theory about unique modes of processing events, unique neurochemical expression, along with subjective experiences along some scaled dimension seems to be a worthwhile research venture. The very role of affect as an information context, and thereby as a determinant of affect specific storage and retrieval strategies, may bring new insight into the essential nature of human information-processing systems.

Other areas that might profitably be pursued with an SDL approach include such issues as the nature of cuing, as it is self-generated or provided by the environment, and its role in recall. Why and how do different cues work at different times and in different states? The issue of cuing is intimately tied to the issue of the nature of the information contained in the memory trace.

Many kinds of brain states in man can and will be examined for possible dissociative effects in the coming years. These include not only the mind-altering drugs but also some of the nonpharmacological subjectively experienced state changes. Some drug and nonpharmacological states, in which an SDL paradigm might be used to test for relative discriminability, will be contrasted with one another, and the robustness of dissociative effects measured among such states. Investigators of drug abuse have become increasingly interested in SDL models both as a tool for probing the etiology of drug abuse and as a potential framework for generating new treatment strategies. These are but a few of the research areas that seem to be reasonable places to use SDL techniques.

A cautionary note is, however, necessary in examining and speculating about potential SDL application. At best, SDL is a model that permits investigators to test a large variety of questions and hypotheses about the way experience is recorded and remembered in different brain states. In itself, SDL has not as yet provided any new theories about memory, clinical states, drug states, or the biology of memory. The SDL techniques themselves are still quite primitive and unwieldy. SDL remains a research arena that is enjoyable, often exciting, but one in which persistent detailed research must proceed as it does in all science. As attractive as it appears to be, it provides no free or simple answers to basic questions.

ACKNOWLEDGMENT

James E. Eich was most helpful in providing a scholarly and thoughtful review of this chapter.

REFERENCES

Bustamante, J. A., Jordan, A., Vila, M., Gonzalez, A., & Insua, A. State dependent learning in humans. *Physiol. Behav.*, 1969, *5*, 793–796.

Bustamante, J. A., Rosello, A., Jordan, A., Pradere, E., & Insua, A. Learning and drugs. *Physiol. Behav.*, 1968, *3*, 553–555.

Darley, C. F., Tinklenberg, J. R., Roth, W. T., & Atkinson, R. C. The nature of storage deficits and state-dependent retrieval under marijuana. *Psychopharmacologia,* 1974, *37,* 139–149.

Deese, J. *The structure of associations in language and thought.* Baltimore: Johns Hopkins Univ. Press, 1965.

Eich, J. E., Weingartner, H., Stillman, R. C., & Gillin, J. C. State-dependent accessibility of retrieval cues in the retention of a categorized list. *J. Verb. Learn. Behav.,* 1975, *14,* 408–417.

Evans, F. J., Gustafson, L. A., O'Connell, D. N., Arne, M. T., & Shor, R. E. Response during sleep with intervening waking amnesia. *Science,* 1966, *152*(2), 666–667.

Evans, F. J., & Kihlstrom, J. F. Posthypnotic amuesia as disrupted retrieval. *J. Abnorm. Psychol.* 1973, *82,* 317–323.

Garner, W. R. *Uncertainty and structure as psychological concepts.* New York: Wiley, 1962.

Goodwin, D. W. Alcoholic blackout and state-dependent learning. *Federation Proc.,* 1974, *33,* 1833–1835.

Goodwin, D. W., Powell, B., Boemer, D., Hoine, H. & Stern, J. Alcohol and recall: State-dependent effects in man. *Science,* 1969, *163,* 1358–1360.

Henry, G., Weingartner, H., & Murphy, D. L. Influence of affective states and psychoactive drugs on verbal learning and memory. *Am. J. Psychiat.,* 1973, *130,* 966–971.

Hill, S. Y., Schwin, R., Powell, B., & Goodwin, D. W. State-dependent effects of marijuana on human memory. *Nature,* 1973, *243,* 241–242.

Ley, P., Jain, V. K., Swinson, R. P., Eaves, D., Bradshaw, P. W., Kincey, J. A., Crowder, R., & Abbiss, S. A state-dependent learning effect produced by amylobarbitone sodium. *Brit. J. Psychiat.,* 1972, *120,* 511–515.

Mandler, G. Organization and recognition. In E. Tulving & M. Donaldson (Eds.), *Organization of memory.* New York: Academic Press, 1972.

Miller, G. A. The magical number seven, plus or minus two: Some limits on our capacity to process information. *Psychol. Rev.,* 1956, *63,* 81–97.

Overton, D. A. Discriminative control of behavior by drug states. In T. Thompson & R. Pickens (Eds.), Stimulus Properties of Drugs. New York: Appleton, 1971.

Overton, D. A. State dependent learning produced by alcohol and its relevance to alcoholism. In B. Kissen, & H. Begleiter (Eds.), *The biology of alcoholism.* Vol. II. *Physiology and behavior.* New York: Plenum, 1972. pp. 193–217.

Overton, D. A. State-dependent learning produced by addicting drugs. *In* Fisher, S., & Freedman, A. H., eds. Opiate Addiction: Origins and Treatment. Washington, D.C., Winston & Sons, 1973.

Overton, D. A. Experimental methods for the study of state-dependent learning. *Federation Proc.,* 1974, *33,* 1800–1813.

Petersen, R. C. Alcohol and cueing. Paper presented at Am. Psychol. Assoc. Symposium, 1975.

Rickles, W. H., Cohen, M. J., Whitaker, C. A., & McIntyre, K. E. Marijuana induced state-dependent verbal learning. *Psychopharmacologia,* 1973, *30,* 349–354.

Stillman, R. C., Weingartner, H., Wyatt, R. J., Gillin, J. C., & Eich, J. E. State-dependent (dissociative) effects of marijuana on human memory. *Arch. Gen. Psychiat.,* 1974, *31,* 81–85.

Stillman, R. C., Weingartner, H., & Eich, J. E. State-dependent scanning and retrieval of visual information (in preparation).

Tarter, R. E. Dissociate effects of ethyl alcohol. *Psychonom. Sci.,* 1970, *20,* 342–343.

Tulving, E., & Pearlstone, Z. Availability versus accessibility of information in memory for words. *J. Verb. Learn. Verb. Behav.* 1966, *5,* 381–391.

Tulving, E., & Thomson, E. Encoding specificity and retrieval processes in episodic memory. *Psychol. Rev.,* 1973, *80,* 352–373.

Weingartner, H., Adefris, W., Eich, J. E., & Murphy, D. L. Encoding-imagery specificity in alcohol state-dependent learning. *J. Exp. Psychol.,* 1976, *2,* 83–87.

Weingartner, H., & Faillace, L. A. Alcohol state-dependent learning in man. *J. Nervous Mental Disease,* 1971, 153, 395–406.

Weingartner, H., Miller, H. A., & Murphy, D. L. Mood-state-dependent retrieval of verbal associations. *J. Abnorm. Psychol.*, 1977, *86*, 276–284.

Weingartner, H., & Murphy, D. L. Brain states and memory: State-dependent storage and retrieval of information. *Psychopharmacol. Bull.*, 1977, *13*, 66–67.

Weingartner, H., Snyder, S., Faillace, L. A., & Markley, H. Altered free associations: Some cognitive effects of DOET (2,5-dimethoxy-4-ethyl-amphetamine). *Behav. Sci.*, 1970, *15*, 296–303.

Wickelgren, W. A. Alcoholic intoxication and memory storage dynamics. *Memory & Cognition*, 1975, *3*, 385–389.

Wright, D. C. Differentiating stimulus and storage hypothesis of state-dependent learning. *Federation Proc.*, 1974, *33*, 1797–1799.

Index

E

F

G

H

I

L

M

N

T

U

V

W

X, Y, Z

A B C 8 D 9 E 0 F 1 G 2 H 3 I 4 J 5